CONSTRUCTION MATERIALS AND ENGINEERING

PIPELINES

DESIGN, APPLICATIONS AND SAFETY

CONSTRUCTION MATERIALS AND ENGINEERING

Additional books in this series can be found on Nova's website under the Series tab.

Additional E-books in this series can be found on Nova's website under the E-book tab.

CONSTRUCTION MATERIALS AND ENGINEERING

PIPELINES

DESIGN, APPLICATIONS AND SAFETY

MIGUEL G. RIVERO
AND
LAUTARO M. MANSILLO
EDITORS

Nova Science Publishers, Inc.
New York

For permission to use material from this book please contact us:
Telephone 631-231-7269; Fax 631-231-8175
Web Site: http://www.novapublishers.com

Library of Congress Cataloging-in-Publication Data

Pipelines : design, applications, and safety / editors, Miguel G. Rivero and Lautaro M. Mansillo.
p. cm.
Includes index.
ISBN 978-1-62100-178-2 (hardcover)
1. Pipelines. 2. Pipelines--Safety measures. 3. Pipelines--Design and construction. I. Rivero, Miguel G. II. Mansillo, Lautaro M.
TJ930.P5695 2011
621.8'672--dc23

2011032397

Published by Nova Science Publishers, Inc. † New York

CONTENTS

PREFACE

In this book, the authors gather topical research in the study of the design, applications and safety of pipelines. Topics discussed in this compilation include drinking water pipeline deterioration; stress corrosion cracking (SCC) behavior in buried pipeline steel; a predictive model for crack growth model of buried pipelines; internal stresses in pipeline coating; evaluation of stressed states of pipelines by ultrasound and aerial altitude gas pipelines. (Imprint: Nova)

Chapter 1 - Maintenance of pipeline networks is a very critical issue, due to the high cost associated and the potential hazard if a catastrophic fail occurs. Deterioration of pipelines is due to a complex group of phenomena, which affects both the condition of the pipeline itself and the composition of the fluid being transported. Metallic pipelines are exposed to soil or atmospheric corrosion externally and aqueous or gaseous corrosion at the inside surface. Inspection and monitoring of pipeline condition is difficult and costly, especially when infrastructure is underground. Internal corrosion is mostly controlled by treatment of the fluid whilst external corrosion requires the use of various strategies such as coatings and/or cathodic protection. On the other hand, statistics analysis of field data and prediction of failure in underground pipeline network can be an advantageous strategy to be considered, as direct inspection involves high costs. One of the most sensible pipeline infrastructures today is that related to drinking water distribution, as this liquid is basic to human being welfare and survival. The most relevant results of several studies related to both main and secondary distribution drinking water pipeline network in a tropical condition are to be presented here. Those studies considered lab experimentation by using pilot systems and also information collected directly from the field. Internal condition of a main network several decades old, was extensively revised in order to characterize the deposits formed, indicating that an appropriate water treatment allows reducing tuberculation and biofilm formation; these phenomena could originate problems such as water quality events, pipeline deterioration or hydraulic operation restriction. Various studies on secondary distribution networks looking to obtain basic information for structuring maintenance programs allows to build adequate methodologies not only for collection of field information but also for analysis of data; review of the most applied standards related to evaluation of pipeline corrosion was another important result. Studies on pilot systems generate information of the effect of pipeline material on water quality changes and biofilm formation; carbon steel, galvanized steel, copper and PVC were evaluated. Finally a wide literature review of studies related to model construction, looking to prediction of lifespan in underground water networks is presented.

Chapter 2 - The failure of pipeline steel is usually catastrophic and causes economic, environmental and conservation losses. About 20% of the failure of pipeline was caused by outer corrosion and 5% by inner corrosion. The first incident of stress corrosion cracking (SCC) on natural gas pipelines occurred in the mid- 1960's [1]. Since that time, there have been hundreds of failures reported in Australia [2], Canada [3], Iran, Iraq, Italy, Pakistan, Saudi Arabia, the former Soviet Union and the United States [4]. In the last 30 years, the pipeline SCC problem has been investigated by different laboratories [5-15]. Most of the early failures were intergranular in nature, whereas, many of the recent failures, such as those that have occurred in Canada are transgranular [16-23].

Chapter 3 - A predictive crack growth model for high pH stress corrosion cracking of pipelines is postulated based on the fundamental understandings of the film rupture mechanism. It is known that the cracking process is governed by the factors in three catalogs: (1) the mechanical properties and microstructure of material that can affect the crack-tip strain rate; (2) the parameters characterizing the external loads and (3) the environmental variables that can alter the kinetics of anodic dissolution and repassivation. The theoretical approaches and/or experimental methods are described for determining the parameters that dominate the crack growth. The experimental validation indicates that the model gives a reasonably good prediction to the effects of important factors relating to materials, environment and loading conditions. Although fatigue damage is negligible in operation of pipelines for gas transportation, the accelerated dissolution at the crack tip due to the cyclic deformation is still required to be considered in the crack velocity estimation. Finally, the procedure for the field application of the new model is outlined.

Chapter 4 - Pipelines are used worldwide for the transportation of oil or gas and must be protected against corrosion over long periods of time to avoid any production failure. An anti-corrosion coating system is generally used in addition to cathodic protection. This coating must fulfil specific conditions such as good mechanical strength and good ageing resistance in corrosive soils or water for instance. Two types of coatings are currently used, a monolayer system or a three layer systems, which have both their advantages and drawbacks. Spontaneous disbonding of the three layer coating was thus observed at the ends of pipelines sections during storage, just after coating application. Moreover, for three-layer systems presenting no apparent defaults, pipelines coating failures have been observed at the epoxy/steel interface after a service period shorter than the expected lifetime. These phenomena were never reported for monolayer systems made with the same epoxy primer, which highlights the influence of internal stresses generated in the thick coating during the process on interfacial strength. The simulation of the process by finite element analysis with adequate material behaviour laws give internal stresses levels in good agreement with experimental measurements and in situ observations. Moreover, most loss of adhesion of the three-layer systems have been observed when the operating temperature was about 50-60°C in wet environments, which suggest wet disbonding. Diffusion phenomena through the different polymer layers must then be taken into account. Depending on primer nature, Fickian or Langmuir water diffusion kinetics show that water molecules diffuse through the thick coatings up to the epoxy-steel interface very rapidly compared to pipelines' service time. The coating failure can then be attributed to epoxy physical or chemical degradation or to interfacial bonds hydrolysis. The diffusion parameters and then, the failure mode in humid environment, strongly depend on temperature, primer nature and fillers proportion and nature. A good correlation can be found for thin coatings between primer disbonding and its saturation time estimated from diffusion data. For thick coatings, wet disbonding occurs before the primer saturation time, which is linked to the presence of internal stresses stored during the process.

Chapter 5 - This chapter contains results of theoretical and experimental investigations of the principal problems of the acoustoelasticity phenomenon application for the nondestructive testing of stressed (strained) state (SSS) of steel pipes and pipelines spent by the author and collaborators for the last 8 years. The foundations of the method of acoustoelasticity for biaxial stress evaluation with the help of bulk waves propagated normally to the plane of stress acting are established. Specific features of stress evaluation in engineering materials by using precise measurements of velocities of elastic waves of a millimeter range of wavelengths are revealed. The results of experimental evaluation of acoustic-elastic coefficients which determine the sensitivity of ultrasonic method to actual stresses in pipe steels used in gas and oil industry are given. The opportunities of precise measurement of phase velocity of ultrasound (principal informative characteristic of the acoustoelastic effect) with use of a pulse-echo method, realized in a special portable device IN-5101A produced by "ENCOTES" Ltd, are discussed. The calculative algorithms are included into the computational module of device and enable one to evaluate in-plane principal stresses in real object. So, the experimental arrangement provides evaluation of uni- or biaxial, tensile or compressive stresses in pipes and constructive details of compressing and pumping stations, including long-term operated materials at different force and climatic actions. The main problems of determination of stressed states of real engineering constructions in a case of the acoustical tensometry, that is monitoring of stresses during manufacturing and exploitation of pipes; and estimation of stresses «in situ», in pre-stressed industrial objects, are considered. The reliability of acoustoelastic manner for biaxial stress evaluation in main pipelines of large diameter was justified experimentally, during the loading of closed pipe by inner pressure of water. The results of the investigations show that observed data are quite close to analytical solution of the problem which was given by Gabriel Lame in 19th century. Practical example of the acoustoelasticity phenomenon application for biaxial stress evaluation is described, namely, measurement of non-design axial stresses in technological pipelines of dust collectors of compressor station. The opportunity of application of "zero-less" ultrasonic tensometry for determination of a stressed state of objects in action has been experimentally confirmed.

Chapter 6 - Design of new cheap aerial pipelines, a large flexible tube deployed at high altitude, for delivery of natural (fuel) gas over a long distance is delineated. The main component of the natural gas is methane, which has a specific weight less than air. The lift force of one cubic meter of methane equals approximately 0.5 kg. The lightweight film flexible pipeline can be located in air at high altitude and, as such, does not damage the environment. This aerial pipeline dramatically decreases the cost and the time of construction relative to conventional pipelines of steel, which saves energy and greatly lowers the capital cost of construction. The article contains a computed project for delivery of 24 billion cubic meters of gas per year.

Chapter 7 - The paper describes experiments to measure characteristics of a gas flow issuing from a finite-volume reservoir to atmosphere through a pipeline of variable length. In view of the continuous variation of flow quantities in time, the process under study could be classed to unsteady processes. Nonetheless, the experimental data were found to be in excellent agreement with the results gained in one-dimensional steady-state calculations performed assuming a constant value of friction coefficient. The latter result points to applicability of quasi-steady calculations to the analysis of complex gas-dynamic processes.

Chapter 8 - Two theoretical models for calculating compressible gas flow with friction through pipeline are considered. The first model, most commonly used in practice, is based on the assumption that the flow is adiabatic, and the total enthalpy of the flow presents a conserved quantity. This approach, however, leads to a noticeable inconsistency between the total-pressure value calculated at pipeline outlet and the friction losses in the pipeline. In the second model, an increase of total flow enthalpy due to the work of friction forces is admitted. Calculations show that, in the latter case, an increase in flow stagnation temperature amounting to several ten degrees is possible, especially for high-velocity flows. Both models were tested for adequacy by comparing the data predicted by these models with available experimental data on the distribution of static pressure over pipeline length. It was shown that no definite conclusion in favor of one of the discussed models can be drawn because of inaccuracies of used experimental data.

Chapter 9 - Effluents emanating from Petroleum reservoirs channeled through steel pipelines comprise of oil, water, gas, and other projectiles suspended in fluid, incessantly flowing in a turbulent manner, in a statistically discernablc continuum. Therefore, Pipelines tend to suffer corrosion and wear having endured continuous impact, particularly at various weld joints within their internal walls. These joints have a subtle, but significant difference in composition from the parent metal, and are actually a concoction of the parent metal alloys with other weld elements in varying amounts, hence its pseudo homogenous nature. The structural integrity and design life of these pipelines invariably depends on these welds possessing equivalent strength and toughness potentials as the parent metal. In order to reduce corrosion and wear to the barest, and geared towards meeting increasing demands of operating pressures and loads, the individual effects of each alloying and weld element must be identified, and a predictable, maneuverable protocol attained. Statistical methods such as the Taguchi Method with Grey Rational Analysis could be applied to optimize the welding process parameters so that elements that increase longevity are incorporated to obtain signature uniquely crafted weld chemical compositions. Tensile, toughness, hardness tests, and micro structure analysis are used to assess weld quality and the veracity of the applied statistical methods is thereafter compared with measured values. The methods applied proved successfully, that the welding process parameters as well as the various types and proportions of the alloying elements that make up the chemical composition of the weld metal can be satisfactorily optimized and improved.

In: Pipelines: Design, Applications and Safety
Editors: M. G. Rivero et al. pp. 1-41
ISBN 978-1-62100-178-2

Chapter 1

DRINKING WATER PIPELINE DETERIORATION

J. Delgado, J. Castano, J. Calderon, and F. Echeverria
Corrosion and Protection Group, University of Antioqua,
Medellin, Colombia

ABSTRACT

Maintenance of pipeline networks is a very critical issue, due to the high cost associated and the potential hazard if a catastrophic fail occurs. Deterioration of pipelines is due to a complex group of phenomena, which affects both the condition of the pipeline itself and the composition of the fluid being transported. Metallic pipelines are exposed to soil or atmospheric corrosion externally and aqueous or gaseous corrosion at the inside surface. Inspection and monitoring of pipeline condition is difficult and costly, especially when infrastructure is underground. Internal corrosion is mostly controlled by treatment of the fluid whilst external corrosion requires the use of various strategies such as coatings and/or cathodic protection. On the other hand, statistics analysis of field data and prediction of failure in underground pipeline network can be an advantageous strategy to be considered, as direct inspection involves high costs. One of the most sensible pipeline infrastructures today is that related to drinking water distribution, as this liquid is basic to human being welfare and survival. The most relevant results of several studies related to both main and secondary distribution drinking water pipeline network in a tropical condition are to be presented here. Those studies considered lab experimentation by using pilot systems and also information collected directly from the field. Internal condition of a main network several decades old, was extensively revised in order to characterize the deposits formed, indicating that an appropriate water treatment allows reducing tuberculation and biofilm formation; these phenomena could originate problems such as water quality events, pipeline deterioration or hydraulic operation restriction. Various studies on secondary distribution networks looking to obtain basic information for structuring maintenance programs allows to build adequate methodologies not only for collection of field information but also for analysis of data; review of the most applied standards related to evaluation of pipeline corrosion was another important result. Studies on pilot systems generate information of the effect of pipeline material on water quality changes and biofilm formation; carbon steel, galvanized steel, copper and PVC were evaluated. Finally a wide literature review of studies related to model construction, looking to prediction of lifespan in underground water networks is presented.

1. Corrosivity of Water in Drinking Water Distribution System

The water distribution system of Medellin delivers drinking water to about three million citizens, supplied by three water treatment plants (Ayura, Manatiales and Villa Hermosa). Each plant process water from a different water source and deliver water to a specific zone of Medellin city. Ayura plant (AY) treats water by coagulation with alum, flocculation and sedimentation, filtration with sand and anthracite and disinfection with Cl_2. Manantiales plant (MN) treats water with pre-oxidation, adsorption with active carbon, alum coagulation, flocculation and sedimentation, manganese oxidation, filtration with sand and anthracite and disinfection with Cl. Finally, Villa Hermosa plant (VH) treats water by coagulation with alum, flocculation, sedimentation and filtration with sand and anthracite.

The pipeline network is mostly made of reinforced concrete (about 54%), ductile iron (about 27%) and steel (10 %). Storage facilities are mainly made of reinforced concrete, and pipeline accessories of ductile iron and steel.

In Medellin city, an undetermined number of ductile iron and carbon steel pipes are in place for long periods of time (in some cases more than 40 years), for which corrosion control is critical to maintain microbial, water quality and pipe integrity.

The corrosion behavior of iron in potable water is undoubtedly important in operating drinking water systems, where carbon steels are applied as piping and tubing materials. Corrosion of system pipes contribute to diminish the reliability of the distribution network, having economic, hydraulic and aesthetic impacts, including water leaks, corrosion product buildup, increased pumping costs, water quality deterioration and high costs of repair and replacement. Corrosion of iron is also the primary factor controlling biofilm growth [1-6]. Interactions between biofilm, humic substances and iron oxide may negatively affect the microbiological quality of the water [7], promote the release of pathogenic microorganisms [8, 9], cause a drastic reduction in the efficiency of disinfectants [8] and induce a chemical decay of the residual chlorine [1, 10, 11].

The corrosion mechanisms of iron in such systems can be electrochemical and/or microbial [12, 13]. However, they have not been fully understood and there are many contradictory reports [1, 6]. Corrosion of iron in the pipes is believed to be quite a complex process which is influenced by many parameters of finished water, including: oxygen concentration, pH, alkalinity, presence of sulfates, chlorides and nitrates, temperature of water, level of disinfectant, concentration of natural organic matter (NOM) and water flow velocity (5, 14-16].

Manipulation of pH, alkalinity and calcium hardness induced the formation of calcium carbonate compounds in the internal wall of water pipes, allowing the removal of color and the control of internal corrosion of a water distribution system [17, 18]. However, in some cases, water quality cannot always explain variations in corrosion behavior; for example, a recent study found that changes in water quality parameters such as pH and alkalinity could not fully account for observed changes in corrosion of iron pipes under stagnant conditions [1].

Studies conducted between 2002 and 2003 in the drinking water system of Medellin City are focused mainly on the measurement of water corrosivity by means of carbon steel corrosion coupons.

Corrosion behavior was studied in the three different subsystems of drinking water network (AY, MN and VH) based on the gravimetric data upon the exposure of carbon steel specimens with the dimensions of 152 mm x 22 mm x 3.2 mm. The specimens were weighed before, and exposed in different sites of drinking water network for 3 months. At the end of exposure time, the corrosion products were removed and the specimens were weighed again to obtain the mass loss and thereby estimate the corrosion rate. The procedure for measuring the corrosivity by this method are described in ASTM D2688 [19] and ASTM G1 [20] standards.

The specimens were installed at the entrance and exit of the three treatment plants. Also, for each subsystem, a tank located as far as possible was selected, in order to evaluate the variation in the corrosivity of water versus distance to the plant. In each tank specimens were installed both inside and in the exit of the tank.

At the exit of treatment plants and tanks, the specimens were installed inside loops, as shown in Figure 1. The loops are derivations of the pipe in which removable elbows and fitting are installed for inserting carbon steel specimens that are exposed to water flow, as well as measuring instruments.

At the entrance of treatment plants and inside tanks, the specimens were fixed on acrylic racks (Figure 2) and placed in open sites where the current flow was not very high.

Table 1 shows the sites chosen for the installation of specimens, as well as the corrosion rates found after 3 months. In each site, four specimens were installed.

The longer distance between a treatment plant and its respective tank occurs between MN plant and Girardota tank (17.4 Km). The distance between AY plant and Belencito tank is 12.1 km, while between VH plant and Limoncito tank is 0.9 Km.

The water is considered as highly aggressive when the corrosion rate is higher than 0.25 mm/y (10 mpy). For moderately aggressive water, the corrosion rate varies between 0.13 and 0.25 mm/y (5-10 mpy), whilst for low aggressive water the corrosion rate is lower than 0.13 mm/y (5 mpy)[21].

Figure 1. (a) Loop at the exit of VH treatment plant; (b) Loop in Limoncito Tank entrance and location of corrosion coupons.

Figure 2. Rack installed in AY plant, before treatment of water.

Table 1. Sites selected and corrosion rates obtained

#	Site	Location	Subsystem	Corrosion rate (mm/y)
1	AY treatment plant	Entrance	AY	0.14
2	AY treatment plant	Exit	AY	0.46
3	Belencito (tank)	Entrance	AY	0.51
4	Belencito (tank)	Inside	AY	0.24
5	MN treatment plant	Entrance	MN	0.26
6	MN treatment plant	Exit	MN	0.37
7	Girardota (tank)	Entrance	MN	0.51
8	Girardota (tank)	Inside	MN	0.24
9	VH treatment plant	Entrance	VH	0.27
10	VH treatment plant	Exit	VH	0.40
11	Limoncito (tank)	Entrance	VH	0.40
12	Limoncito (tank)	Inside	VH	0.36

At the entrance of AY plant, the corrosion rate is the lowest of all registered (0.14 mm/y), corresponding to moderately aggressive water, but near to the threshold between moderately and highly aggressive. At the entrance of the other two plants, the corrosion rate corresponds to a high level of aggressiveness. In all plants, the water experiences an increase in aggressiveness after treatment, being more aggressive at the MN plant exit (0.46 mm/y). However, the aggressiveness of treated water does not appear to be related to the aggressiveness of the effluent. On the other hand, comparing the corrosion rates obtained from plants, it was found that the water at the exit of AY plant was the most aggressive.

The corrosion rates at the entrance of tanks are always higher than those calculated for the respective plant effluent. For all subsystems, the highest corrosion rates were obtained at the entrance of tanks. The greater the distance between the treatment plant and the tank, the higher the increase in corrosion rate. The highest difference between the corrosion rate in the plant exit and in the entrance to the respective tank was found in MN subsystem, while the lowest difference was found in VH subsystem, where the Limoncito tank is located only at 0.9 km. The visual assessment of the corrosion products of steel specimens allowed to observe some morphological differences related to the water flow in the exposure site. In the loops, the specimens were exposed to a relatively high flow rate, and the corrosion products were oriented in the direction of flow (Figure 3). On the other hand, for the specimens installed in relatively low flow rate (in treatment plant influents), the morphology was more irregular (Figure 4). The morphology and color was similar for corrosion products found both in the specimens installed in loops as the inner wall of pipelines.

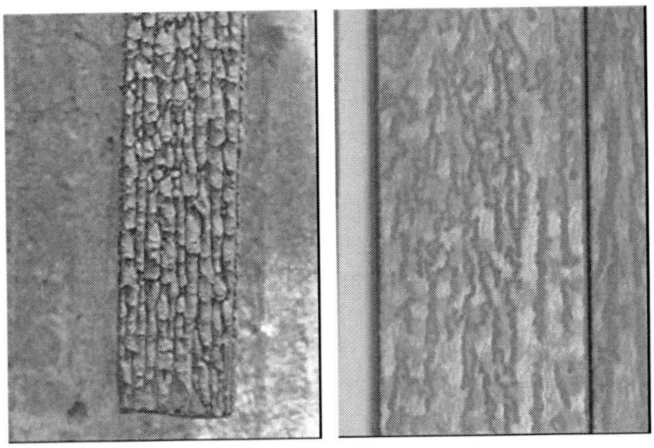

Figure 3. Specimen installed in the loop of VH treatment plant, before and after removal of corrosion products.

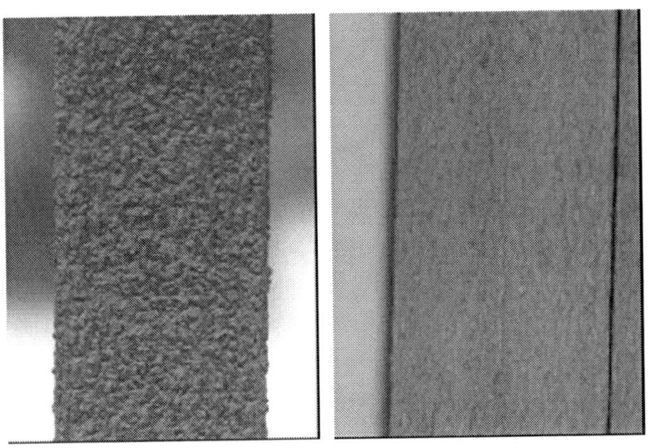

Figure 4. Specimen installed in VH treatment plant influent, before and after removal of corrosion products.

Table 2 shows the variation in some water quality parameters (minimum and maximum) at the exit of the plants, based on historical data (between 2002 and 2003).

Table 2. Variation in water quality parameters at the exit of the plants

	MN	AY	VH
pH	7.76-9.27	7.51-8.25	6.73-7.95
Alkalinity (ppm $CaCO_3$)	9.31-26.96	13.21-19.75	14.33-23.87
Hardness (ppm $CaCO_3$)	7.79-28.84	12.04-15.74	13.38-24.53

By comparing these values with the aggressiveness of water evaluated using specimens, it is observed that water from MN plant, evaluated as less aggressive water, has the higher pH values. Water from AY plant, evaluated as the most aggressive, has lower pH values than MN, although in AY plant there are maximum values higher than 8, which is recommended to minimize pipeline corrosion. Weight loss is generally found to increase with increasing pH in the range 7 to 9 [1]. However, one study found that both weight loss and iron concentration decreased as pH was raised from 8.5 to 9.2 [22], which could explain the lower aggressiveness of MN water.

In the three treatment plants, alkalinity and hardness are lower than recommended. Increasing both parameters generally leads to lower corrosion rates [22, 23,] and it is recommended that alkalinity is maintained at greater than 60 ppm as Ca CO_3 [24], and hardness higher than 37.5 ppm as $CaCO_3$ [25] . In MN water the highest alkalinity and hardness values recorded could have influenced their less aggressiveness.

2. Characterization of Deposits Formed inside Pipelines

Minerals, organic matter and bacterial biomass are the main constituents of drinking water deposits [8, 26], and water quality can be also affected by the existence of such deposits [8, 10]. In addition to corrosion products, particulate matter is other source of deposits in water distribution systems. Particulate matter is transported by water, microbial activity and physicochemical reactions within the water bulk [8, 26, 27].

For each plant one sampling site, as close as possible to the plant exit, was selected for the characterization of deposits formed. Additional sampling sites, in which water quality problems have been reported, were selected as follows: two for the subsystems AY and MN and one for subsystem VH, the smaller subsystem. Samples were taken from the interior of the pipeline network and from the inside of the nearest tank (Table 3). Deposit samples were collected randomly at each sampling site, using either a stainless steel or a plastic spatula for hard and soft deposits. Historic data (years 2000 to 2003) of water pH was provided by the service company; in addition water samples were taken at both the plant exit and a distant tank for each subsystem and analysed regarding free chlorine residuals.

Deposit samples were dried during 48 hours at 40°C. Then, samples were grinded using an agate mortar until approximately 98% passed a No. 325 mesh. XRD, FTIR, ESEM and EDS analyses were carried out on the samples.

Three main deposits were found across the water distribution system under study: Brown deposits (formed everywhere in the system), tubercle deposits (formed on steel surfaces) and white deposits (found only at some places). Deposits with similar characteristics to brown and tubercle deposits have been reported in other studies [7, 9, 28, 29].

FTIR and XRD analyses of the three main deposits are resumed in Table 4. FTIR analyses could be used as an indication of the formation or presence of organic material in the sample, and therefore, as an indicator of the presence of biofilm. The presence of absorption bands around 2950 and 2880 cm^{-1} can be a good indication of the presence of the organic matter, and hence for biofilm, because these bands are not affected from interference from bands of inorganic material (silicates, hydroxides and carbonates) [30].

Results of XRD analysis of brown deposits reveal important variations according to the colour intensity: Dark deposits appear to be amorphous whilst the lighter brown ones contain crystalline compounds, quartz as the main constituent mixed with an aluminium silicate hydroxide, most probably kaolinite.

Table 3. Sites selected for sampling of deposits in the main water network of Medellin city

Sub system	Sampling site		Distance from plant (Km)	Material
	Pipeline	Tank		
AY	X		3.2	Reinforced concrete
	X		10.3	Reinforced concrete
		X	10.3	Concrete
	X		12.1	Reinforced concrete
		X	12.1	Concrete
MN	X		1.9	Reinforced concrete
	X		17.4	Reinforced concrete
		X	17.4	Concrete
	X		7.9	Reinforced concrete
		X	7.9	Concrete
VH	X		0.2	Steel
	X		0.9	Steel
		X	0.9	Concrete

Table 4. Results of the FTIR and XRD analyses of the different deposits found

	Brown deposit	Tubercle deposit	White deposits
FTIR	Aluminosilicates Organic matter	Goethite Magnetite Lepidocrocite C-C compounds Adsorbed carbonates	Calcite Complex silicate, either containing water or hydroxide groups Phyllosilicate (kaolin/smectite)
XRD	Goethite Magnetite Kaolinite	Goethite Magnetite Lepidocrocite	Aluminium silicate hydroxides Calcite Quartz

Similar compounds in potable water deposits were reported elsewhere [8, 31]. Tubercle deposits formed at different sites in the water system showed mainly the presence of magnetite and goethite, with minor amounts of lepidocrocite XRD analysis of a brown deposit sample formed on a metallic surface reveals mixtures of magnetite, goethite and alluminosilicates. XRD analysis of various white deposits reveals that soft material is made of varying amounts of quartz, calcite and basic aluminosilicates whilst the compact deposit is principally made of calcite. On the other hand, Table 5 shows EDS results for brown deposits, revealing a heterogeneous morphology and composition. The main components in most samples are C, O, Al, Si, Mn, Fe, Ca and Mg.

Table 5. Most typical values (in percentage) of several samples of brown deposits obtained from EDS analysis

Element	Relative percentage		
	Median (at.%)	Minimum (at.%)	Maximum (at.%)
C	11	8	37
O	59	47	63
Al	10	6	15
Si	8	4	12
Mn	3	2	6
Fe	2	1	13
Mg	1	1	3
Ca	1	0	3

Figure 5. Cross-section of a tubercle showing inner soft part (I), hard shell (S) and outer part (O). Also, EDS analysis and SEM images of shell and outer part are shown.

The results of ESEM and EDS analyses indicate that tubercle deposits are morphologically and chemically homogeneous across the distribution system. A tubercle cross section reveals the existence of three different parts: a hard shell in the middle of soft, loose inner and outer materials. No big differences in chemical composition were found by EDS (Figure 5). However, morphology of middle shell is compact and its colour is black, which are typical features of magnetite. On the other hand, inner and outer materials show globular crystals typical of goethite [32, 33].

White deposits showed no morphological distinctive characteristics and EDS analysis of a deposit collected from subsystem AY revealed O and Al as major components whereas a similar deposit found in subsystem VH was mainly made of C, O, Ca and Si.

Analysis of historic data reveals that water pH values at plant exits are: 7.8 ± 0.3 for subsystem AY, 7.9 ± 0.6 for subsystem MN and 6.8 ± 0.4 for subsystem VH. This data also indicates that water pH measured at tanks is nearly the same than at plant exit for subsystems MN and VH whereas for subsystem AY it was 7.2 ± 0.5. The formation of goethite and lepidocrocite is favoured in solutions with pH from 5 to 7, whilst pH values above 8 privilege magnetite formation [34].

Therefore, higher amounts of goethite and lepidocrocite will be expected to be found in tubercle deposits collected from subsystem VH, while magnetite formation will more viable in subsystem MN and at the plant exit of subsystem AY. Free chlorine residuals at the plant exits is between 20 to 27% higher than at distant tanks; therefore, it might be possible, that high free chlorine residuals content in water favours the formation of magnetite over goethite, as observed in atmospheric corrosion [35]. In any case, higher amounts of magnetite in pipeline corrosion products reduce iron release into water [29, 36].

In Table 6, general conditions favouring either magnetite or goethite and lepidocrocite are presented. Goethite and lepidocrocite are favoured in VH subsystem, whereas magnetite is favoured in AY and MN subsystems [30].

Magnetite formation is enhanced under the low oxidation conditions at the core of the tubercle [35, 37]. Microbial activity may also encourage magnetite formation inside the tubercle [38]; this is supported by the morphology of the tubercle samples studied here [36, 38]. White deposits were most probably formed as a result of sedimentation of suspended particles carried by the water flow [8, 31] or by local changes of physico-chemical conditions [39, 40].

Table 6. Some conditions that favour the formation of goethite/lepidocrocite or magnetite in tubercle deposits

Goethite and lepidocrocite	Magnetite
pH between 5 and 8	pH higher than 8
Inside tanks	Distribution system
Stagnant conditions	Higher water flow
Less chlorine content	Higher chlorine content
Higher carbonate ions content	Less carbonate ions content
Higher iron release to water	Less iron release to water

3. External Corrosion on Underground Pipeline Network: Selection of the Main Variables that Take Play in the Corrosion of the Structure

The corrosion of buried pipes is an electrochemical phenomenon in which the anodic reaction is oxidation of the metal, while the cathodic reaction is the reduction of oxygen. Due to the high heterogeneity of soil, the permeation of oxygen from the atmosphere to the metal is different and usually present differential aeration cells, establishing a localized corrosion process in which regions of the metallic structure in contact with a low permeation soil, such as clay, are formed in anodic areas and those regions of the structure in contact with a soil of greater permeation of oxygen, such as sand or selected backfill material are cathodic areas. A large buried structure, such as a pipe, metal corrosion happens in both anodic and cathodic areas; however, the corrosion in the anodic area is limited by diffusion of oxygen in water-saturated clay soil, while in the sandy soil corrosion is limited by the precipitation of corrosion products on metal surface.

Soil heterogeneity makes the corrosive attack on buried structures as a complex phenomenon in which a large number of variables are involved. The diffusion of oxygen into the metal depends on the structure and composition of the soil, these properties, in turn; define other important factors for the corrosion process, as the resistivity, the bacterial population, presence of sulfate or chloride salts, pH value and moisture content, among others. All these factors determine the corrosiveness of the soil and it is now possible to classify the different types of soil, according to their corrosiveness, evaluating these factors [41]. For steel, ductile or gray iron pipes durability depends more on the corrosivity of the soil that the nature of the metal [42]. This makes important to assess in advance the risk of corrosion of a pipeline by evaluation of that soil variables and use the recommended methods for their control [43].

Because is not always possible to measure all the variables that define the intensity of corrosive attack, for practical purposes it is useful to determine which of these variables have the major impact on determining the corrosivity of the soil. Thus, assess the risks of corrosion by measure a minimum number of variables, with adequate reliability. In the present case of study two groups of variables were taken account in 15 observation units: a first group relate with the characteristic of the soil, been considered the resistivity, redox potential, pH and temperature. The second types of variables were related with the inherent condition of the pipe, been considered the structure-soil electrochemical potential and the thickness of the pipe-wall. The multivariate analysis allows a factorial classification and leads to the transformations of a set of correlated variables into a smaller set of not correlated variables. Making the corresponding correlations or "weights" of each original variable with the respective adjacent factors, it is possible to reduce the number of initial variables to two new factors.

These new factors satisfactorily explain the corrosivity of the soil. The first new factor collects the information of the soil and is composted mainly by the resistivity, redox potential and pH values. The second new factor is related to the material activity and is explained only by the structure-soil electrochemical potential [44].

Once the multivariate analysis is carryout, is useful to make a cluster analysis of the observation units using the new factors as input variables and the Euclidian distance respect to the centroids (sample means). This methodology allows to classifier the observation units into groups with similar corrosivity. One advantage of this methodology is that depending on the euclidean distance already taken, other subgroups could be observable, in which the observation units have even greater similarity. Applying analysis of variance (ANOVA) to the observation units for the new factors 1 and 2, it is possibly to verify if there are significant differences between the two groups of the observation units.

The result for the current case shows a P-value of 0.0003 for the factor 1, indicating significant difference between the two groups; while the analysis of variance according to the factor 2 shows a P-value of 0.3754 indicating no significant difference between the groups. This means that the population of the observation units must be studied by analyzing of the two groups or subpopulations that are different between them and the factor 1 (which relate to soil parameters: resistivity, redox potential and pH values) is the best factor able to explain the variability of the gressivity between one soil and another, and thus is the best factor able to classify the different types of soil according to their corrosivity [44].

There are various methodologies to explore soil aggressiveness; a rating system developed by Peabody [45], which determines the degree of corrosivity based only in the measurement of resistivity of the soil under study, a classification system based in the AWWA C-105 standard and the methodology described in the DIN 50-929 part 3 standard.

Comparison of soil aggressiveness evaluation indicates close agreement between results obtained applying Peabody rating system [45] and that depicted by AWWA C-105 standard [46]. The evaluation indicated in the DIN 50-929 standard is more complete than the other two methodologies mentioned here, as it considers up to twelve different parameters (both physical and chemical) of the soil sample.

When direct inspection of the pipeline is possible, other methodologies have been proposed to assess the condition of a pipeline network [43, 46]. In those cases some of the variables considered are: Pipe–soil potential, loss in pipe wall thickness, loss of thickness due to pitting, presence of internal coating and number of failures occurred in a particular site.

When results of evaluation of soil aggressiveness are compared with those of direct assessment of pipe condition it has been reported that the former appears to predict better pipe condition that the actual state of the pipe [46]. This fact can be explained as corrosion is an accumulative process and this is more precisely assessed by direct inspection of the pipe, whereas the aggressiveness of the soil is a onetime measurement.

4. Evaluation of Internal Corrosion of Metals Used in Drinking Water Distribution Networks

4.1. Corrosion of Metals Used in Water Distribution Networks

Between the widely varied of metals used in drinking water distribution systems and building networks, iron, carbon steel, copper and stainless steel are the most employed. Nevertheless, these materials could undergo some kind deterioration by interaction with water; the magnitude of that deterioration depends on the hydraulic operation conditions, nature of the metal and the physicochemical characteristics of the water.

Pipe corrosion promotes metal dissolution, loss of water quality (red or blue water, bad flavor and smell problems, formation and growth of pathogenic microorganisms), reduction of hydraulic load capacity, and loss of water, among others disable operation incidents [47,48].

Drinking waters have close-neutral pH and low values of some physicochemical parameters such as: conductivity, concentration of anions (mainly chloride, sulfate and nitrate), concentration of total solids, alkalinity and total hardness; these characteristics may water to become a corrosive media.

There are different methods to correlate water physicochemical parameters with their corrosivity. These methods provide an indication of the tendency of water to dissolve or precipitate calcium carbonate ($CaCO_3$). Because a film of calcium carbonate provide corrosion protection to the metal pipe by providing a diffusion barrier that decreases the rate of dissolved oxygen reduction, the ability of the water to carbonate film formation are normally related to water corrosivity.

The tendency to formation of $CaCo_3$ is called as "corrosivity index" and is used to determine whether the water is aggressive towards the metal pipe. Some of the most corrosivity index used to evaluate corrosion in natural water are Langelier Index (LSI), Ryznar Index (RSI), Puckorius Index (PSI), Saturation Index (SI) [49], these indices are briefly explained below.

Langelier Indexe (LSI): Langelier Index showed the rate of calcium carbonate saturation for predicting the tendency of water to form deposits or to be incrusted, in which a positive value of the index indicates that the water is supersaturated with respect to $CaCO_3$ and incrustation tendency, a negative result indicates the ability to dissolve $CaCO_3$ and possible corrosive tendency, and a score of zero indicates a water in chemical equilibrium, with no tendency to form deposits or dissolve $CaCO_3$. The follow equations are used to calculate LSI index from physicochemical parameters of water:

$$LSI = pH_A - pH_S,$$

where, pH_A is the actual pH value, pH_S is the pH value of saturation or pH value at which equilibrium $H_2O/CaCO_3$ is achieved.

$$pH_S = (9.3 + A + B) - (C + D)$$

where,

$A = (\text{Log}\ [TDS] - 1)/10$

$B = -13.12 \times \text{Log}\ (°C + 273) + 34.55$

$C = \text{Log}\ [Ca^{+2}\ \text{as}\ CaCO_3]$

$D = \text{Log}\ [\text{Alcalinidad as}\ CaCO_3]$

TDS = Total dissolved Solids

Ryznar Index (RSI): the Ryznar Index allows an approximate measurement of the amount of deposits that can be expected, as is showed in Table 7. It is calculated as follows:

$$RSI = 2(pH_S) - pH_A$$

Table 7. Values and significance of Ryznar Index

RSI Value	Meaning
4.0 – 5.0	Strongly encrusted
5.0 – 6.0	Slightly incrusted
6.0 – 7.0	Slightly corrosive
7.0 – 7.5	Significantly corrosive
7.5 – 9.0	Strongly corrosive
> 9.0	Higher corrosive

Puckorius Index (PSI): this index determines the ability of water flowing through the network system to precipitate or dissolve calcium deposits. The index predicts a trend of the pipe to be incrusted when the value is less than 4.5, a tendency to corrosion if it is greater than 6.5, the optimum conditions without corrosion is achieved when the PSI is in the range between 4.5 and 6.5 values. PSI is calculated as follows:

$$PSI = 2\ (pH_{EQ}) - pH_S$$

where, $pH_{EQ} = 1.465 \times \text{Log [alkalinity]} + 4.54$

Saturation Index (SI): it is one of the most commonly used to determine the tendency of water to precipitate or dissolve $CaCO_3$, predicts the supersaturation of water with respect to calcite when the value is positive, or a water undersaturated with respect to calcite if the value is negative. This index is calculated as [50]:

$$SI = pH - pH_S$$

where, $pH_S = (pK_2 - pK_s) + pC_a + pAlk$

$pK_2 = -\log(H^+)(CO_3^{2-})/(HCO_3^-)$

$pK_s = -\log(Ca^{2+})(CO_3^{2-})$

$pC_a = -\log(Ca^{2+})$

$pAlk = -\log(\text{total alkalinity})$

The total alkalinity is the concentration (mol/l) of acid, need to neutralize dissolved alkaline ions by titration according to:

$$\text{Total alkalinity} = (H^+)_{tit} = 2(CO_3^{2-}) + (HCO_3^-) + (OH^-)$$

4.2. Pilot Systems to Assess Internal Corrosion in Drinking Water Network

There are different device used to assess water corrosivity in laboratory studies. Robbins device is used to analyze the growth of microorganisms [51,52] and monitor the biofilm development of bacteria which induce corrosion in pipelines [53]. The Pederson device employed with slight modifications to study biofilm formation in different metallic and nonmetallic materials [54, 55]. The annular reactor system, which is composed of two cylinders, the external one is stationary and the internal one is rotational, is used to varied analyses. By changes in the rotational speed of the inner cylinder it is possible to modifier the shear stress on the external cylinder wall, corresponding to the condition of flow to be simulated. This system is rather used in order to investigate biofilm growth, in which plates or coupons of the study material are located on the external surface of the inner cylinder [56] or in the internal surface of the external cylinder [57]. Furthermore, it is useful to analyze the changes in the physicochemical water parameters by taking water samples at the output of the reactor [58]. However, the ratio of surface area / volume in the annular reactor system is higher than in real pipes, making that the retention time in these systems corresponds to longer retention times in a real network system [59]. Taking account the limitations of the annular reactor, nowadays it is more common the use of pilot distribution system, because it allows obtain more realistic conditions of a network operation, being the hydraulic and surface area / volume ratio more close to the real operation of the network [60]. Pilot distribution systems have several features that allow its use in current studies and are the reactors more commonly used for corrosion studies. The pilot have been used to simulate water distribution systems, either alone or in conjunction with test coupons or coupled with pilot plant. Among their advantages over other reactors there are: the variables affecting corrosion can be systematically monitored and evaluated; provide data about the corrosion rates of the material, effects of fluctuations or changes in water quality or alternative treatment. Pilot systems may also be used to microbial growth studies, biofilm analysis and assess water quality problems [61, 62].

As practice example we present the results of a research conducted to assess the corrosivity of the drinking water of the Medellín City (Colombia, South America). In this study a pilot system of four pipe loops (Figure 6.) was constructed with the purpose to determine the evolution of water quality, biofilm growth, formation of deposits on the inner wall surface of the pipe and corrosion rates. In this work only the corrosion results are present. These loops allowed obtaining similar operational conditions to the water distribution system of the city. Three metal pipes were evaluated: ASTM 500 carbon steel, pure copper and galvanized steel (50 μm of coating thickness). An additional pipe loop of PVC was used as reference. Each loop consisted of a test pipe with 3.0 m length and 0.0254 m of diameter. These dimensions were the same for all systems. The pipes used are type M copper tube and 1.0 in schedule 40 pipe of carbon steel and galvanized steel, which was connected at ends with a PVC pipe to form a closed pipe loop. Each loop system operated with flow-through in order to provide real contact between inner surface material and water, and with recirculation flow to provide a much longer retention time to evaluate the changes in the water parameters. The adequacy of the system to operate in any configuration is accomplished through a series of valves. Each system had a pump, flow meter, sampling valve and a set of valves to adjust the operation of the system. The impeller of pump is made of brass and the pump is made of cast iron. The exposed area of both metals is 0.005 and 0.0116 m^2 respectively. The metallic gate valve that is located at the pump discharge is made of bronze and its inner area is 0.0013 m^2. These surfaces represent less of 4% of the total exposed area and its influence is negligible. The feeding water has low hardness and alkalinity. The physicochemical parameters of the feed water used in this study are presented in Table 8. The physical characteristics of the pipe loops are shown in Table 9. The evaluation time was six months.

Figure 6. Pilot system of four pipe loops (carbon steel, galvanized steel, copper and PVC) to evaluation of water quality changes, biofilm growth, formation of deposits and corrosion rates.

Table 8. Physical-chemical parameters of the feeding water during the test

Parameters	Average value	Parameters	Average value
Temperature	19.9 ± 0.8 °C	Chlorite ion	5.87 ± 0.63 mg/L
pH	7.06 ± 0.21	Sulfate ion	7.05 ± 0.54 mg/L
Conductivity	66.0 ± 4.3 μS/cm	Total Hardness, CaCO3	30.1 ± 11.9 mg/L
Dissolved oxygen	6.0 ± 0.5 mg/L	Total solids	61 ± 10 mg/L
Residual Chloride	0.87 ± 0.15 mg/L	Total dissolved solids	52 ± 13 mg/L
Turbidity	0.86 ± 0.50 NTU	Heterotrophic bacteria	5 CFU/100mL
Alkalinity, as CaCO3	15.0 ± 1.6 mg/L	Total Coliforms	0 CFU/100mL

Table 9. Physical characteristics of the pipe loops used in the pilot system

Parameters	Copper	Carbon Steel	Galvanized steel
Pipe length evaluated (m)	3.0	3.0	3.0
Total volume of water (m^3)	0.0076	0.0075	0.0075
Surface area of metal pipe (m^2)	0.23	0.23	0.24

Coupons of the pipe materials were placed into each loop to evaluate corrosion rate. Metallic plates of carbon steel (ASTM-A-36), copper (>99% purity), galvanized steel (80 μm coating thickness) are used in the respective systems of metal pipe, and stainless steel (AISI 304) coupons for the reference system (PVC loop). Coupons were prepared and installed following the ASTM standard D-2688 [63] to evaluate the corrosion rate by mass loss method. Corrosion rate was determined at 1, 3 and 6 months of exposure. Mass loss was determined in triplicate and the median value was used to calculate the final corrosion rate.

Table 10. Estimation of the corrosivity index of the drinking water used in the test

Index	Month						Interpretation of the results
	1	2	3	4	5	6	
Langelier Index (LSI)	-2.14	-2.09	-1.53	-1.79	-1.87	-1.88	< 0, the water will dissolve CaCO3.
Ryznar Index(RSI)	11.2	11.0	10.1	10.8	10.9	10.9	> 9.0, extremely corrosive
Puckorius Index (PSI)	11.9	11.6	11.0	11.7	11.9	11.9	> 6.5 corrosive tendency
Saturation Index (SI)	-2.28	-2.23	-1.67	-1.93	-2.00	-2.02	< 0, corrosive tendency

With the purpose of predicting the aggressiveness of the drinking water, the corrosivity indexes were determined taking into account the physicochemical of water parameters, Table 4 shows the Langelier, Ryznar, Puckorius and saturation index, calculated for supply drinking water prior the loops entrance. According to the values of the indexes the drinking water exhibits a corrosive tendency and it does not has a predisposition to form fouling or deposits, because the water is soft and has low alkalinity.

Corrosion rates calculated at 1, 3 and 6 months of exposition are shown in Table 11. Corrosion rate values for the first 30 days of exposition are higher compared to values reported in the literature for other kind of waters, as for example fresh waters [64]. High corrosion rate could result as consequence of low hardness and low alkalinity of the water [50, 65] as well as the amount of dissolved oxygen in water. Normally, metals immersed in potable water present higher corrosion rate because the high concentration of oxygen and low alkalinity which inhibit the formation of protective layers (carbonates). This is more significant for carbon steel.

As can be seen in Table 11, corrosion rates of carbon steel are practically constant during the six month of exposition, this means that not protective corrosion layer of carbonate are formed on the carbon steel surface. For the other metal tested a decreasing of the corrosion rates with the exposure time are observed. Galvanized steel, copper and stainless steel shows a decreasing in their corrosion rates with time, possibly due to the formation of protective corrosion products.

Table 11. Corrosion rates at 1, 3 and 6 months of exposition

Metal	Corrosion Rate (μm/year)		
	Month		
	1	3	6
Carbon steel	454 ± 27	453 ± 11	451 ± 14
Copper	52 ± 4	38 ± 1	27 ± 1
Galvanized steel	138± 17	68± 3	35±4
Stainless Steel	0.32 ± 0.06	0.10 ± 0.04	0.08 ± 0.03

In galvanized steel pipe, zinc could form a protective layer consisting of zinc oxide, hydroxides and various basic salts. Copper presents a lower corrosion rate which could be caused by the formation of copper oxide product, this compound is stable and little soluble in water [66]. Stainless steel displays a good corrosion resistance; the high concentration of dissolved oxygen could have stabilized the protective passive layer on the steel and improved its corrosion resistance. Additionally, there are few chloride ions (5.87 mg/L) in the tested water, which reduce the possibility of localized attack.

4.3. Pilot System Study at Ayura Treatment Plant

A pilot system was designed, built and operated at the exit of AY treatment plant. One of the activities in the pilot was to obtain samples of material deposited and characterize such deposits. The pilot installed included three pipe sections (pipe loop reactors) to simulate the effect of different materials in the drinking water distribution system under the same hydraulic conditions and with the same influent quality (Figure 7). These materials were: carbon steel (CS), ductile iron coated with mortar (IM) and PVC. Each pipe section had removable nipples for periodic evaluation of the inner walls.

Hydraulic conditions established for the pilot operation are shown in Table 12. After 130 days, the flow was suddenly increased to see if there was detachment of the biofilm or the materials deposited in the pipe walls and their possible influence on water quality.

Water was continuously conducted by the pilot system and the flow was interrupted only during samplings, which were performed at 38, 65, 86 and 114 days. At each sampling, one nipple of each section was removed and replaced with a sanitized nipple. In each removed nipple, the inner walls were scraped in the laboratory (Figure 8), and the material obtained was analyzed by XRD, FTIR and ESEM.

Figure 7. General view of pilot system.

Table 12. Hydraulic conditions established for the pilot operation

Parameter	Values
Velocity (m/s)	0.4-0.65.
Flow (m^3/h)	Section 1 (CS): 11.53-15.8. Section 2 (DI): 10.2-14.03 Section 3 (PVC): 16.5-21.2
Shear stress, (τ, N/m^2)	2-3

Figure 8. Sampling at nipples for analysis of deposits.

In all the samplings, the amount of deposited solids was higher in carbon steel than in the other two materials. In PVC, the amount of solids was lower because the interaction of this material with water is very low. On the other hand, the adherence of films formed on the wall or sedimented material was lower in PVC, due to its relatively low roughness. It is noteworthy that the three sections operated at shear stress of the same order. In addition, the amount of material deposited is lower in IM than in CS, because the mortar with which the pipe is coated, protects it from corrosive attack on the ferrous material. Anyway, mortar is not free from attack by aggressive waters, which can solubilize the silico-aluminates that are part of these materials.

Table 13 show the compounds found by XRD and FTIR, in the three materials studied. In CS section, in all the samplings, common constituents in the steel corrosion in aqueous medium were found: goethite (α-FeOOH), lepidocrocite (γ-FeOOH) and magnetite (Fe3O4). Since the first sampling, goethite was the major constituent found.

Table 13. Results of the FTIR and XRD analyses of the deposits found in the pilot

	CS	IM	PVC
XRD	Fe_3O_4 α-FeOOH γ-FeOOH	$Ca_3Al_2(SiO_4)_2(OH)_4$ $CaAl_2Si_2O_8.4H_2O$ $Fe_3Si_2O_5$ $CaCO_3$ $NaO.3Al_4.Si_6O_{15}(OH)_6.4H_2O$ SiO_2 Al_2SiO_5	$Ca_3(PO_4).xH_2O$ SiO_2 MnO_2
FTIR	Fe_3O_4 α-FeOOH C=O	$CaCO_3$ $NaO.3Al_4.Si_6O_{15}(OH)_6.4H_2O$, $CaAl_2Si_2O_8.4H_2O$, $Fe_3Si_2O_5$	$Ca_3(PO_4).xH_2O$ SiO_2 C=O

In FTIR analysis, significant amounts of organic components of the carboxyl group were found, associated with the presence of organic constituents. In steel pipelines, the ESEM analysis shows cotton-ball type structures, no larger than 5 μm, characteristic of goethite (similar than those shown in Figure 5), as well as flowery and plate crystals, characteristic of lepidocrocite (Figure 9) [32, 33, 86].

(a) (b)

Figure 9. Iron corrosion products found in steel section of pilot system: (a) tubercles (b) ESEM image of lepidocrocite.

In the IM section, the same compounds were found in all the samplings. Most of them are silicates and compounds that can be originated in the coating. There was also kaolinite, which may have been transported from the influent.

In the PVC section, hydrated calcium phospate is the major constituent. Also, manganese compounds were found, some with a degree of complexity, as hydroxyphosphate iron, manganese and calcium (jahnsite).

Finally, under the pilot hydraulic conditions evaluated, there are no effect of pipe materials in water quality. Water retains its properties despite the presence of biofilm. However, drastic changes in the hydraulic conditions can remove deposits and biofilm, generating poor water quality events.

5. Electrochemical Techniques for Evaluation of Metal Corrosion in Drinking Water

Metallic corrosion in drinking water is an electrochemical process, in which pipe material is the cathode and anode electrodes (different areas on the same material) and water is the electrolyte media. The dissolved oxygen reduction is commonly assumed as the cathodic reaction [67, 68] and metal dissolution is the anodic reaction. Because the electrochemical nature of the corrosion processes, the electrochemical techniques are powerful tools for evaluation of both aggressiveness of the water and the corrosion resistance of different metallic materials used to construction drinking water networks [69, 70, 71]. The main electrochemical techniques employed to assess corrosion and metallic dissolution in drinking water are potentiostatic and linear sweep polarization and electrochemical impedance spectroscopy (EIS). Polarization gives a general idea of the kinetic of both anodic metallic dissolution and dissolved oxygen cathodic reduction. Mass transport influence could be also assess by polarization techniques; however due to low conductivity of the drinking water some care must be taken account and correction of the ohmic droop is necessary in order to obtain real current—voltage curves of the electrochemical system [67, 68]. Another problem that researcher must be confront is the distortions of the polarization curves due to the film of the corrosion products formed on the metal surface. Corrosion products can be formed quickly, depending mainly on the nature of the metallic material, and the film formed can control the overall kinetic of the metallic corrosion [68– 72]. Because diffusion processes relate with the dissolved oxygen and metallic ions normally happen, hydrodynamic electrodes are recommended to use. Electrochemical impedance spectroscopy (EIS) is an useful technique which could help to resolve the problems already mentioned of the low conductivity and formation of the corrosion products, because EIS measurements give explicitly information of the electrolyte resistance [67, 68] and the existence of the corrosion products film [68, 69, 72].

5.1. Carbon Steel and Cast Iron Pipes

Carbon steel and cast iron materials show similar behavior when are in contact with drinking water. Polarization curves show that the kinetic of the corrosion of the mild steel is a time dependant process; this indicates the presence of a surface phenomena. At early states of immersion the kinetic is controlled by activation energy of the cathodic process (oxygen reduction). For longer exposition time the kinetic control is due to diffusion of oxygen and to the formation of the no protective porous layer of the corrosion products [68, 72]. The film formed onto iron surface is constituted by a bilayer of oxides and oxi-hydroxides of iron. In drinking water with low content of carbonates the outer layer of corrosion products is constitute mainly by some conduct compounds like magnetite (Fe_3O_4) and is more compact than the inner layer. The inner layer is constituted by no-conductive compounds like lepidocrocite (γ-FeOOH) and goethite (α-FeOOH). The electric resistance of the outer layer increases with the immersion time and it is not influenced by hydrodynamic condition of the system; on the other hand, the electric resistance value of the inner film decreases slightly with immersion time, which indicates that the film is not protector and iron pipe surface is always active. Because of this, the corrosion rates of carbon steel exposed in drinking water by 6 month as showed in Table 11 are high and constant values. The instantaneous corrosion rates of carbon steel estimated by potentiostatic polarization in drinking water of the Medellín city was 150 μm/year, this value is in the same magnitude order of the corrosion rate calculated by mass loss method as showed in Table 11. The formation of an arrangement of three different rust layers has also to be reported [73], but this arrangement depends on the presence or absence of free-chlorine in the water.

During the potentiodynamic polarization of carbon steel immersed in drinking water the anodic current density increases continuously with the increasing of anodic potentials; that occurs because the oxide layer formed onto steel does not protect the metal against corrosion [74]. Similarly, the OCP strongly increases with the increasing of hydrodynamic condition, possibly indicating a greater interaction between a non-protective oxide layer and the boosted supply of dissolved oxygen with higher levels of forced convection. This may explain the time delay to reach steady state conditions and low adherence of the corrosion products layer formed onto metal surface. The chemical composition of the iron oxide layer could be modified by continuous supply of dissolved oxygen, given by high hydrodynamic condition, and chemically unstable oxide layer could be formed upon the iron surface.

Electrochemical impedance of carbon steel immersed in drinking and natural waters displays a high frequency tail which is usually related to a dielectric behavior of a surface film of insoluble corrosion products acting as a discontinuous dielectric layer [72]. At lower frequencies, a flattened loop appears, possibly formed by two constant times related to diffusion and faradic process. According to the evidence showed by EIS measurements (Figure 10), the overall kinetic corrosion of cast iron and carbon steel immersed in drinking water could be explain by the redox process happening in a porous electrode, elucidated by Levie theory [75]. Impedance modeling made by equivalent circuit develop by Frateur [73] allowed the assessment of experimental noise and the removal of non-stationary data, making possible to extract the anodic charge transfer resistance from the fitting procedure. This method provides a reliable value of the corrosion current and rate from simple linear polarization technique.

5.2. Copper Pipes

Copper pipe is generally considered as corrosion resistant because it develops a protective oxide layer when exposed to neutral or slightly alkaline drinking water. Copper corrosion in drinking water, like carbon steel, is a time dependant process, this mean that corrosion rate of copper immersed in that electrolyte depend on the oxide evolution onto copper surface. Copper immersed in drinking water reach quickly a stable open circuit potential (OCP) and the evolution of OCP goes to positive values, indicating a anodic control of the corrosion process by the formation of a film of corrosion products [69]. The film of corrosion products is constituted by cuprous, cupric oxides, and cupric hydroxide during initial period, but with increasing immersion time, the copper surface is covered by stable cuprous oxide [69, 76, 77]. A more detailed description of the copper behavior was obtained by electrochemical impedance. EIS measurement carried out at OPC showed that the corrosion process of copper in drinking water is related with three time constants: formation and growth of copper oxide film, diffusion of cuprous ion through the oxide film and the charge transfer of the metal dissolution [76, 77]. At large time of exposition, the overall corrosion process is controlled by the diffusion of cuprous ions [77] and the polarization resistance is larger. Hydrodynamic condition of the pipe also affects copper behavior in drinking water; it was observed that high flow velocity reduces the thickness of the protective cuprous oxide film and as consequence, the polarization resistance decreases and the instantaneous corrosion rates increases [69, 78]. The instantaneous corrosion rates of copper estimated by potentiodynamic polarization in drinking water of the Medellín city was 20 μm/year, this value is in agreement with corrosion rates estimated by mass loss method of coupons of copper exposed in pipe loop systems during 6 months, as showed in Table 11.

Potentiodynamic polarization curves of copper made in drinking water showed strong increase of current density in low anodic polarization, indicating that copper is very sensitive to small anodic over-potential from the PCA. Small anodic over-potential in a copper pipeline of a drinking water network can occur due to improper grounding of external electrical service devices, galvanic corrosion when copper is in electrical and ionic contact with a dissimilar metal, population of bacteria in non conducting layers on the metal surface that could increase electrochemical potential for differential oxygen cells and some other events. Copper pipe sitting in an underground service line would tend to take on the temperature of the soil, whereas the other end of the same tube within an exterior-interior wall of a home might be 20 °C colder or warmer. Likewise, adjacent-connected hot and cold water lines naturally lead to sustained temperature gradients. Thermogalvanic corrosion resulting from temperature gradients may cause the early failure of pipes exposed to a sustained temperature gradient. Experiments of thermogalvanic current carried out in copper pipes exposed to the stable temperature gradient (warm on top, cold on bottom) showed that anodic current could reach to values between 14 μA and 8 μA. Positive currents indicated that the cold section of pipe is anodic to the warm section of pipe [79].

According to that investigation if the amount of extra copper corroded at 10 μA were completely released to the water during stagnation, an additional 4 mg/l of copper would be released to the water each day. Those quantities of copper are not negligible for environmental agencies. Additionally, if the metal dissolution occurs as localized form, the copper pipe will be completely drilled in short time. Electrochemical impedance of copper immersed in drinking water made at low anodic over-potentials exhibit at least four time-constants (Figure 11), which can be associated to the double layer capacitance, the relaxation of two intermediate species and a diffusion process. The charge transfer resistance decreases and an inductive arc are evident with increasing anodic potential, which is related to a much less protective oxide layer and the release of adsorbed species [69].

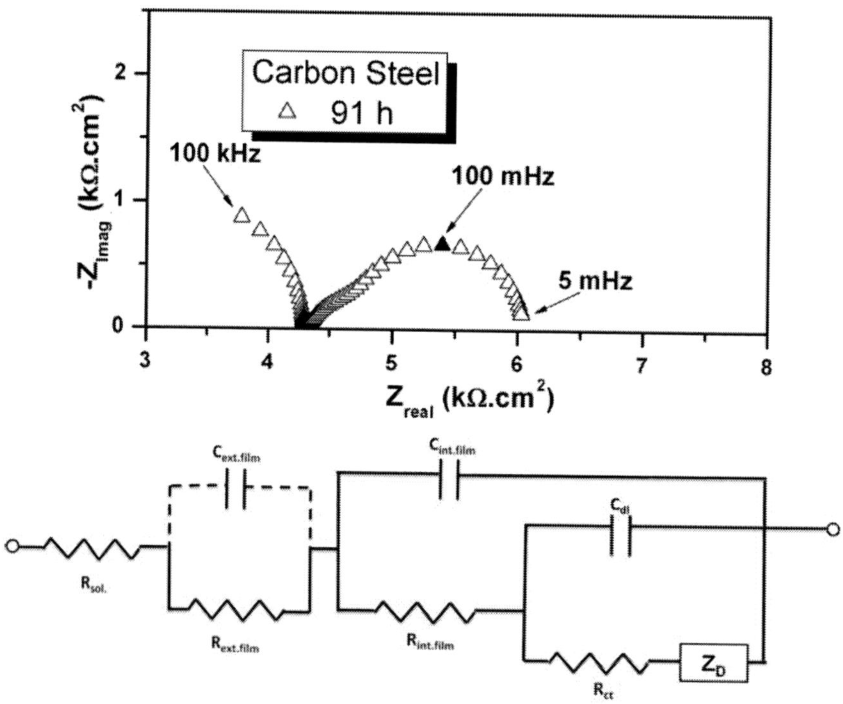

Figure 10. Typical electrochemical impedance diagram of carbon steel immersed in drinking water and electric circuit commonly used to fit impedance data.

5.3. Stainless Steel Pipes

Stainless steel is commonly used in drinking water pipeline systems because of its resistance to corrosion. Stainless steel can develop a thin, uniform and stable chromium oxide layer that protects the metal against corrosion.

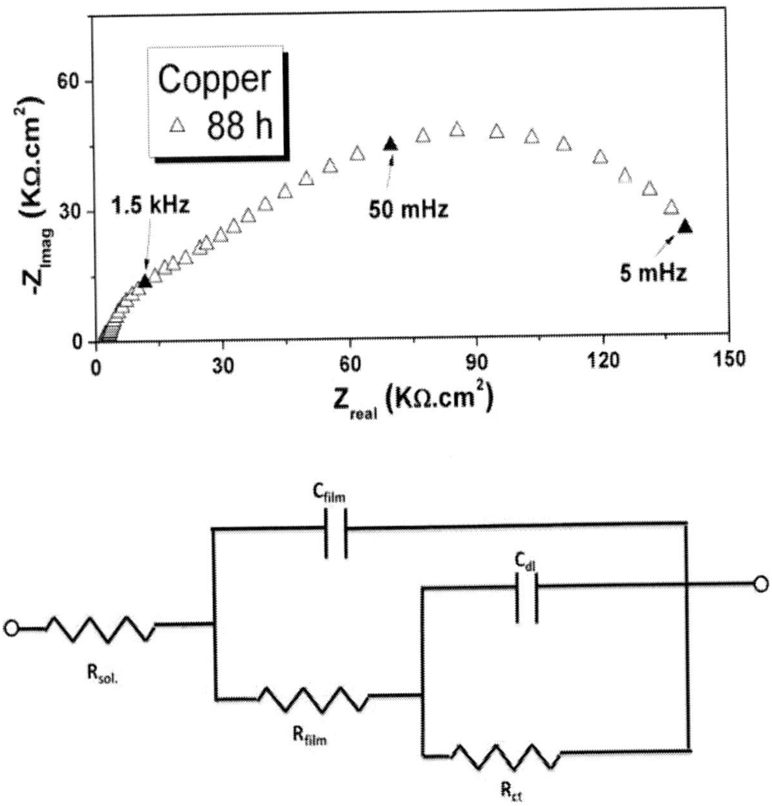

Figure 11. Typical electrochemical impedance diagram of copper immersed in drinking water and electric circuit commonly used to fit impedance data.

The corrosion behavior of stainless steel depends mainly on the composition of stainless steel. The classification of stainless steel alloys when used in drinking water pipeline systems normally is made by the evaluation of the corrosion behavior of metal and by the assess of amount of metal species leaching into the drinking water. Leaching tests are frequently used in qualification of materials used for water distribution systems and allow the evaluation of the amount of "heavy-metal ions" migrating into the water in contact with the metal under test [80]. For metallic materials used in drinking water systems, the European standard test method is to expose the test samples to a real drinking water. Then, the stagnation profile that reflects the usage of a domestic installation is followed for at least 26 weeks and the amount of leached corrosion products in the test water is determined [81].

Stainless steel immersed in drinking water reach a stable open circuit potential (OCP) after 5 h of immersion and the evolution of OCP goes to positive values. The change of the OCP from initial immersion time to 24 h of immersion is close to 150 mV.

This behavior indicates the formation of a stable oxide layer within the first hour on the metallic surface [82]. Potentiodynamic polarization curves of stainless steel in drinking water show both the cathodic and anodic branches of the polarization curves typical Tafelian behaviors, which indicated activation control of the corrosion mechanisms at low overpotentials. Consistently with the activation control, the polarization curves at low cathodic and anodic over-potential appeared to be roughly independent from hydrodynamic condition. This indicates that the surface passivation happens by formation of an oxide film in the solid state and not by reactions of formation-precipitation of corrosion products in aqueous media; case in which it would be expected to be sensitive to hydrodynamic conditions, by inducing shear forces. At high anodic over-potential stainless steel can experiment localized corrosion, the capacity to repassivation depend on the metal composition [82].

The instantaneous corrosion rates of stainless steel estimated by potentiodynamic polarization in drinking water of the Medellín city was 0.3 μm/year, this value is close to corrosion rates estimated by mass loss method of coupons exposed in pipe loop systems during 6 months, as showed in Table 11. Stainless steel displays high corrosion resistance, the high concentration of dissolved oxygen could stabilize the protective passive layer on the stainless steel and improved its corrosion resistance. Additionally, the drinking water used to the test has few chloride ions (5.87 mg/L), which reduce the possibility of localized attack on the metal.

The passive state of stainless steel immersed in drinking and pure water is commonly studied by electrochemical impedance spectroscopy at OCP [82, 83]. Nyquist plots of stainless steel electrode displays only a single open capacitive loop at OCP (Figure 12).

The impedance modulus strongly increase with immersion time and then sccm to saturate for times longer than two days, suggesting a rapid passive film strengthen. The impedance spectra recorded for different immersion time commonly is associated with the existence of two time constants according to the modified Randles-Ershler equivalent circuit [82, 83], where resistances and capacitances are either related to the bilayered character of passive layer formed on the stainless steel surface or assigned to the oxide-electrolyte interface and the redox processes occurring in the passive layer [84, 85].

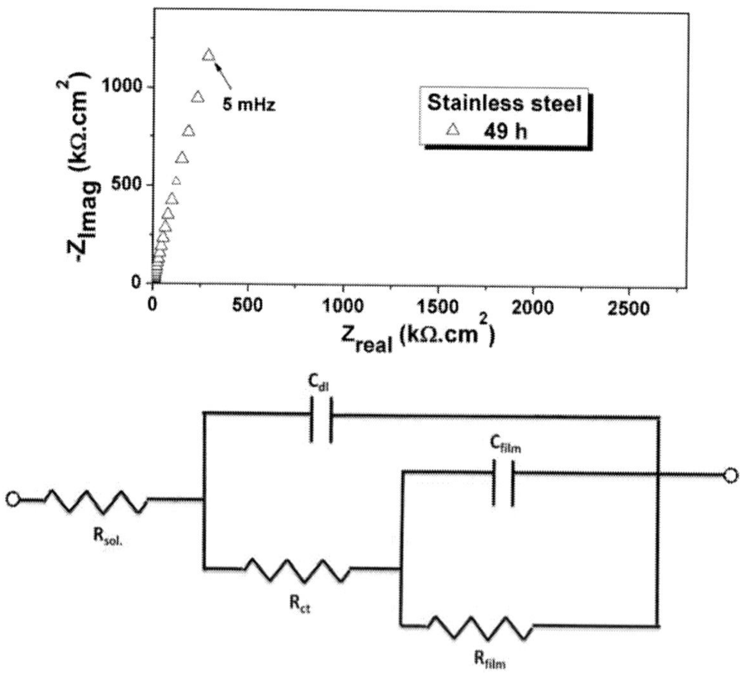

Figure 12. Typical electrochemical impedance diagram of Stainless steel immersed in drinking water and electric circuit commonly used to fit impedance data.

The capacitance of the passive layer decreases with increases of immersion time showing values between 10 and 4 μF.cm^{-2}. Whereas electric resistance of the passive layer increases, displaying values between 0.1 and 4 MΩ.cm^{2}, depending on the hydrodynamic condition.

6. Prediction of Lifespan in Underground Distribution Water Networks

In addition to economic losses, failures in drinking water networks also generate social negative impacts for both local authorities and service companies [87]; those failures are mainly due to pipeline corrosion, malfunction in valves and other accessories as well as ground displacements. Regular inspection of underground infrastructure is restricted and therefore, it is required to have tools able to predict network condition and consequently be able to prioritize maintenance well before operation and people become affected.

It is considered that water distribution systems are unique when talking on failure patterns; even more, for the same network, important differences can be exist between individual pipes and therefore the element employed for analysis must be as small as possible [88].

Direct inspections of pipelines are only useful for collecting information of its present condition, however little can be used for prediction of failures; breaks are related to factors which are of local nature and can be interrelated. Consequently, the information obtained during inspections at some particular locations does not help to forecast pipeline condition.

Important information about failure patterns is contained in burst historical records, which can be analyzed together with all the available data of both pipeline network (material, water pressure, diameter, depth, etc.) and other related factors (soil, road traffic, water table, etc.). The results of such an analysis are a valuable tool for financial planning in maintenance departments. In addition, when coupling the results to a geographic information system, critical points can be detected and a more comprehensive maintenance strategy can be defined.

In the present work, firstly it will be reviewed the factors which more influence underground network deterioration such as: material, pressure, soil characteristics, age, diameter, loads, length and depth. Then after, the main part of this work will be deal with the methodologies and models employed to assess or predict the condition of the pipeline network: Regression models, probabilistic models and finally models applying artificial neural network concepts.

6.1. Factors Influencing Pipeline Deterioration

Material: Within the conventional materials employed in distribution water networks it can be found steel (mostly galvanized), ductile iron, cast iron, polymeric (mainly PVC and polyethylene), concrete and asbestos cement. A report based in a 1990 survey in Sweden indicates that steel and ductile iron in most cases fails due to corrosion whereas for the other materials failure causes were either no related to the material or it could not be determined [89]. Concrete and asbestos cement pipeline failure mechanism has been described to occur mainly due to various chemical processes affecting the cement matrix [90]. Metallic pipes deteriorates mostly by pitting corrosion though at inside of the pipelines other processes can be present (tuberculation, crevice and erosion) [91-92]. For PVC and polymeric pipes the mechanism of deterioration are not yet properly understood as these materials present slower deterioration rates and they are only been commercially used for about 40 years [93].

In general terms, mechanism of pipeline failure involve mainly three aspects: (i) pipe properties, interaction soil-pipe and quality of installation; (ii) internal and external loads and (iii) internal and external deterioration [93].

Network age: In general it is accepted that pipeline age is one important factor to be considered when determining network vulnerability to failure [94]; however, in most cases besides pipe age, network conditions have to be also considered [87, 95]. It is clear that, as in most cases in material deterioration, every case is a case and failure mechanism cannot be explained by a single variable. In addition, it has to be considered that network failure is not always evident and it is only observed when water losses due to minor leaks are considered [94]. On the other hand, it should be considered that failures related to construction or design defects are not possible to evaluate and may induce deviations in failure time prediction [87].

Soil: As stated before metallic pipes mostly fail due to corrosion processes and in order to determine soil aggressiveness various methodologies have been formulated based in the evaluations of soil physicochemical characteristics. One of the most widely used method to asses soil corrosivity is depicted in the AWWA C105 standard; this considers resistivity, pH, redox potential, draining capacity and sulfur content. A more comprehensive test is contained in the DIN 50-929 part 3 standard; it test type of soil, resistivity, water content, pH, buffer capacity, sulfite, sulfate and chloride contents, ground water presence, soil homogeneity and soil-pipe potential. AWWA standard only allows to found whether soil is or not corrosive whereas the DIN method is able to classify soil as virtually not aggressive, weakly aggressive, aggressive and strongly aggressive. Some authors have used AWWA standard to obtain soil aggressiveness levels [96-98].

Diameter, length and depth: It has been found that in smaller diameter failure appears to occur at higher rates [87, 95]; however, failure consequences of pipes with diameter ranging from 30 to 300 mm are considered to be low, while burst of pipes with larger diameter is related to severe economic, social and environmental consequences [99]. Short length pipes are in general related to urban areas where soil is more variable and consequently failure rates are higher than in rural conditions where soil is more homogeneous trough a given length [95]. Some consider that a new burst is more likely to take place close to a failed pipe and when several burst are observed in the same area this must be a good enough reason to do rehabilitation works [100]. Investigations of the effect of network depth on failure rate indicates that up to about 5.5 m failures diminish, but from that point burst occurrence start to increase; 90% of incidents take place at depths of 4 m or less [90].

Loads: From the mechanical point of view a pipeline should resist both internal operational pressures and external loads; performance of pipes will vary mostly according to material and pipe wall thickness. Metallic and polymeric materials are considered to be flexible, whereas concrete and asbestos cement are rigid [93, 95, 97].

Pipelines are exposed to several loads: vertical and horizontal soil pressures, pipeline weight, internal pressure, static and dynamic road loads and seismic loads. Prediction of pipeline durability based on mechanical behavior have de disadvantage of being intensive in data requirements but some consider those models as the ultimate goal [93]. Analysis of mechanical performance of pipeline installed in seismic areas in USA and Japan indicates that number of failure decreases as pipe diameter increases regardless earthquake intensity and pipe material; of course, flexible materials present less failure than rigid ones [101].

Burst historic data: A rigorous documentation of burst events is essential for service companies to be able to use mathematical modeling of such data looking to design network maintenance programs [102]. Data should contain information on material, diameter and pipe wall thickness, loads affecting the pipeline, burst characteristics and geographic location, network age and soil characteristics; depth, temperature, network pressure, water table and geotechnical data could be important too [102]. In absence of historic failure data, probabilistic models may give good results to predict failure probabilities [99].

6.2. Analysis of Deterioration in Pipeline Infrastructure

The bathtub curve is widely used for describing the life cycle of a buried pipeline (see figure 13) and often used for modeling. The first stage, burn-in stage starts with a high burst frequency which rapidly decreases to become relatively constant at low frequencies during the normal operation stage. The initial high number of failures is considered due to installation errors and pipe manufacture defects.

Failures in the second stage are rare and mostly due to unsystematic events. Then after during the wear-out stage burst frequency start to increase due to deterioration and aging of the infrastructure. Increase in failure frequency at the final stage may vary according to maintenance programs.

6.2.1. Statistical Models

Statistical methods have been employed to analyze historical data in order to be able to identify burst patterns in water pipeline systems; this information is then after used to predict future network breaks. According to the classification adopted by Kleiner et al. [103], when reviewing the several models reported that explain, quantify and predict failures on water pipeline networks, these models can be either regression or probabilistic; regression models allow to calculate burst number as a function of time whereas probabilistic models use a survival function to determine the probability of a pipe to stay without any failure after a given period as a function of several variables [103]. In regression models, the population under study has to be divided into groups, which are homogeneous regarding the factors (operational, environmental and related to the pipe) that affect the breakage rate of the pipeline; careful statistical analysis should be done when grouping the data in order to obtain representative results [104]. Regression models can be subdivided into linear and exponential time dependent models; in general terms a time linear model has the following form:

$$Y = \beta_0 + \Sigma_i \beta_i x_i$$

where β_0 and β_i are unknown constants to be estimated and Y is a linear function of the explanatory variables x_i. On the other hand, a time exponential model in its more basic form can be expressed as a Y non- linear function considering a combination of unknown regression parameters β_0 to β_i, and explanatory variables x_i. In table 1 various models reported in the literature are resumed together with the parameters involved in each case.

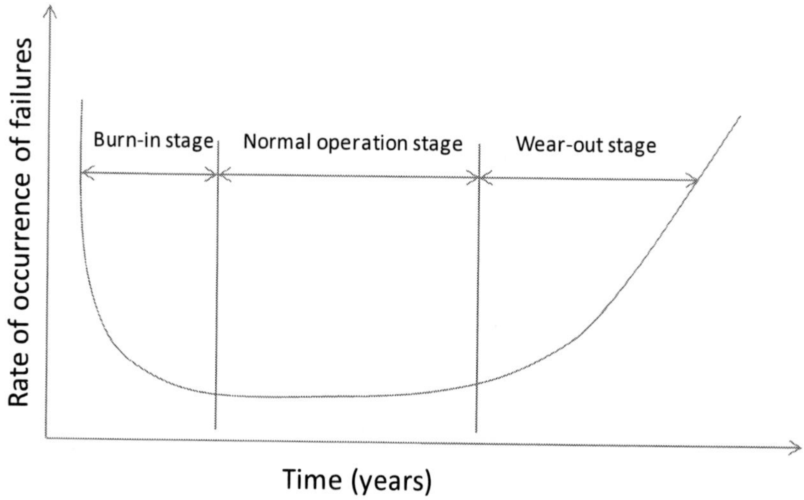

Figure 13. Bathtub curve depicting the variation of the rate of occurrence of failures in the life cycle of buried pipe.

Table 14. Regression models reported in the literature for prediction of deterioration of water pipeline networks

Function	Parameters and variables	Ref.
Linear models		
$N = k_0 A$	N = Number of breaks per year k_0 = Regression parameter	[105]
$A = 0.028 \times SR - 6.33 \times pH - 0.049 \times r_d$	A = Pipe age to the first break (years) SR = Resistivity of saturated soil (Ω.cm) pH = Soil pH r_d = Redox potential (mV)	[106]
$P = a_0 + a_1 L + a_2 A$	P = Reciprocal of the probability of a day with no breaks L = Pipe length A = Pipe age a_0, a_1, a_2 = Regression coefficients	[107]
$P = a_0 + a_1 L + a_2 A$	P = Reciprocal of the probability of a day with no breaks L = Pipe length A = Pipe age a_0, a_1, a_2 = Regression coefficients	[107]
$N = \beta_0 D + \beta_1 AC + \beta_2 CI + \beta_3 CSC + \beta_4 DI + \beta_5 PVC + \beta_6 STL + \beta_7 L + \beta_8 Y + \beta_9 P + \sum_{j=10}^{20} \beta_j LU_j + \sum_{k=21}^{25} \beta_k ST_k + \beta_{26} TEMP + \beta_{27} RAIN + \beta_{28} SMAX + \beta_{29} MX \cdot MN + \beta_{30} PC1 + \beta_{31} PC2 + \beta_{32} PC3$	N = Number of breaks per six months per pipe. D = Diameter of the pipe segment in inches. AC, CI, CSC, DI, PVC, and STL = Binary variables denoting asbestos cement, cast iron, concrete steel cage, ductile iron, polyvinyl chloride, and steel, respectively.	[108]

	L = Length of pipe in feet. Y = Year of installation of the pipe. P = Operating pressure of the pipe in pounds per square inch. LU = Land use above the pipe. ST = Type of soil around the pipe. TEMP = Average monthly temperature over each 6-month period. RAIN = Total rainfall measured in hundredths of an inch in each 6-month period. PC1, PC2, PC3 = Three principal components obtained after doing a principal components analysis (PCA) on six soil corrosivity covariates.	
Exponential models		
$N(t) = N(t_0)e^{A(t+g)}$	t = Time from present to a given past break (years) N(t) = Number of breaks per unit length per year ($km^{-1}yr^{-1}$) $N(t_0)$ = N(t) at installation year. g = Pipe age at time t. A = Breaking rate coefficient (yr^{-1})	[109]
$N(t) = C_1C_2N(t_0)e^{A(t+g)}$	Same as above, in addition: C_1 = Ratio between [break frequency for cast iron with previous breaks] and [overall break frequency for cast iron]. C_2 = Ratio between [break frequency for pit cast pipes 500 mm diameter] and [overall break frequency for pit cast pipes].	[110]
$NY = x_1 + x_2D + x_3P + x_4I + x_5RES + x_6LH + x_7T$ $REP = y_1e^{y_2t}e^{y_3T}e^{y_4PRD}e^{y_5DEV}SL^{y_6}SH^{y_7}$	x_i, y_i = Regression parameters. NY = years from installation to the first repair. D = Pipe diameter. P = Absolute pressure inside the pipe I = % of pipe overlain by industrial development. RES = % of pipe overlain by residential development. LH = Length of pipe in highly corrosive soil T = Type of material (1:metallic, 0:reinforced concrete) REP = Number of repairs PRD = Differential pressure t = pipe age from first break DEV = % length of pipe in low and moderately corrosive soils. SL = Area of the pipe surface in low corrosivity soils SH = Area of the pipe surface in high corrosivity soils.	[111]
$Y = \beta_0\exp(\beta_1(t) + \beta_2(yi))$	Y = Number of breaks per year β_0 = Coefficient of regression β_1 = Coefficient of time since last break β_2 = Coefficient of year of installation t = Time since last break yi = Year of installation	[108]

By means of probabilistic models, a survivor function is estimated for each individual pipe, which provides the probability that a pipe will survive without breaks beyond a specified time, as a function of a number of independent variables related to environmental and operational conditions and pipeline characteristics [112]. Probabilistic multivariate models can account the effect of most variables, making them more accurate when predicting breaking rates and additionally the use of these methods diminish the need of partitioning of the data; however significant expertise is required as the mathematics employed is complex [103]. In addition to these models there are models employing probabilistic processes on grouped data to predict probabilities of breakage which are named probabilistic single-variate models; a resume of both type of probabilistic models existing in the literature is gathered in table 15. Multivariate models are mostly based on the proportional hazard model proposed by Cox [113], single-variate models can be cohort survival, Bayesian diagnostic and semi-Markov chain models, within the most frequently reported; a critic discussion on existing models was published by Kleiner and Rajani [103].

Table 15. Probabilistic models reported in the literature for prediction of deterioration of water pipeline networks

Function	Parameters and variables	Ref.
Multivariate models		
$h(t,Z)=h_0(t)e^{b^T Z}$ $h_0(t)=2\times10^{-4}-10^{-5}t+2\times10^{-7}t^2$	T = Time to next break $h(t,Z)$ = Hazard function $h_0(t)$ = Baseline hazard function Z = Vector of variables b = Vector of coefficients to be estimated	[114]
Early stage: Same as in reference [28] (above). Late stage: $h=\lambda=e^{b^T Z}$	Hazard is constant for the late stage.	[115] [117]
$h(t,Z)=h_0(t)e^{b^T Z}$ $h_0(t)=\lambda\beta(\lambda t)^{\beta-1}$	t = Time to the next break $h(t)$ = Hazard function λ,β = Scale and shape parameters of the Weibull distribution (correspondingly).	[118]
$H(t)=(t/\theta)^{\beta}$ $\theta=\theta_0 e^{\alpha Z}$	t = Pipe age $H(t)$ = Number of break per unit of length at the age t θ,β = Scale and shape parameters, respectively θ_0 = Baseline value α = Vector of coefficients to be estimated by regression Z = Vector of variables affecting breakage rate	[119] [120]

Function	Parameters and variables	Ref.
$\ln(T) = \mu + x^T \beta + \sigma Z$ $\Rightarrow T = f(\mu, \sigma Z) e^{x^T \beta}$	T = Time to next break x = Vector of explicative variables Z = Random variable distributed as Weibull σ = Parameter to be estimated β = Vector of parameters to be estimated Z = Random variable distributed as Gumbel	[121] [122]
$d_m(t) = k(t - t_0)^{\alpha}$ $v_m(t) = k'(t - t_0)^{\alpha'}$ $\alpha' = \alpha - 1 < 1.0$ $k' = k\alpha$	d_m = Maximum pitting depth t_0 = Time of corrosion initiation k = Proportional constants α = Exponential factors v_m = Pitting rate	[123]
Single-variate models		
$f(t) = \frac{(a+1)be^{b(t-c)}}{[a+e^{b(t-c)}]^2}$ $S(t) = \frac{a+1}{a+e^{b(t-e)}}$ $h(t) = \frac{be^{b(t-c)}}{a+e^{b(t-c)}}$	$f(t)$ = Probability density function $h(t)$ = Hazard function $S(T)$ = Survival function t = Useful lifetime of pipe a = Aging factor (year^{-1}) b = Failure factor (year^{-1}) c = Resistance time of pipeline (years)	[124] [125] [126]
$P = \frac{P_{c/f} P_f}{P_{c/f} P_f + P_{C/nf}(1 - P_f)}$	P = Probability of failure with specific characteristics p_f = Failure probability of the system $P_{C/f}$ = Probability of founding specific characteristics in a failed segment $P_{c/nf}$ = Probability of founding the same characteristics in a no failed segment	[127]
t_1: Generalized Gamma distribution $t_i(i > 1)$: Exponential distribution	t_i = Time between *(i-1)*th and the *i*th pipe break	[128]
$P(x) = \frac{m^x e^{-m}}{x!}$ $m = m(s,t)$	m = Mean value of subsequent breaks within the cluster domain x = Number of subsequent breaks within the cluster domain s = Distance from the first break in the cluster t = Time after the first break in the cluster	[129] [130]

6.2.2. Mechanical Models

From the mechanical point of view pipe breakage involve the following factors: intrinsic mechanical characteristics of the pipeline (pipe material and structural properties, pipe-soil interaction and quality of installation), extrinsic mechanical variables (internal and external loads) and deterioration phenomena (chemical/electrochemical/biochemical interactions with the internal and external environment) [93]. The models reviewed here intent to improve the understanding of the behavior of a buried network and the relationships with the above mentioned factors. Table 16 shows some of the mechanical models reported in the literature and reviewed in detail by Rajani and Kleiner [93].

Table 16. Mechanical models reported in the literature for calculation of failure parameters in water pipeline networks

Function	Parameters	Ref.
$p_s = \sum_{i=0}^{N_T} \frac{\beta h_f^i}{\left[\left(\frac{1}{K_{tip}}\right) + \left(\frac{B_d}{k_s d_f^i}\right)\right]} H(s - d_f^i B_d)$	p_s = Frost load at any point s d_f = Frost depth i = Time step number N_T = Number of time steps h_f = Total frost heave B_d = Trench width K_{tip} = Stiffness of elastic half-space of unfrozen soil mass below freezing front β = Attenuation factor k_s = Backfill–sidewall shear stiffness H = Boussinesq function s = Location where frost load is calculated.	[131]

Function	Parameters	Ref.
$\sigma_\theta = \frac{P_i r}{t} + \frac{6k_m E_p t r}{E_p t^3 + 24 k_d P_i r^3}\left(C_d \gamma B_d^2 + \frac{I_c C_t F}{A}\right) + \sigma_f$	σ_θ = Total in-plane stress C_d, C_t = Earth and surface load coefficients k_m = Bending moment coefficient E_p = Pipe elastic modulus P_i = Internal pressure r = Mean pipe radius t = Wall thickness k_d = Deflection coefficient I_c = Vehicle impact factor F = Wheel load on surface σ_f = Stress due to pipe ring deflection B_d = Trench width γ = Soil backfill unit weight A = Effective length of pipe.	[132]
$\sigma_x = x_1 E_p \frac{du}{dx} + x_2 P_i - x_3 E_p \alpha_p \Delta T$ $\sigma_\theta = \frac{P_i D}{t}\left(\frac{D}{t}\frac{E_p}{E_s} v_p k_s K\right)$	σ_θ, σ_x = Longitudinal response of pipeline to changes in internal and external pressures and temperature changes. E_p = Pipe elastic modulus X = Axial displacement in longitudinal direction x ν_p = Pipe Poisson ratio α_p = Coefficient of pipe thermal linear expansion P_i = Internal pressure ΔT = Temperature differential E_s = Elastic soil modulus D = Pipe diameter t = Wall thickness	[133]

Function	Parameters	Ref.
	k_s = Pipe–soil reaction modulus $\chi_1,\chi_2,\chi_3,\kappa$ = Constants as function of soil and pipe properties.	
$p_0(d,l) = \frac{hs_y}{r}\left(\frac{1-A/A_0}{1-A/(MA_0)}\right)$	p_0 = Pressure in the pipe at failure (pitting corrosion) d = Maximum depth of corrosion defect A = Cross-sectional area of metal lost in the corroded region projected onto the longitudinal axis of the pipe. A_o = Original cross sectional area of the corroded region l = Maximum total length of corrosion defect r = Pipe radius	[134]
	h = Pipe wall thickness s_y = Pipe yield strength M = Folias factor	
$\sigma_n = \frac{\alpha K_q}{\beta\left[d/t\sqrt{a_n}\right]^s}$	σ_n = Nominal tensile stress β = Geometric factor dependent on the dimensions and shape of the corrosion pit a_n = Lateral dimension of corrosion pit size K_q = Provisional fracture toughness d = Pit depth α, s = Constants obtained from experimental tests t = Pipe wall thickness	[135]
$p=K_nK_a(10-pH)^n\rho^{-n}t^nA_a$	p = Average pit depth a,K_n,K_a = Empirical constants A_a = Pipe surface area exposed to corrosion pH = Soil pH ρ = Soil resistivity n = Soil aeration constant t = Time (years).	[136]
$\rho = \left(\frac{t}{P_e+P_i}\delta\right) - t$	ρ = Remaining life t = Age of water main δ = Thickness of original pipe wall P_e = External pit depth P_i = Internal pit depth	[137]
$\ln(PD) = \frac{e^{a+bIA}}{1+e^{a+bIA}}$ $IA = \sum_{j=1}^{6}\alpha_j\frac{LnX_j - \hat{\mu}_{LnX_j}}{S_{LnX_j}}$	PD = Pit depth IA = Index of aggressiveness a, b = Regression constants α_j = Factor score coefficients X_j = Variables considered (soil humidity, pH, redox potential, soil–pipe potential, soil resistivity and sulphide soil content) μ = Corresponding medium value S = Corresponding standard deviation	[138]

References

[1] McNeill, L.S.; Edwards, M. *J. AWWA,* 2001; Vol. *88*, pp 88–100.

[2] Edwards, M. *Water Sci. Technol.,* 2004; Vol. *49*, pp 1–8.

[3] Volk, Ch.; Dundore, E.; Schiermann, J.; Lechevallier, M. *Water Res.,* 2000; Vol. *34*, pp 1967-1974.

[4] Kleiner, Y.; Rajani, B. *Urban Water,* 2001; Vol. *3*, pp 131-150.

[5] Shams El Din, A. M. *Desalination,* 2009; Vol. *238*, pp 166–173.

[6] Imran, S. A.; Dietz, J. D.; Mutoti, G.; Taylor, J. S.; Randall, A. A.; Cooper, C. D. *J. AWWA,* 2005; Vol. *97*, pp 93-100.

[7] Zacheus, O. M.; Lehtola, M. J.; Korhonen, L. K.; Martikainen, P. J. *Water Res,.* 2001; Vol. *35*, pp 1757- 1765.

[8] Gauthier, V.; Portal, J. M.; Yvon, Y.; Rosin, C.; Block, J. C.; Lahoussine, V.; Benabdallah, S.; Cavard, J.; Gatel, D.; Fass, S. *Water Sci. Technol.,* 2001; Vol. *1*, pp 89-94.

[9] Lechevallier, M. W.; Babcock, T. W.; Lee, R. G. *Appl. Environ. Microbiol.* 1987; Vol. *53*, pp 2714-2724.

[10] Lehtola, M. J.; Nissinen, T. K.; Miettinen, I. T.; Martikainen, P. J.; Vartiainen, T. *Water Res.,* 2004; Vol. *38*, pp 601-610.

[11] Cheng, X.; Peterkin, E.; Burlingame, G. A., *Water Res.,* 2005; Vol. *39*, pp 3781–3790.

[12] Chandy, J. P.; Angles, M. L. *Water Res.,* 2001; Vol. *35*, pp 2677–2682.

[13] Sander, A.; Berghult, B.; Broo, A. E.; Johansson, E. L.; Hedberg, T. *Corrosion Sci.,* 1997; Vol. *39*, pp 77–93.

[14] Kuch, A. *Corrosion Sci.,* 1988; Vol. *28*, pp 221-231.

[15] Nawrocki, J.; Raczyk-Stanistawiak, U.; Swietlik, J.; Olejnik, A.; Sroka, M. *Water Res.,* 2010; Vol. *44*, pp 1863-1872.

[16] Takasaki, S.; Yamada, Y. *Corrosion Sci.,* 2007; Vol. *49*, pp 240–247.

[17] Rocklidge, S. J.; Ketchum Jr., L. H. *Water Res.,* 2002; Vol. *36*, pp 2689-2694.

[18] Kirmeyer, G.; Friedman, M.; Clement, J. *Guidance Manual for Maintaining Distribution System Water Quality*. AWWA: Denver (CO), 2000; pp 105.

[19] ASTM D-2688. *Standard Test Methods for Corrosivity of water in the Absence of Heat Transfer (Weight Loss Methods)*, American Society for Testing and Materials: West Conshohocken (PA), 2011.

[20] ASTM G-1. *Standard Practice for Preparing, Cleaning, and Evaluating Corrosion Test Specimens*, American Society for Testing and Materials: West Conshohocken (PA), 1999.

[21] Boffardi, B. P. Standards for corrosion rates [E-text type]. http://www.flozone.com/flozone/white_papers/qc/Standards%20for%20Corrosion%20 Rates.pdf. 2011.

[22] Kashinkunti, R.D.; Metz, D.H.; Hartman, D.J.; DeMarco, J. In *Proceedings of the 1999 AWWA Water Quality Technology Conference*; AWWA: Tampa (FL), 1999.

[23] McNeill, L. S. *Water Quality Factors Influencing Iron and Lead Corrosion in Drinking Water*; PhD Thesis; Virginia Polytechnic Institute and State University: Blacksburg (VA), 2000; pp 1-12.

[24] Horsley, M.B.; Northrup, B.W.; O'Brien, W.J.; Harms, L.L. *Proceedings of the 1998 AWWA Water Quality Technology Conference*; AWWA: San Diego (CA), 1998; Paper 5C-3.
[25] EMWD (2006). Cuadro de la calidad del agua 2006 [E-text type]. http://www.emwd.org/water service/pdf/ccr_2006-table-esp.pdf.
[26] Gauthier, V.; Geârard, B.; Portal, J. M.; Block J. C.; Gatel D. *Water Res.* 1999; Vol. *33*, pp 1014-1026.
[27] Rao, T.S.; Sairam, T.N.; Viswanathan B.; Nair K.V.K. *Corrosion Sci.,* 2000, Vol. *42*, pp 1417-1431.
[28] Sly, L.I.; Hodgkinson, M.C.; Arunpairojana V. A. *Appl. Environ. Microbiol.,* 1990; Vol. 56, pp 628-639.
[29] Tang, Z.; Hong, S.; Xiao, W.; Taylor, J. *Corrosion Sci.,* 2006; Vol. *48*, pp 322-342.
[30] Echeverria, F.; Castaño, J. G.; Arroyave, C.; Peñuela, G.; Ramirez, A.; Morato, J. *Ingeniare. Rev. Chil. Ing.,* 2009; Vol. *17*, pp 275-281.
[31] Lin, J.; Coller, B.A.W. *Water Res.,* 1998; Vol. *32*, pp 1019-1026.
[32] Castaño, J. G.; Botero, C. A.; Restrepo, A. H.; Agudelo, E. A.; Correa, E.; Echeverría, F. *Corrosion Sci.,* 2010; Vol. *52*, pp 216–223.
[33] Antunes, R. A.; Costa, I.; de Faria, D. *Matéria,* 2003; Vol. *8*, pp 27-34.
[34] Cornell R.M.; Schwertmann, U. *The iron oxides*; VCH: Weinheim, Germany, 1996; pp 573.
[35] Rossman, L. A. *Water Res.,* 2006; Vol. *40*, pp 2493-2502.
[36] Sarin, P.; Snoeyink, V.L.; Bebee, J.; Jim, K. K.; Beckett, M. A.; Kriven, W. M.; Clement, J. A. *Water Res.,* 2004; Vol. *38*, pp 1259-1269.
[37] Parshutin, V. V.; Revenko, V. G.; Sholtoyan, N. S.; Chernova, G. P.; Bogdashkina, N. L. *Prot. Met. Phys. Chem. Surf.,* 2010; Vol. *46*, pp 489–495.
[38] Lin, J.; Ellaway, M.; Adrien, R. *Corrosion Sci.,* 2001; Vol. *43*, pp 2065-2081.
[39] Zhang, Y.; Shaw, H.; Farquhar, R.; Dawe, R. *J. Petrol. Sci. Eng.,* 2001; Vol. *29*, pp 85-95.
[40] Andritsos, N.; Karabelas, A.J. *Int. J. Heat Mass Transfer*., 2003; Vol. *46*, pp 4613-4627.
[41] ANSI/AWWA C105, *American Water Works Association*, Denver (Co).
[42] M. Romanoff, *"Results of National Bureau of Standards Corrosion Investigation in Disturbed and Undisturbed Soils*: Technical Bulletin No. 86, 12th Annual Appalachian Underground Corrosion Short Course, 1967.
[43] Spickelmire, B., *Corrosion considerations for ductile iron pipe*. Mater. Perform., 2002; Vol. 41(7), pp 16–23.
[44] Calderón, J.A.; Delgado, J.; Peñaranda, S.; Restrepo, A. *"Statistic assessment of the factors that take part in the corrosion on buried pipe"*. Revista Facultad de Ingeniería, septiembre, 2005; Vol. 34, pp 23-34.
[45] Peabody, A.W. *NACE basic corrosion course*, Chap. 5, 2nd Edn. NACE, Houston (TX), 1971.
[46] Restrepo, A.; Delgado, J.; Echeverria, F. *Evaluation of Current Condition and Lifespan of Drinking Water Pipelines*, J. Fail. Anal. and Preven., 2009; Vol. 9, pp 541–548
[47] Rushing, J.C.; Edwards, M. *Corros. Sci*., 2004; Vol. 46, 8, pp 1883.
[48] Sarin, P.; Snoeyink, V.L.; Bebee, J.; Jim, K.K.; Beckett, M.A.; Kriven, W.M. Clement, J.A. *Water Res*., 2004; Vol. 38, 5, pp 1259.

[49] Imran S. A.; Dietz J. D.; Mutoti G.; Taylor J. S.; Randall A. A. *Modified Larsons Ratio Incorporating Temperature, Water Age, and Electroneutrality Effects on Red Water Release.* Journal of Environmental Engineering, 2005; Vol. 131(11), pp 1514-1520.

[50] Jones, D. A. *Principles and Prevention of Corrosion, 2nd Ed.* Prentice-Hall, New York, 1996; pp 361.

[51] T. Schwartz, S.; Hoffmann, U. *Obst. "Formation and bacterial composition of young, natural biofilms obtained from public bank-filtered drinking water systems".* Water research, 1998; Vol. 32, pp 2787- 2797.

[52] Nickel, J. C.; Ruseska, I.; Wright, J. B.; Costerton, J. W. *"Tobramycin Resistance of Pseudomonas aeruginosa Cells Growing as a Biofilm on Urinary Catheter Material".* Antimicrobial agents and chemotherapy, 1985; Vol. 27. pp 619-624.

[53] Costerton, J. W.; Lashen, E. S. *"Influence of biofilm on efficacy of biocides on corrosion-causing bacteria".* Material Performance, 1984; Vol. 23, pp 13-17.

[54] Momba, M. N. B.; Makala, N. *"Comparing the effect of various pipe materials on biofilm formation in chlorinated and combined chlorine-chloraminated water systems".* Water S.A, 2004; Vol. 30, pp 175 - 182.

[55] Momba, M. N. B.; Makala, N. *"Combining chlorination and chloramination processes for the inhibition of biofilm formation in drinking surface water system models".* Journal of Applied Microbiology, 2002; Vol. 92, pp 641-648.

[56] Gagnon, G. A.; Rand, J. L.; O´Leary, K. C.; Rygel, A. C.; Chauret, C.; Andrews, R. C. *"Disinfectant efficacy of chlorite and chlorine dioxide in drinking water biofilms".* Water Research, 2005; Vol. 39, pp 1809 - 1817.

[57] Gagnon, G. A.; Huck, P. M. *"Removal of easily biodegradable organic compounds by drinking water biofilms: analysis of kinetics and mass transfer".* Water Research, 2001; Vol. 35, pp 2554-2564.

[58] Volk, C.; Dundore, E.; Schiermann, J.; Lechevallier, M. *"Practical evaluation of iron corrosion control in a drinking water distribution system".* Water Research, 2000; Vol. 34, pp 1967-1974.

[59] Ndiongue, S.; Huck, P. M.; Slawson, R. M. *"Effects of temperature and biodegradable organic matter on control of biofilms by free chlorine in a model drinking water distribution system."* Water Research, 2005; Vol. 39. pp 953-964.

[60] Boe-Hansen, R. *"Microbial growth in drinking water distribution systems".* Ph.D. Thesis. Environment and Resources DTU, Technical University of Denmark, 2001.

[61] Eisnor, J. D.; Gagnon, G. A. *"A framework for the implementation and design of pilot-scale distribution systems".* Journal of Water Supply: Research and Technology—AQUA, 2003; Vol. 52, pp. 501-519.

[62] Rios, J. F.; Calderón, J. A.; Echeverría, F.; Peñuela, G. *"Diseño de un sistema piloto para el estudio de la corrosión del material de tubería y su contribución en el deterioro de la calidad del agua potable".* Rev. Fac. Ing. Univ. Antioquia, 2008; N.° 43, pp 102-113.

[63] ASTM D2688 – 11. *Standard Test Method for Corrosivity of Water in the Absence of Heat Transfer (Weight Loss Method).* ASTM International, 2011.

[64] Talbot, D.; Talbot, J. *Corrosion Science and Technology.* CRC Press : Boca Raton, 1998.

[65] Boulay N.; Edwards M. *Role of temperature, chlorine, and organic matter in copper corrosion by-product release in soft water.* Water Research, 2001 Vol. 35(3), pp 683 - 690.

[66] Shim J. J.; Kim J. G. *Copper corrosion in potable water distribution systems: influence of copper products on the corrosion behavior.* Materials Letters, 2004; Vol. 58(14), pp 2002 - 2006.

[67] Duprat, M.; Lafont, M.; Dabosi, F.; Moran, F. *Study of the corrosion and inhibition processes of a carbon steel in a low conductivity medium by electrochemical methods.* Electrochim. Acta, 30, (1985) 353-365.

[68] M. Sfaira, A. Srhiri, H. Takenouti, M. Marie de Ficquelmont-Loïzos and A. Ben Bachir, and M. Khalakhil. Corrosion of Mild Steel in low Conductive Media Simulating Natural Waters. J. Applied Electrochemistry, 2001; Vol. 31, pp 537-546.

[69] Rios, J. F.; Calderón, J. A.; Nogueira, R. P. *Electrochemical Behavior of Copper in Drinking Water: Evaluation of Dissolution Process at Low Anodic Overpotential.* J. Braz. Chem. Soc., 2011; Vol. 22, No. 7, pp 1362-1370.

[70] Shim, J.J.; Kim, J.G. *"Copper corrosion in potable water distribution systems: influence of copper products on the corrosion behavior".* Materials Letters, 2004; Vol. 58, pp 2002-2006.

[71] Mikó, A.; Erning, J.W. Schlerkmann, H.; Mathiesen, T. *"Comparative investigation of stainless steels used in drinking water distribution systems".* Electrochim. Acta., 2009; Vol. 54, pp 7507-7513.

[72] Bousselmi, L.; Fiaud, C.; Tribollet, B.; Triki, E. Electrochim., 1999; Acta. 44, 24, pp 4357.

[73] Frateur, I.; Deslouis, C.; Orazem, M.E.; Tribollet, B. *Modeling of the cast iron/drinking water system by electrochemical impedance spectroscopy.* Electrochim. Acta, 1999; Vol. 44, pp 4345-4356.

[74] Choi, Y.-S.; Shim, J.-J.; Kim, J.-G. *"Corrosion behavior of low alloy steels containing Cr, Co and W in synthetic potable water".* Materials Science and Engineering A, November 2004; Vol. 385 N° 1-2, pp 148-156.

[75] De Levie, R. Advances in Electrochemistry and Electrochemical Engineering,1967; Vol. 6, pp 329.

[76] Shim, J.J.; Kim, J. G. *Copper corrosion in potable water distribution systems: influence of copper products on the corrosion behavior.* Materials Letters, 2004; Vol. 58, pp 2002– 2006.

[77] Valcarce, M. B.; De Sánchez, S. R.; Vázquez, M. *A comparative analysis of copper and brass surface films in contact with tap water.* J MATER SCI, 2006; Vol. 41, pp 1999–2007.

[78] Feng, Y.; Teo, W. K.; Siow, K. S.; Tan, K. I.; Hsieh, A.K.; Corros. Sci., 1996; Vol. 38, pp 369.

[79] Rushing, J.C.; Edwards, M. *The role of temperature gradients in residential copper pipe corrosion.* Corrosion Science, 2004; Vol. pp 1883–1894.

[80] American Society for Testing and Materials, Annual Book of ASTM Standards, 11. 04, 2001; (D 3987-85).

[81] EN 15664-1: Influence of metallic materials on water intended for human consumption—dynamic rig test for assessment of metal release—Part 1. Design and operation.

[82] Mikó, A.; Erning, J.W.; Schlerkmann, H.; Mathiesen, T. *Comparative investigation of stainless steels used in drinking water distribution systems.* Electrochimica Acta, 2009; Vol. 54, pp 7507 – 7513

[83] Krakowiak, S.; Darowicki, K.; Slepski, P. *Impedance investigation of passive 304 stainless steel in the pit pre-initiation state.* Electrochim. Acta, 2005; Vol. 50, pp 2699 – 2704.

[84] Jüttner, K.; Lorenz, W.J.; Paatsh, W. *Corrosion Science*, 1989; Vol. 29, pp 279.

[85] Hitzig, J.; Jüttner, K.; Lorenz, W.J. Electrochemical Society, 1986; Vol. 133, pp 887.

[86] Ma, Y.; Li, Y.; Wang F. *Mater. Chem. Phys.*, 2008; Vol. *112*, pp 844–852.

[87] Berardi, L.G.; O. Kapelan, Z. Savic, D. A. *Development of pipe deterioration models for water distribution systems using EPR.* Journal of Hydroinformatics, 2008; Vol. 10(2), pp 113-126.

[88] Andreou S. A.; Marks D. H.; Clark R. M. *A new methodology for modelling break failure patterns in deteriorating water distribution systems: Theory.* Adv. Water Resources, 1987; Vol. 10 (3), pp 2-10.

[89] Levlin, E. *Water and waste Pipes.* Water Resources Engineering. Royal Institute of Technology, 2004; pp 1-12.

[90] Davies, J.P.; Clarke, B. A.; Whiter, J. T.; Cunningham, R. J. *Factors influencing the structural deterioration and collapse of rigid sewer pipes.* Urban Water, 2001; Vol. 3(1-2), pp 73-89.

[91] Srikanth, S.; Sankaranarayanan, T.S.N.; Gopalakrishna, K.; Narasimhan, B.R.V.; Das, T.V.K.; Das, S.K., *Corrosion in a buried pressurised water pipeline.* Engineering Failure Analysis, 2005; Vol. 12(4), pp 634-651.

[92] Makar, J.M. *A preliminary analysis of failures in grey cast iron water pipes.* Engineering Failure Analysis, 2000; Vol. 7(1), pp 43-53.

[93] Rajani, B.; Kleiner, Y. *Comprehensive review of structural deterioration of water mains: physically based models.* Urban Water, 2001; Vol. 3(3), pp 151-164.

[94] Avila-Rangel H.; Saldarriaga-Valderrama J.G. *Renovación y Rehabilitación de Redes de Distribución de Agua Potable y Alcantarillados*, in XX Congreso Latinoamericano de Hidráulica. Cuba, 2002.

[95] Boxall J.B.; O'Hagan A.; Pooladsaz S.; Saul A.J.; Unwin D.M. *Estimation of burst rates in water distribution mains.* Proceedings of the Institution of Civil Engineers, Water Management, 2007; Vol. 160 (2), pp 73-82.

[96] Restrepo, A.; Delgado, J.; Echeverría, F. *Evaluation of Current Condition and Lifespan of Drinking Water Pipelines*, J Fail. Anal. and Preven., 2009; Vol. 9 pp 541–548.

[97] Spickelmire, B. *Corrosion considerations for ductile iron pipe.* Materials performance, 2002; Vol. 41(7), pp 16-23.

[98] Dillon, C.P. *Corrosion Control in the Process Industries.* 2nd Ed. 1994, Washington: MTI by NACE International.

[99] Moglia, M.; Davis, P.; Burn, S. *Strong exploration of a cast iron pipe failure model.* Reliability Engineering and System Safety, 2008; Vol. 93(6), pp 885-896.

[100] Hadzilacos, T.K., D.; Preston, N.; Melbourne, P.; Camarinopoulos, L.; Eimermacher, M.; Kallidromitis, V.; Frondistou-Yannas, S.; Saegrov, S. *UtilNets: a water mains rehabilitation decision-support system.* Computers, Environment and Urban Systems, 2000; Vol. 24(3), pp 215-232.

[101] Conner, R.C. *Nuevos avances en materiales de hierro dúctil, XXV Congreso Interamericano de Ingeniería Sanitaria y Ambiental, AIDIS*. México, (D.F), 1996; pp 1-7.

[102] Deb, A.; AWWARF; Hasit, Y.; Grablutz, F. *Prioritizing Water Main Replacement and Rehabilitation*, Denver(CO), AWWA Research Foundation and American Water Works Association. 2002; pp 223.

[103] Kleiner, Y.; Rajani, B. *Comprehensive review of structural deterioration of water mains: statistical models*. Urban Water, 2001; Vol. 3(3), pp 131-150.

[104] Kleiner, Y.; Rajani, B. *Using limited data to assess future needs*. Journal of AWWA, 1999; Vol. 91(7), pp 47-62.

[105] Kettler A. J.; Goulter I. C. *An analysis of pipe breakage in urban water distribution networks, Canadian Journal of Civil Engineering*, 1985; Vol. 12, pp 286-293.

[106] McMullen, L. D. *Advanced concepts in soil evaluation for exterior pipeline corrosion. In Proceedings of the AWWA*. Annual Conference, Miami 1982; pp 45-59.

[107] Jacobs, P.; Karney, B. *GIS development with application to cast iron water main breakage rates*. In Proceedings, 2nd Int. Conf. on Water Pipeline Systems, BHR Group Ltd., Edinburgh (Sc), 1994; pp 53–62.

[108] Yamijala S.; Guikemab, S.D.; Brumbelowc, K. *Statistical models for the analysis of water distribution system pipe break data*, Reliability Engineering and System Safety, 2009; Vol. 94(2), pp 282-293.

[109] Shamir, U.; Howard, C.D.D. *An Analytic Approach to Scheduling Pipe Replacement*, Journal of the AWWA, 1979, Vol. 71(5), pp 248 258.

[110] Walski, T. M.; Pelliccia, A. *Economic analysis of water main breaks*, Journal of the AWWA, 1982; Vol. 74(3), pp 140–147.

[111] Clark R. M.; Stafford C. L.; Goodrich J. A. *Water distribution systems: a spatial cost evaluation*. Journal of Water Resources Planning and Management Division ASCE, 108(WR3), 1982; pp 243 – 256.

[112] Yamijala, S. *Statistical Estimation of Water Distribution System Pipe* Break Risk, Master Thesis, Texas AandM University, 2007.

[113] Cox D.R. Regression models and life tables. Journal of Royal Statistic Society 34(B), 1972; pp 187–220.

[114] Marks, D. H.; et al. *Predicting urban water distribution maintenance strategies: a case study of New Haven, Connecticut*. Report to the Environmental Protection Agency by the Department of Civil Engineering, Massachusetts Institute of Technology, Cambridge, Mass, 1985.

[115] Andreou S.A.; Marks D.H.; Clark R.M. *A new methodology for modelling break failure patterns in deteriorating water distribution systems*, Advances in Water Resources, 1987; Vol. 10(1), pp 2-10.

[116] Andreou S.A.; Marks D.H.; Clark R.M. *A new methodology for modelling break failure patterns in deteriorating water distribution systems*: Applications, Advances in Water Resources, 1987; Vol. 10(1), pp 11-20.

[117] Marks, H. D.; Andreou, S.; Jeffrey L.; Park, C.; Zaslavski, A. *Statistical models for water main failures*, US Environmental Protection Agency (Co-operative Agreement CR810558), M.I.T Office of Sponsored Projects No. 94211. Boston, Mass. 1987.

[118] Brémond, B. *Statistical modelling as help in network renewal decision.* European commission co-operation on science and technology (COST), Committee C3 – diagnostics of urban infrastructure, Paris (Fr), 1997.

[119] Constantine, A. G.; Darroch J. N.; In S. Osaki, D. N. P. Murthy (Eds.), Pipeline reliability: Stochastic models in engineering technology and management. Singapore: World Scientific., 1993.

[120] Constantine, A. G.; Darroch, J. N.; Miller, R. *Predicting underground pipe failure.* Australian Water Works Association, 1996.

[121] Lei, J. *Statistical approach for describing lifetimes of water mains – Case Trondheim Municipality.* SINTEF Civil and Environmental Engineering, Report No. 22F007.28, Trondheim, Norway, 1997.

[122] Eisenbeis, P.; Rostum, J.; Le Gat, Y. *Statistical models for assessing the technical state of water networks – Some European experiences.* In Proceedings of the AWWA Annual Conference, Chicago, 1999.

[123] Caleyo, F.V.; Velazquez J. C.; Valor, A.; Hallen, J. M. *Probability distribution of pitting corrosion depth and rate in underground pipelines:* A Monte Carlo study. Corrosion Science, 2009; Vol. 51(9), pp 1925-1934.

[124] Herz R.K. *Ageing process and rehabilitation needs of drinking water distribution networks.* Journal of Water SRT – Aqua, 1996; Vol. 45 (5), pp 221–231.

[125] Herz R.K. *Exploring rehabilitation needs and strategies for water distribution network.* Journal of Water SRT – Aqua, 1998; Vol. 47 (6), pp 275–283.

[126] Deb A.K.; Hasit Y.; Grablutz J.F.M.; Herz R.K. *Quantifying future rehabilitation and replacement needs of water mains*, AWWA Research Foundation, Denver (CO), 1998.

[127] Kulkarni, R.B.; Golabi, K.; Chuang, J. *Analytical techniques for selection of repair-or-replace options for cast iron gas piping systems* – Phase I. as Research Institute, PB87-114112, Chicago (IL), 1986.

[128] Gustafson, J.M.; Clancy, D. V. *Modelling the occurrence of breaks in cast iron water mains using methods of survival analysis.* In Proceedings of the AWWA Annual Conference, Chicago, 1999.

[129] Goulter I.C.; Kazemi A. *Spatial and temporal groupings of water main pipe breakage in Winnipeg.* Canadian Journal of Civil Engineering, 1988; Vol. 15(1), pp 91–97.

[130] Goulter I.C.; Davidson J.; Jacobs P. *Predicting water-main breakage rate.* Journal of Water Resources Planning and Management, ASCE, 1993; Vol. 119(4), pp 419–436.

[131] Zhan, C.; Rajani, B.B. *On the estimation of frost load in a trench: theory and experiment.* Canadian Geotechnical Journal, 1997; Vol. 34(4), pp 568–579.

[132] Watkins, R.K.; Spangler, M.G. *Some characteristics of the modulus of passive resistance of soil – a study in similitude.* Highway Research Board Proceedings, 1958; Vol. 37, pp 576–583.

[133] Rajani, B.; Zhan, C.; Kuraoka, S. *Pipe-soil interaction analysis for jointed water mains.* Canadian Geotechnical Journal, 1996; Vol. 33(3), pp 393–404.

[134] Kiefner, J. F.; Vieth, P. H. *Project PR-3-805: a modified criterion for evaluating the remaining strength of corroded pipe.* Pipeline Corrosion Supervisory Committee of the Pipeline Research Committee of the American Gas Association, 1989.

[135] Rajani, B.; Makar, J.; McDonald, S.; Zhan, C.; Kuraoka, S.; Jen, CK.; Viens, M. *Investigation of grey cast iron water mains to develop a methodology for estimating service life.* AWWA Research Foundation, Denver (CO), 2000.

[136] Doleac, M. L.; Lackey, S. L.; Bratton, G. N. *Prediction of time-to failure for buried cast iron pipe.* Proceedings of AWWA Annual Conference, Denver (CO), 1980, pp 21–28,

[137] Randall-Smith, M.; Russell, A.; Oliphant, R. *Guidance manual for the structural condition assessment of the trunk mains.* Swindon (UK), Water Research Centre, 1992.

[138] Restrepo, A.; Delgado, J.; Echeverria, F. *Evaluation of Current Condition and Lifespan of Drinking Water Pipelines,* J. Fail. Anal. and Preven., 2009; Vol. 9, pp 541–548.

In: Pipelines: Design, Applications and Safety
Editors: M. G. Rivero et al. pp. 43-68
ISBN 978-1-62100-178-2

Chapter 2

SCC BEHAVIOR IN BURIED PIPELINE STEELS: REVIEW ARTICLE

A. Torres-Islas[1]*, S. Serna[1] and B. Campillo[2]

[1]UAEM, Centro de Investigación en Ingeniería y Ciencias Aplicadas-FCQeI, Av. Universidad 1001, Col. Chamilpa, 62210-Cuernavaca, Mor., México

[2]Facultad de Química, Instituto de Ingenieria-UNAM, Circuito Exterior S/N, Cd. Universitaria, 04510, D.F. México

1. INTRODUCTION

The failure of pipeline steel is usually catastrophic and causes economic, environmental and conservation losses. About 20% of the failure of pipeline was caused by outer corrosion and 5% by inner corrosion. The first incident of stress corrosion cracking (SCC) on natural gas pipelines occurred in the mid- 1960's [1]. Since that time, there have been hundreds of failures reported in Australia [2], Canada [3], Iran, Iraq, Italy, Pakistan, Saudi Arabia, the former Soviet Union and the United States [4]. In the last 30 years, the pipeline SCC problem has been investigated by different laboratories [5-15]. Most of the early failures were intergranular in nature, whereas, many of the recent failures, such as those that have occurred in Canada are transgranular [16-23]. It is now recognized that there are at least two forms of external SCC on underground pipelines. The intergranular form is referred to as high-pH, classical SCC, while the transgranular form is referred to as low-pH, non-classical, or near-neutral-pH SCC.

Now China plans to build a natural gas pipeline from the Xinjiang Uygur Autonomous Region in Northwest China to Shanghai in East China. It is a 4,200-kilometre pipeline, which is expected to wind its way through six provinces and one autonomous region. The total cost of the gigantic project is expected to be more than 140 billion yuan (US$ 16.86 billion) and it will be China's second largest infrastructure project next to the Three Gorges Dam Project. Thus, it is essential and urgent to design the pipeline soundly and continue research into SCC

* E-mail: atimarquis93@ yahoo.com.mx; Tel: +52777 397084; fax: + 52777397084.

to provide against possible troubles or predict the consequences. Moreover, with the constructions of the Sino-Russia and the Sino-Kazakstan transnational gas and oil pipeline imminent, it is very important to continue to research failure causes of pipeline.

2. Near-neutral pH and High pH SCC in Pipelines

There are several differences between near-neutral pH SCC and high pH SCC, and further details are given below.

2.1. High pH SCC

Although pipelines are coated for protection against corrosion when they are put into the ground, there is always the risk that the steel pipe could become exposed to the surrounding environment. The pipe would then be vulnerable to corrosion. Since the corrosion is an electrochemical reaction, an electric current is passed through the soil to the pipe to effectively prevent corrosion. This process of applying a voltage to the pipe through the soil gives the pipeline a cathodic potential and is referred to as cathodic protection.

High pH SCC occurs only in a relatively narrow cathodic potential range in the presence of a carbonate /bicarbonate environment and at a pH greater than 9. In the cathodic potential range and environment required for high pH SCC, a protective film forms on the steel surface [24]. This film is a thin oxide layer that forms from the electrochemical reaction that takes place.

If the protective film on the pipe surface is not broken, SCC cannot start because the film acts as a barrier between the pipe surface and the environment. But if the steel is subjected to a strain that stretches the metal until it is permanently deformed, the film, being brittle, will crack and bare metal will be exposed to the environment. This process of rupturing the film to expose the metal is what creates the opportunity for SCC to initiate.

2.2. Near-neutral pH SCC

Near-neutral pH SCC can develop in pipelines under normal operating conditions when a coating is broken down it lacks cathodic protection of the pipeline surface and ground water contacts the outside surface. Because the characteristics of the coating are different and one may be more prone than another to debonding or forming holiday, they have different effects. In the presence of a cathodic current, the solution exhibits higher pH, then high pH SCC will occur.

In comparison with high pH SCC, there are only about ten years of research on near-neutral pH SCC. Because of the differences between the two types of cracking the research is generally not transferable. Consequently, additional knowledge must be developed about near-neutral pH SCC. The characteristics of near-neutral pH and high pH SCC are compared in Table l.

Table 1. Comparisons between high pH and near-neutral pH SCC of pipeline [27-29]

Pipeline	High pH SCC	Near-neutral pH SCC
Location	a) Typically within 20 km downstream of compressor b) Typically where discharge gas temperature exceeds 35 °C	a) Not restricted to any particular location. b) No dependence on gas discharge temperature
Corrosive environment	Laboratory solutions typically used to simulate cracking are concentrated 1 N carbonate/bicarbonate, of pH near 9.5	Solution found associated with cracking in the field are dilute, low conductivity, near neutral pH trapped waters
Crack morphology	Intergranular	Transgranular
Crack orientation	Longitudinal	Longitudinal
Numerous surface cracks	Yes	Yes
Linking of crack	Yes	Yes
Patches of cracks	Yes	Yes
Corrosion of crack faces	No	Yes
Corrosion of pipe	Usually not	Sometimes
Width of potential range	Narrow (< 100 mV)	Probably > 100 mV
P/S potential, CCS	-722 mV	Native potential
Temperature dependence	Arrhenius behavior	Not established
Crack appearance	Cracks show little evidence of secondary corrosion	Cracks are with evidence of substantial secondary corrosion

Parkins cautioned that the way in which near-neutral pH SCC initiates and then develops is not yet completely understood, and so what we discuss here should be taken in that light [25]. In this submission, Parkins discussed how dissolution and hydrogen are believed to be factors in the growth of near-neutral pH SCC. He conceived that the mechanism of crack growth involves dissolution and the ingress of hydrogen into the steel; the hydrogen facilitating crack growth by promoting reduced ductility. While it is clear from evidence of corrosion on the sides of cracks, developed in service or laboratory tests, that dissolution occurs within crack enclaves, it is doubtful that growth can be accounted for entirely in terms of a dissolution process. That is because at high stress or strain observed growth rates are markedly greater than that can be accounted for by rates of dissolution in near-neutral pH environments. However, the evidence in support of hydrogen playing a role in the overall growth process is circumstantial rather than direct [26]. Parkins [24] indicated that the factors contributing to the development of near-neutral pH cracking would appear to include the following:

Cracks are probably initiated at pits on the steel surface wherein a localized environment is generated that has a pH low enough to produce atomic hydrogen in the pit.

The presence of carbon dioxide in the groundwater assists in creating near-neutral pH levels. Some of the discharged atomic hydrogen enters the steel, degrading the mechanical properties locally so that cracks are initiated or grown by a combination of dissolution and hydrogen embrittlement. Continuing anodic dissolution in the crack is necessary for crack growth, assisted by hydrogen entry into the steel. The plastic stress level necessary to produce cracking may not be related solely to fracturing the embrittled steel. It may also contribute by rupturing the protective film, allowing hydrogen to reach and then penetrate the steel.

Thus, Parkins [30] is of the view that near-neutral pH SCC crack growth involves dissolution and hydrogen, and he is supported in that view by Leis [30], Wilmott and Jack [30], Lambert and Plumtree [30] and Beavers [30].

2.3. High pH vs Near-neutral pH SCC Crack Characteristics

The majority of instances of SCC in high pressure gas pipelines at high pH has been associated with the propagation of intergranular cracks (IGSCC), but several instances of near neutral pH SCC have been found in the form of transgranular cracking (TGSCC). High pH SCC generally produces intergranular cracking, where the cracks grow around or between the grains in the steel. In IGSCC, the very small amount of iron carbonate that sometimes is present is incorporated in the thin magnetite films that invariably form. These films are strongly sticking to the crack sides and effectively prevent any lateral dissolution on the sides. As a result, IG cracks are very tight, narrow or fine. Conversely, TGSCC, where the cracks follow a path across or through, is often associated with the formation of a relatively large amount of white iron carbonate between the coating and pipe surface. TG crack sides suffer significant lateral dissolution, with appreciable amounts of loosely adherent corrosion products forming in the crack enclaves. The side walls of the cracks corrode and the cracks appear much wider than high pH SCC cracks. However, the crack generally becomes narrower as the crack deepens.

The different nature of the crack propagation is the result of the different effects of the environments and the susceptibility of the steel. In the high pH SCC, the grain boundaries are more susceptible to dissolution than the grains and so that is where the cracks form. Parkins viewed that transgranular cracking can also occur in high pH SCC when the cracks become relatively deep or are subjected to relatively high stress levels or high fluctuating stress [32]. The significance of this point is that the presence of a transgranular crack by itself is not enough to be certain that a crack is near-neutral pH SCC.

3. SCC Factors

For SCC to occur on an engineering structure, three conditions must be met simultaneously; a specific crack-promoting environment must be present, the metallurgy of the material must be susceptible to SCC, and tensile stress above some threshold value must be present. Each of these factors is discussed below for high pH and near-neutral pH SCC of pipelines.

3.1. Metallurgy

Metallurgy can affect SCC through composition and microstructure. However, pipeline steels, and certainly the conventional steels that have historically been used in the last 50 years, do not contain elements found in similar carbon-manganese steels used in literally hundreds of construction applications without reports of SCC.

More recently, the yield strength of pipeline has gradually increased by the addition of micro-alloying elements such as vanadium, columbium and/or titanium. The addition of these elements tends to produce a controlled rolling and cooling of the steel plate used to manufacture pipe and has resulted in finer grained bainite steel microestructures.

A number of research investigations involving small scale, laboratory reproduced SCC and using both high and low pH environments have been conducted without achieving a meaningful correlation between steel chemistry and susceptibility to SCC. Danielson and Jones (2001)[33] discuss the high-pH SCC testing of six different heats of X52, as well as three heats (X65, X70, X80) of modern steels. Their paper concludes: "In general, the microstructure/microchemistry had a small effect on the SCC behavior."

Nevertheless, certain batches of pipeline steel have been found to be much more susceptible to SCC than other batches with similar compositions and microstructures (Beavers and Harper 2004)[34] A full understanding of this remains to be developed, but current research suggests that other characteristics of the steel, such as creep response to cyclic loading, may be important.

In this way, our results (A. Torres et al. 2008)[35] working with a recently developed microalloyed pipeline steel, indilute $NaHCO_3$ solutions between 0.1- 0.005M and four different microstructures achieved by heat treatments included the as-received condition, have found that as a general trend X- 70 steel is resistant to SCC in concentrated $NaHCO_3$ solutions, but it was highly susceptible to SCC in the most dilute concentration, at 50°C, regardless the heat treatment. However, the mechanism of SCC in the most dilute solution was dominated by film rupture and anodic dissolution in the steel with major ferrite and perlite presencewhereas for minor ferrite presence and martensitic microstructure, the SCC mechanism is dominated by hydrogen embrittlement.

M. A. Arafin and J. A. Szpunar (2011)[36] have found thatX-80 and X-100 pipeline steels are highlysusceptible to hydrogen induced cracking (HIC)in contact with near neutral pH carbonate-bicarbonate solutions, and that a bainitc lath boundary separation takes place due to hydrogen trapping in these areas.

In fact Lu B. T. and Luo J. L. (2006)[37] found that pipeline steels ranging from X52 to X100 steels and the weldments of X70 and X65 with various heat treatments and as a consequencedifferent microstructures and strength levels shownthat the near-neutral pH SCC resistance of pipeline steel is reduced, generally, with an increase in the strength level, but the strength dependence of SCC resistance is heavily affected by the microstructures of the pipeline steels. The steels with a fine-grained, bainite-ferrite structure possess a much better combination of strength and SCC resistance than those with a ferrite plus pearlite structure. However, the introduction of the welding process will significantly degrade SCC resistance in the steels containing a bainitic-ferrite structure. This degradation effect is caused mainly by the decomposition of the bainitic-ferrite structure into a separate microstructural entity. On the other hand, an increase in the pearlite content in the microstructure has a detrimental effect on the SCC resistance of pipeline steels with a ferrite plus pearlite structure.

Therefore, due to results in actual researches, it is clear that each one of the different microstructures play an important role in the inherent SCC resistance.

On the other hand, pipeline is manufactured using one of four processes: seamless pipe, electric resistance welded pipe made from steel coils, flash welded pipe from plate and submerged arc-welded pipe made from steel plate. The vast majority of line pipe for gas transmission service is produced by one of the three seam-welded processes. Pipe in grades X60 and higher achieves some of its strength from the controlled rolling procedures used to reduce the thickness of the original cast slabs to the final pipe wall thickness. These procedures include not only the hot work but also the cooling rate of the plate or strip after hot rolling. The microstructure may contain varying amounts of ferrite, pearlite and bainite with wide variation in the crystalline grain size. There is strong evidence that the items above either promote or inhibit SCC. The first reported cases of SCC exhibited intergranular cracking (high-pH SCC) in steels with a microstructure that consisted of grains of low-carbon ferrite and higher carbon colonies of pearlite. Typically, the more recently detected near neutral-pH SCC has occurred on slightly higher yield strength steels with a much finer grain size and a higher toughness. About this, J.T. Bulger et al. (2006)[38] concludes that pipeline steels with a microstructure of fine-grained bainite+ferrite have a better combination of strength and SCC resistance than those with ferrite+perlite structures. However, there are a number of cases of near neutral-pH SCC in older, large grained, low-strength steel and cases of high-pH SCC in newer, fine-grained steel. Thus, regarding reported cases of SCC, no generalizations regarding metallurgy can be made.

3.2. Mechanical Properties

The mechanical properties of highest interest for most gas transmission piping are the yield strength and the toughness. Generally, the best economics result from selecting the highest strength pipe material available for the design of a new pipeline system. As improved manufacturing procedures are being developed, higher grades of pipe are being purchased. There is no strong evidence that increasing strengths up to and through grade X70 increases susceptibility to SCC initiation or growth.

About mechanical behavior of pipeline steel W. Chen et al. (2007)[39] found that the effect ofresidual stresses on the occurrence of pitting and stress corrosion cracking formation in pipeline steel exposed to neutral pH aqueous environments can be readily blunted due to plastic deformation (room temperature creep) and/or extensive anodic dissolution. As a result, a high positive tensile residual stress gradient is necessary for continued growth of SCC in pipeline steels exposed to this neutral pH environment. The tensile residual stress represents a large mechanical driving force for crack nucleation and short crack growth. Active cracks may become dormant as the near-surface residual stress gradient changes, due to self-equilibration, from highly tensile to a lower tensile state or to a compressive state. The change in residual stress level can occur within 1 mm of the surface, resulting in a large proportion of dormant SCC.

In addition, generally plastic prestrain reduces the SCC resistance in various welded zones Baotong Lu (2009)[40] determined that the SCC susceptibility and behavior of the welded test materials can be put in the following order: heat-affected zone (HAZ) > weld metal (WM) > base metal (BM). Fractographic analysis indicates that there are two cracking modes during SSRT tests, mode I and mode II. Mode I cracks propagate along the direction perpendicular to the maximum tensile stress, and mode II cracks lie in planes roughly parallel to the plane where the maximum shear exists. The SCC of the BM is governed by mode I cracking and fracture of the HAZ, and the WM is dominated by mode II cracking. In damage analysis it has been shown that the detrimental impact of plastic prestrain on the residual SCC resistance cannot be evaluated with the linear superposition model. A plastic prestrain sensitivity, a material constant independent of plastic prestrain, is proposed to characterize the susceptibility of SCC resistance to plastic prestrain, and it increases with the SCC susceptibility of the steels. It was also found that the enhanced SCC susceptibility caused by plastic prestrain may be related to an increase in yield strength and the correlation of the ratio of the reduction in area in NS4 solution to that in air (RA_{SCC}/RA_{air}) with the yield strength is microstructure dependent.

Another interesting analysis is related to electrochemical-stress relation in X70 pipeline steel. About this behavior in U-bend specimens M.C. Li (2008)[41] found employing scanning localized electrochemical impedance spectroscopy(LEIS) technique, that the deformation-induced stress, if not sufficiently high, has an inhibitive effect on corrosion reaction, pitting occurrence and crack initiation in pipe steel under high-pH condition. Such an inhibitive effect is due to the enhanced generation of carbonate product and the resultant surface block effect at the stressed zones. The tensile and compressive stresses have identical effect on inhibition of dissolution and pitting of the steel. However, tensile stress enhances the steel dissolution more significantly than compressive stress, and thus, generates more carbonate product, resulting in higher localized impedance. Pits easily occur around the neutral axis of the U-bend specimen, where the steel deformation and the resultant stress are ignorable. For pipelines encountering non- uniform stress distribution, the role of stress in crack initiation is critically different.

In this way as a final point, in our last SCC study ofmicroalloyed pipeline steel A.torres-Islas et al. (2011)[42] in the longitudinal and transverse steel rolling directions classical SSC dissolution and embrittlement cracking mechanisms were identified in both rolling directions. Some evidence seems to indicate that the HELP (Hydrogen Enhanced Local Plasticity) mechanism is present in the SCC steel process. This mechanism interacts with a Widmastaten ferrite promoting hydrogen ingress minimizing their mechanical properties.

Therefore mechanical properties in microalloyed pipeline steels are directly related with microstructure, type of stress, stress level and environment conditions.

3.3. Pipeline Operating Conditions

As previously discussed, SCC requires three conditions to be satisfied simultaneously: 1) a tensile stress above the threshold stress, 2) an appropriate environment at the steel surface, and 3) a susceptible material.

Below some value of tensile stress, referred to as the threshold stress, crack initiation does not occur. The threshold stress is difficult to accurately define but, depending on the range of stress fluctuation, is on the order of 40 to 100 percent of the yield strength for classical SCC. A threshold stress for near neutral-pH SCC has not been established (Beavers 1999)[43]

The operator has at least some control of the applied tensile stress when it is strictly the result of internal pressure in the system. Unfortunately, residual tensile stresses from manufacture, bending stresses from pipe movement, overburden loads from soil, dents or gouges, or from heavy equipment can cause as much or more tensile stress as that caused by internal pressure, all of which is beyond the control of the operator.

Note that in Canada some 10 to 20 percent of the SCC reported is oriented in a circumferential direction, i.e., the dominant stress affecting the crack is oriented axially to cause crack growth (Canadian Energy Pipeline Association CEPA 1997)[44]. The direct longitudinal stress caused by pressure can be up to half of the hoop stress. However, pipe flexure will result in additional stress, and the resultant value can exceed the hoop stress, with the maximum/minimum values at the extreme fibers of bending. The SCC cases reported by CEPA are associated with undulating terrain where pipe loading resulted from soil creep or localized bending. Localized bending may also occur at dents resulting in higher axial stresses in the local region.

Pipe that has been cycled into the plastic range multiple times sometimes may experience a condition known as cyclic softening. This appears as a loss of yield strength and can significantly reduce the threshold stress. Again, the operator has little control over cyclic softening. About this Zheng (2008)[45] found that SCC under static stressing conditions confirms that the cracking is indeed a true SCC process, although the rate of which is low without dynamic loading. In contrast to the high pH pipeline stress corrosion cracking in the carbonate–bicarbonate solution, this form of cracking in dilute near neutral environment takes a much longer time to initiate. Once initiated, the crack growth rate is highly sensitive to the loading rate of the applied mechanical force. In addition, the operator has little control over the pH of the groundwater and is unable to control the aggressiveness of the environment, except for new construction by installing a premium coating system. Unfortunately, these coating systems may not be considered suitable for recoating in the ditch. Note also that the pH of the groundwater will be modified by the electrochemical reaction at the pipe surface.

The operator does have some control over the operating temperature. For example, some operators have installed cooling towers to help control SCC.

3.4. Coating

Coating type and condition (sometimes a function of the installation procedure, associated quality control or lack thereof, and weather conditions at the time of installation) have a profound effect on SCC. This is especially true when the coating has a tendency to disbond (i.e. the coating comes away from the pipe but does not break), or forms holidays (i.e., breaks or gaps in the coating). This is true for tape coatings, such as the polyethylene-backed tapes used predominantly in the early 1960s to 1980s. These tapes are spirally wrapped around the pipe with an overlap at the helix line. "Tenting" occurs between the pipe surface and the tape along the ridge created by longitudinal, spiral, and girth welds. Tenting also occurs at the overlap between the helix of the wrap.

When the tape disbonds from the pipe, moisture can accumulate beneath the tape surface. The tape itself has fairly high electrical insulation properties, thus preventing cathodic current from reaching entrapped moisture beneath the tape at the pipe surface. In Canada, about three-quarters of reported near neutral-pH SCC-related service incidents have occurred under these tape coatings. The cracks tend to occur at or near the toe of the seam weld where stress is concentrated and water has access, as well as where the coating has been damaged or disbonded (Canadian National Energy Board NEB 1996).

Asphalt and coal tar coatings are relatively thick and can be brittle. The coatings can disbond, especially due to poor surface preparation. Over time, the volatiles can disperse, leaving the coating relatively brittle. Unlike tape coatings, when these coatings disbond, they usually, but not always, become saturated with moisture and conduct cathodic current, thus protecting the pipe. If the coating is brittle, it may break into pieces, also allowing a path for the cathodic current protection. SCC might still occur when the soil is so resistive that the cathodic current cannot reach the pipe. For these coating types, there is no preferential location, but SCC might occur wherever disbondment or holidays occur (NEB 1996).

It is generally agreed that fusion-bonded epoxy (FBE) coatings, which are often the coating of choice for newly installed pipelines in the United States, are an effective protection against SCC. Extruded polyethylene, because the coating system is monolithic, also appears to be effective, except possibly at tape-wrapped girth welds.

Beavers(1994)[46]describes the role that coatings play in external SCC of natural gas pipelines based on the relationship between the coatings and the three controlling parameters in SCC (metallurgy, stress, and environment). Surprisingly, the coating type can influence all three controlling parameters: there are clear differences in the performance of the different coatings. The FBE coatings exhibit a number of good characteristics including resistance to disbonding and the ability to conduct CP current at disbonded areas. The practice of grit blasting the surface prior to application of the FBE coating also improves resistance to SCC if properly performed. On the other hand, the polyethylene-tape coatings exhibit a number of properties that are detrimental to SCC resistance. These include poor disbonding to resistance, shielding of CP current and poor resistance to the diffusion of corrosive gases.

On the other hand a mathematical model has been developed by Thomas R. Jack (2004)[47] to predict the generation and evolution of the environment under a disbonded permeablecoating as a consequence of the action of CP. The model couples the electrochemical reactions on the surface of the pipe to the transport of species to and from the pipe surface through the permeable coating and the surrounding soil. The model is structured to use available field data (such as soil and groundwater data and information from CP surveys) to predict conditions on the pipe surface.

Latest research, as in the case of L. Niu and Y.F. Cheng (2008)[48], has focused in energy development activities in the Arctic area, which impose serious challenges for maintenance of integrity of the oil/gas transmission pipelines; so the optimization of selecting and evaluating coatings for feasible use in northern pipelines is of great importance with respect to the extreme environments to which the coated pipe may be subjected. In Cheng research, various protective coatings for integrity maintenance of oil/gas pipelines were reviewed. In particular, the high performance composite coats (HPCC) that is specifically designed for the northern pipeline operation, and found that HPCC provides an effective alternative for pipeline integrity maintenance.

HPCC's improved resistance to abrasion, flexibility and cathodicdisbondment, and the high electrochemical impedance have enabled the long-term pipeline protection in the harsh northern environment. The innovative coating technology is essential to the maintenance of pipeline integrity, not only in the northern area and not only for corrosion attack, but also stress corrosion cracking of pipelines that is due to the dynamic soil condition must be prevented through the coating application. Figure 1 shows the development of coating technology in oil/gas pipeline in Canada.

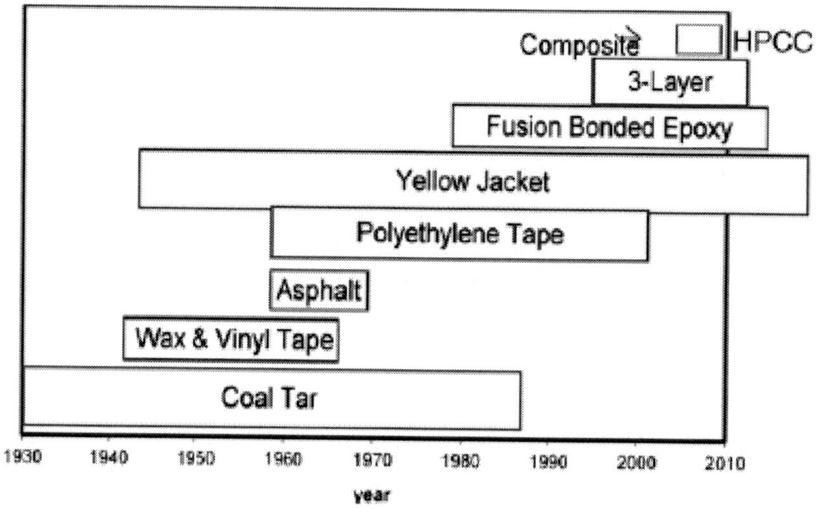

Figure 1. Development of coating technology in oil/gas pipeline in Canada.

3.5. Soil Conditions

In 1973, Wenk described results of analyses of soil and water extracts (from the soil) taken from high-pH locations (Wenk 1974)[49]. While supporting data were not provided, it was stated that SCC had occurred in a wide variety of soils, covering a range in color, texture, and pH. No single characteristic was found to be common to all of the soil samples. Similarly, the compositions of the water extracts did not show any more consistency than the physical descriptions of the soils, according to Wenk. On several occasions, small quantities of electrolytes were found beneath disbonded coatings near locations at which stress corrosion cracks were detected. The principal components of the electrolytes were sodium carbonate and bicarbonate. Sodium bicarbonate crystals were also found on pipe surfaces near some SCC colonies (Fessler 1973)[50]. Based on the presence of the sodium-based carbonates and bicarbonates, it is likely that these were high-pH SCC sites. Therefore, it is not surprising that these results are not consistent with the results of the Trans-Canada Pipeline (TCPL) studies performed in the 1980s and 1990s, when near neutral-pH SCC was found.

Mercer described the results of a field study conducted by British Gas Corporation in 1979[51]. Soil data from both the UK and U.S. were collected and analyzed. As in the study by Wenk, detailed information on the soil analyses was not provided, but it was concluded that soil chemistry had no obvious direct influence on high-pH SCC. The moisture content of the soil, the ability of the soil to cause coating damage, and localized variation in the level of CP were the primary soil-related factors identified.

Delanty and O' Beirne (1991,1992)[52,53] reported on the results of more than 450 investigative excavations performed on TCPL's system in the mid to late 1980s. In the tape-coated portions of the system, near neutral-pH SCC was found in allof the various types of terrains and soils (e.g., muskeg, clay, silt, sand, and bedrock) present on the system. There was no apparent difference in the soil chemistry for the SCC and non-SCC sites. However, the SCC was predominantly located in imperfectly to poorly drained soils in which anaerobic and seasonally reducing environmental conditions were present.

In the same system, near neutral-pH SCC was found in the asphalt-coated portions of the system, predominantly (83 percent) in extremely dry terrains consisting of either sandy soils or a mixture of sand and bedrock. There was inadequate CP in these locations, based on pipe-to-soil potential measurements or pH measurements of electrolytes found beneath disbonded coatings. The remainder of the SCC sites on the asphalt-coated portions of the system had localized areas of inadequate CP, based on pH measurements of electrolytes.

Delanty and Marr (1990, 1992)[54,55]developed an SCC severity rating model for near neutral-pH SCC for the tape-coated portions of TCPL's system in eastern Canada. The predictors in that model were soil type, drainage, and topography. The soil classifications were based on method of deposition. The most aggressive soil types were lacustrine (formed by deposits in lakes), followed by organics over glaciofluvial (formed by deposits in streams fed by melting glaciers), and organics over lacustrine. The prevalence of SCC in glaciofluvial soils was about 13 percent of that in lacustrine soils, and about 17 percent of that in soils with organics over glaciofluvial or lacustrine. Very poorly or poorly drained soils were found to be the most aggressive, while level-depressed soil was found to be the most aggressive topography. The SCC model did not contain parameters associated with soil chemistry because the results of previous geochemical projects were inconclusive.

As described above, neither the early field studies conducted on high-pH, nor the later field studies conducted on near neutral-pH SCC, detected a correlation between the occurrence of SCC and soil chemistry. On the other hand, high-pH SCC was not reported where the extensive field study of near neutral- pH SCC was performed in Northern Ontario (Delanty and O'Beirne1991, 1992)[52, 53], suggesting that the soil conditions were not conducive to this form of cracking. Furthermore, no near neutral or high-pH SCC was found in Northern Ontario where elevated pH electrolytes were detected, possibly because the soil conditions could not support the development of concentrated carbonate-bicarbonate solutions, even when the CP conditions were conducive to such development. These observations suggest that a further analysis of field soil data might provide insight into the role of soil/groundwater chemistry on the occurrence of SCC Beavers and Garrity (2001)[56].

Near neutral-pH SCC may be associated with local topographical depressions, e.g., at the base of hills or streams, where the groundwater either channels along the pipe or across it. Flowing water may help to maintain the near neutral-pH environment by supplying CO_2 to the electrolytic solution in a disbonded area. The majority of laboratory investigation has been performed in NS4 electrolyte solution containing 5 percent CO_2. NS4 is simulated trap water that is typical of liquids found beneath disbonded polyethylene tape coatings at locations where near neutral-pH SCC was found. Research shows that the crack growth rate increases with increasing CO_2 concentrations, and that the cracking becomes dormant in CO_2 free environments Beavers et al. (2001)[54].

On the other hand A. Benmoussa et al. (2006)[55] using potentiodynamic polarisation and EIS methods in soil parameters on pipeline steel corrosion behaviour exposed in near-neutral pH soil simulating solution has demonstrated that corrosion phenomenon is accentuated by soil resistivity, pH, temperature, moisture content and chemical composition of electrolytes contained in the soil. Almost all results showed that the steel corrosion increases and corrosion current density increases with temperature in the range from 20 to 60°C. Impedance curves showed that the charge transfer resistance (Rt) increases with increasing immersion duration. Parameters such as corrosion current density (Icorr), polarization resistance (Rp), and soil resistivity (q) can serve as the parameters for evaluation of soil corrosivity.

4. CATHODIC PROTECTION

Cathodic protection (CP) is closely related to the high-pH cracking process. The CP current collecting on the pipe surface at disbondments, in conjunction with dissolved CO2 in the groundwater, generates the high-pH SCC environment. CP can also place the pipe-to-soil potential in the potential range for cracking. The potential range for cracking generally lies between the native potential of underground pipelines and the potential associated with adequate protection (-850 mV CSE) (Fessler 1973, Parkins 1977)[6,50]. Based on field pH measurements of electrolytes associated with near neutral-pH SCC colonies, it has been concluded that this form of SCC occurs in the absence of significant CP, either because of the presence of a shielding coating or high-resistivity soils that limit CP current to the pipe surface (Delanty, 1991)[52].

Based on the available laboratory and field data, it can be concluded that the polarized pipe-to-soil potential of pipeline segments that are potentially susceptible to high-pH SCC should be maintained above (i.e. more negative than) –850 mV CCS. Potentially susceptible segments can be assessed using ASME B31.8S Appendix A3 for gas pipelines, which considers historical information, coating type, operating temperature, age, operating stress, and distance downstream from the compressor station. For liquid pipelines, the distance downstream of the pump station can be used in the ASME assessment (NACE 2004)[47]. The other CP criteria (100mV polarization or 850 mV with CP applied) should not be used on potentially susceptible segments. Consideration should be given to seasonal fluctuations in the potential to minimize the likelihood that the pipe falls into the cracking range on a seasonal basis.

Our results (A. Torres-Islas 2008)[35] in stress corrosion cracking study of microalloyed pipeline steels in dilute $NaHCO_3$ solutions showed that according to NACE recommended criterion the protection potential has to be ~-0.79 V_{SCE} in this type of solution. Thus, applying -500 and -600 $mV_{SCE}$to the steel X-70, leads to a cathodic protection lowering their SCC susceptibility.

By examining the Pourbaix E-pH diagram of iron in aqueous solution at -500 and -600 mV_{SCE} (100 and 200 mV) cathodic potentials, for 0.005 M $NaHCO_3$ solution with pH 8.35, hydrogen reduction is thermodynamically impossible. Thus, in this potential range the quenched steel and the quenched + tempered steel, that were prone to hydrogen embrittlement in SSRT tests with hydrogen pre-charged specimens, do not present evidence of embrittlement.

The scenario was quite different at -600 and -1600 mV_{SCE}cathodic potentials, in which the steel area reductions were considerably lesser, presenting brittle fracture surfaces. By analyzing some of the fracture surfaces and transverse cracks it could be determined that the main failure mechanism is hydrogen embrittlement.

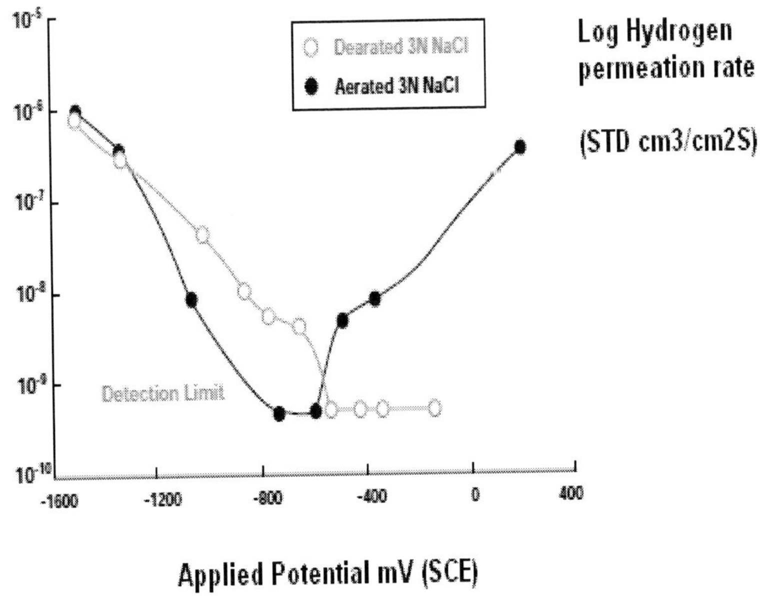

Figure 2. Effect of potential on hydrogen entering steel.

Furthermore at this cathodic potential range a great amount of hydrogen can be evolved due to water reduction that can diffuse into the steel making it brittle, no matter the steel microstructure. In accordance with the above Ping Liang (2009)[56] in a stress corrosion cracking susceptibility of X80 steel study, under applied cathodic potentials in a simulated soil solution using SSRT, found that no apparent change of reduction in area was found at -775 mV_{SCE} in contrast to the open circuit potential, and many dimples were visible onfracture surface. However, when the applied potentials were lower than -1000 mV, the SCC susceptibility increased as a result of evolution hydrogen, which diffuses into the steel. Pitts were found to be an important factor in the initiation of cracks.

Near neutral-pH SCC is most prevalent on pipelines with shielding coatings (e.g. tape) and has occurred where the pipeline is apparently protected based on CP information. Nevertheless, it is worthwhile to maintain adequate protection to avoid SCC and corrosion at or near holidays. Effective CP also will minimize the occurrence of near neutral-pH SCC with non-shielding coatings considering that application of cathodic potentials relatively near to Ecorr in all conditions of steel diminishes the susceptibility to SCC. However, at extremely cathodic potentials, the failure is due to HE, rather than SCC.

5. HYDROGEN EMBRITTLEMENTAND THE EFFECT OF ELECTRODE POTENTIAL

The electrochemical potential of the alloy can have a marked influence on the tendency for SCC to occur. For hydrogen embrittlement of high strength steel a more negative potential will tend to increase the rate of hydrogen evolution, and thereby the susceptibility to hydrogen embrittlement. It is less obvious how it happens (Shreir, LL, Jarman, RA and Burstein, GT 1994)[57] but more positive potentials than the typical free corrosion potential may also increase the entry of hydrogen. Figure 2 shows the amount of hydrogen permeating through a steel membrane as a function of the applied potential.

SCC processes that do not involve hydrogen typically occur over a limited range of electrode potential. It is often found that cracking occurs in the transitional potential regions between active and passive or between passive and pitting (Figure 3). In these regions the surface of the component will be in the passive region, while the crack tip will be in the active or pitting state.

In service the electrode potential is not usually controlled directly, with applied cathodic protection being the main example. Rather the potential is determined indirectly by the composition of the environment, particularly the presence of oxygen and other cathodic reactants. Thus modification of the oxygen content can often have a profound influence on SCC susceptibility. On the other hand hydrogen dissolves in all metals to a moderate extent. It is a very small atom, and fits in between the metal atoms in the crystals of the metal.Consequently it can diffuse much more rapidly than larger atoms.

For example, the diffusion coefficient for hydrogen in ferritic steel at room temperature is similar to the diffusion coefficient for salt in water.

Hydrogen tends to be attracted to regions of high triaxial tensile stress where the metal structure is dilated. Thus, it is drawn to the regions ahead of cracks or notches that are under stress. The dissolved hydrogen then assists in the fracture of the metal, possibly by making cleavage easier or possibly by assisting in the development of intense local plastic deformation. These effects lead to embrittlement of the metal; cracking may be either inter- or transgranular. Crack growth rates are typically relatively rapid, up to 1 mm/s in the most extreme cases. The bcc (body-centred cubic) crystal structure of ferritic iron has relatively small holes between the metal atoms, but the channels between these holes are relatively wide. Consequently, hydrogen has a relatively low solubility in ferritic iron, but a relatively high diffusion coefficient.

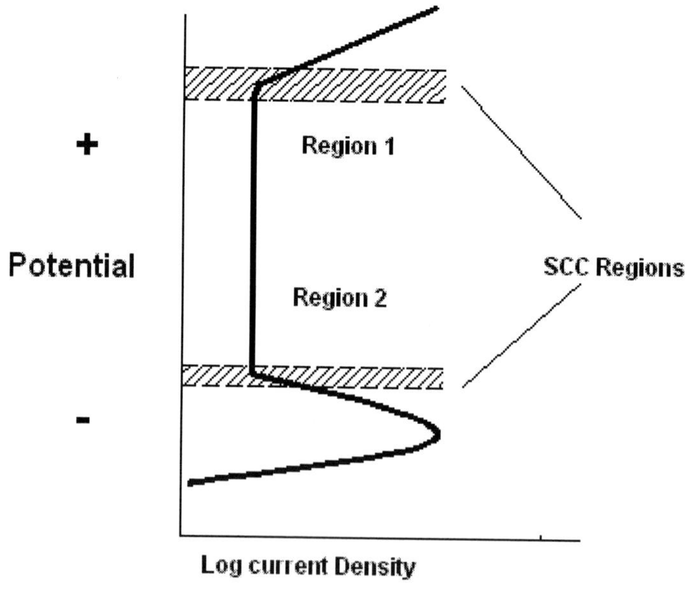

Figure 3. Effect of potential on SCC susceptibility.

In contrast the holes in the fcc (face-centred cubic) austenite lattice are larger, but the channels between them are smaller, so materials such as austenitic stainless steel have higher hydrogen solubility and a lower diffusion coefficient. Consequently, it usually takes much longer (years rather than days) for austenitic materials to become embrittled by hydrogen diffusing in from the surface than it does for ferritic materials, and austenitic alloys are often regarded as immune from the effects of hydrogen.

In addition all steels are affected by hydrogen, as is evidenced by the influence of hydrogen on corrosion fatigue crack growth, and the occurrence of hydrogen-induced crackingunder the influence of very high hydrogen concentrations. However, hydrogen embrittlement under static load is only experienced in steels of relatively high strength. There is no hard-and-fast limit for the strength level above which problems will be experienced, as this will be a function of the amount of hydrogen in the steel, the applied stress, the severity of the stress concentration and the composition and microstructure of the steel. As a rough guide hydrogen embrittlement is unlikely for modern steels with yield strengths below 600 MPa, and is likely to become a major problem above 1000 MPa. The hydrogen may be introduced into the steel by a number of routes, including welding, pickling, electroplating, exposure to hydrogen-containing gases and corrosion in service.

As mentioned earlier high strength pipeline steels are currently being considered for new construction and line extensions for economic reasons. These steels are designed for high strength and high toughness for fracture resistance. Jeffrey Xie (2009)[58] working with high strength steel (X100) under permafrost condition, with the comparison to the regular pipeline steel (X65 steel), found that temperature is an important parameter influencing the polarization resistance, solution resistance, hydrogen generation and the effect of hydrogen on the mechanical properties. For both X100 and X65 steels, polarization resistance is low and similar in values at temperatures above the freezing point, while increasing more than ten (10) times with only 4°C difference from 1°C to -3°C. In general these two steels behave similarly in terms of electrochemical properties for all the temperature spectra. Under freely corroding conditions, the generation rate of hydrogen is small. However the application of cathodic protection increases the generation of hydrogen, especially at temperatures above the freezing point. As a result, both steels became less ductile with increasing cathodic polarization. Though for the condition of permafrost or temperature lower than 20°C, the ductility loss of X100 steel due to hydrogen permeation is comparable to that of X65 steel. This suggests that the use of X100 steel pipe should not result in additional problems associated with hydrogen embrittlement or cracking as a result of cathodic protection application.

On the other hand C. F. Dong et al. (2009)[59] in their study "The effect of inclusions in the steel crack initiation" have shown that the amount of hydrogen-charging into the X100 steel specimen increases with the charging time and charging current density. Hydrogen-charging will enhance the susceptibility of the steel to HIC. The cracks initiate primarily at inclusions, such as aluminum oxides, titanium oxides and ferric carbides, in the steel. The diffusivity of hydrogen at room temperature in X100 steel isdetermined to be 1.04 x10 $^{-8}$ cm^2/s.

In the same way, the study of hydrogen-charging process of pipeline steels X52 Julien Capelle (2009)[60] has shown that this steel is sensitive to hydrogenating in deoxygenated NS4 solution of near-neutral pH value. Applied stress, equal to hoop stress in pipe wall under operating internal pressure 70 bars, significantly accelerates hydrogen charging of steel and hydrogen concentration in loaded steel can be even five times higher than in unloaded steel. It was found that critical hydrogen concentration exists that causes significant reduction of steel local fracture resistance. This value might be applied as an important engineering parameter for reliability assessment of pipelines in exploitation.

Concerning pipeline welded areas, corrosion behavior of welded X70 pipeline steel in near-neutral pH solution was studied by G. A. Zhang and Y.F. Cheng (2009)[61]. They found that the microstructure of weld metal consists of acicular ferrite and grain boundary ferrite, while that of heat-affected zone is a mixture of acicular ferrite, bainitic ferrite and a few martensite/austenite microconstituents. (The microstructure of base steel is typically ferrite and pearlite). Electrochemical corrosion mechanism of welded X70 steel does not experience change upon hydrogen-charging, or stressing, or both. Hydrogen-charging is capable of enhancing the local anodic dissolution of the steel. The resistance of corrosion product layer l decreases with hydrogen-charging, and the heat-affected zone has the largest dissolution current upon hydrogen-charging. The increase of applied stress enhanced the anodic dissolution of welded X70 steel, especially the heat-affected zone, in near-neutral pH solution. Maximum current is observed in the heat-affected zone, and increases with the increase of applied stresses. The total synergistic effect of hydrogen-charging (10 mA/cm2) and applied stress (550 MPa) on anodic dissolution of welded X70 steel in near-neutral pH solution is determined to be within the range of 5.7 and 6.5, with a maximum value when encountering the heat-affected zone. In agreement, J. Capelle (2010)[62] found that in X52, X70 and X100 pipeline steel in NS4 solution, efficiency of hydrogen permeation in metal is quite low and depends on time of exposure. Applied tensile stress, which is equivalent to gross hoop stress in pipe wall under operating conditions, can accelerate the hydrogen absorption in several times. For studied steels the resistance to hydrogen absorption decreases with decreasing steel strength.

More studies have obtained similar results such as that of Ron Wang (2009)[63] inX70 pipeline steel with hydrogen pre-charging and dynamic hydrogen charging in 0.5 mol/L H_2SO_4 solution under slow strain rate tensile testing. The results show that under the hydrogen pre-charging, the fracture toughness decreased in a linear relationship with the hydrogen concentration as the hydrogen concentration was more than 1 ppm in weight. The fracture surfaces were characteristic of dimples. Under the dynamic hydrogen charging, the fracture toughness for hydrogen-induced cracking decreased linearly with logarithm of the hydrogen concentration without stress and the hydrogen-induced fracture had the appearance of cleavage facets. In a related study A. Alhussein (2010)[64] mentioned that the cracks, in the pipeline API 5L X52 steel charged with hydrogen, propagate following the porosity path without any distinct direction and the absorbed hydrogen atoms placed inside the crystalline sites of steel cause the embrittlement of material so that a small effort is sufficient to create cleavage.

Finally, Toribio's (2011)[65] consideration of the material as a live entity immersed in the surrounding environment has an important consequence: the concept of material is strongly linked with the existence of superficial or internal defects or geometrical flaws such as cracks or notches (from the macroscopic point of view) or imperfections in the microstructure such as lattice defects, dislocations, micro-voids, etc. (from the microscopic point of view). Thus the classical approach in mechanical and structural engineering in which the material is totally defined by its constitutive equation – or, even worse, merely by its elastic properties – turns to new approaches (materials science and fracture mechanics approaches) according to which material behaviour depends not only on the intrinsic characteristics of the material itself, but also on the circumstance, i.e., on extrinsic factors such as load history (load magnitude, kind of loading, loading rate, etc.) and environment (temperature, humidity, corrosive agents, etc.), which make previous defects grow. It is important to notice, therefore, that a given material does not have a behaviour per se, but can exhibit one or another behaviour depending on the circumstance, i.e. on the specific working conditions (mechanical and physico-chemical environment).

As a final remark, it is essential to continue research into SCC, because many of the basic questions about SCC have not yet been answered. The SCC mechanism, which accounts for why IGSCC is intergranular and why TGSCC is transgranular, should be understood. There is also a need to continue to develop mitigative measures to deal with SCC. Most notably, the development of a fully reliable SCC in-line inspection tool will significantly improve the industry ability to detect SCC.

Overall, further focused research will enhance knowledge of SCC and will contribute to the development of measures to protect the public and the environment from the consequences of pipeline failures due to SCC.

6. Recent Investigations of the Authors in the Field

Advances in alloy design technique and the adoption of new controlled rolling processes leads to a new generation of pipeline steel with higher strength and toughness. Fine-grained acicular and bainitic microstructures and, most recently, martensitic microstructures were developed by rapid cooling after controlled rolling.

In the present work, near neutral SSC study in NS4 test solution at room temperature was carried out on recently developed microalloyed pipeline steel for oil and gas transport. This study involves Polarization Curves (PC), Electrochemical Impedance Spectroscopy (EIS) and Slow Strain Rate Testing (SSRT).Some microalloyed steel SSRT samples were hydrogen precharged to study their effect on the SCC in the longitudinal and transverse steel rolling directions.

Classical SSC dissolution and embrittlement cracking mechanisms were identified in both rolling directions. Some evidence seems to indicate that the HELP (Hydrogen Enhanced Local Plasticity) mechanism is present in the SSC steel process. This mechanism interacts with a Widmastaten ferrite promoting hydrogen ingress minimizing their mechanical properties.

Figure 1(a) illustrates the behavior of the steel in the potentiodynamic polarization test; as can be seen the open circuit potential (OCP) or corrosion potential (E_{corr}) was -224 mV_{SCE} and the i_{corr} obtained by the Tafel extrapolation method using only the anodic curve was of 0.2 mA/cm^2.

In this case the E_{corr} value obtained is more anodic than other values reported in the specialized literature for these steels in contact with acid solutions, which fall in the -600 to -400 mV_{SCE} range [71-73]. Regarding the obtained i_{corr} value this was increased by an order of magnitude.

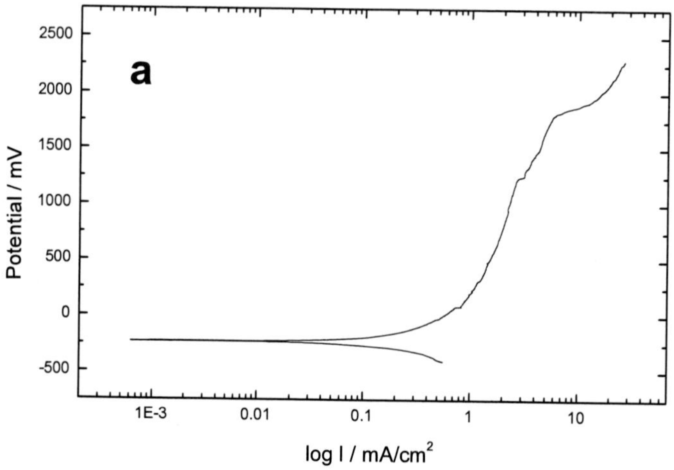

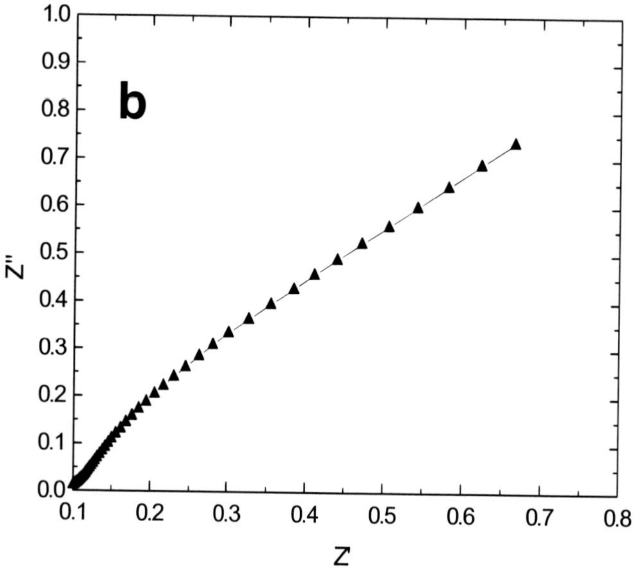

Figure 1. Electrochemical corrosion steel behavior in NS4 test solution, a) polarization curves, b) EIS Nyquist plot.

Table 1. Microalloyed steel chemical composition

C	Si	Mn	P	S	Cr	Mo	Ni	Al	Co	Cu	Nb	Ti	V
.0319	.2355	1.031	.003	.0026	.4243	.1674	1.300	.0520	.0043	.0106	.0235	.0149	-

Table 2. Microalloyed steel mechanical properties

E (GPa)	YS (MPa)	UTS (MPa)	Elong (%)	Deformation (mm)	Rolling Direction
170.5	515	681	14.588	7.294	Longitudinal
163.35	501	663	14.132	7.066	Transversal

E= Young Modulus, YS = Yield Strength, UTS=Ultimate Tensile Strength, Elong= Elongation.

On the other hand Figure 1(b) presents electrochemical EIS results on a Nyquist diagram; the impedance in the complex plane has the tendency to form a straight line with an angle of 45°. This behavior indicates that there was mainly a diffusion mechanism on the surface of the steel. This fact is confirmed taking into consideration the asymptotic form of the polarization curve that is indicative of a continuous anodic dissolution. However the polarization in the anodic curve shows some variation in their continuity, this behavior is related eventually with the presence of a barrier between the metal and the solution. Although it is not a protective film, it affects the i_{corr}, as shown in the polarization curve at 1829 mV_{SCE} the i_{corr} increases suddenly which indicates that the barrier at this point suffers some type of damage or breakdown.

The results in Figure 2 show that regardless of the longitudinal or transverse direction of the steel in SSRT testing, in air and air with hydrogen preload is a ductile fracture. On the other hand in the NS4 and NS4 solution with hydrogen preload a fragile fracture was present. This behavior indicates that the test solution and the E_{corr} play an important role in the steel hydrogen embrittlement and the SCC susuceptibility. Furtheremore, the high energy of hydrogen dissociation related to the steel deformation in a stress-strain field facilitates the penentration of the hydrogen in the steel crystal lattice [66].

The displacement of the E_{corr} to more positive values, indicates that the cathodic reaction was predominant before reaching equilibrium, therefore the presence of the hydrogen increases migrating to the interior of the microstructure as molecular hydrogen. The hydrogen migration thermodynamic behavior has been widly reported, [66-74] finding invariably that the fracture mechanism and susceptibility to the SCC was directly related to the passive film rupture, crack tip anodic dissolution, and hydrogen embrittlement However, analyzing our results the polarizacion curves indicate that steel doesn't generate a passive layer in contact with the NS4 solution, which would be according to the E_{corr} presenting a corrosion behavior dominated by an anodic dissolution mechanism.

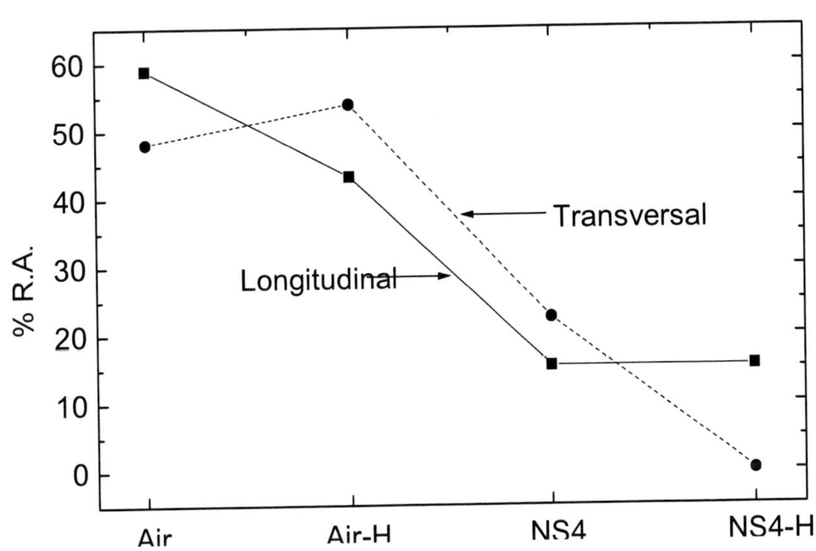

Figure 2. Steel performance under SSR test under different experimental conditions: Air, Air-H, NS4 and NS4-H.

On the other hand, the hydrogen preload in air tests does not generate significant embrittlement; by contrast in the transverse steel it confers greater ductility.This behavior is most likely related to a mechanism known as Hydrogen-Enhanced Localized Plasticity (HELP) [75]. Such mechanism suggests that the stress level decreases due to the addition of hydrogen, which promotes the slipping of dislocations, leading to a failure caused by a ductile process.

The SSR tests without hydrogen preload in contact with the NS4 solution don't present susceptibility to form a passive film as shown in Figure 1. Thus the fragilization observed was consequent of the hydrogen evolving during the corrosion process. For the transversal rolling direction steel, the necking before the fracture ocurr as a result of the transition from a uni-axial stress to a tri-axial stress state due to localized plastic flow. Also, crack initiation in the inclusion-matrix interfaces contribute to the necking behavior in the central sample region. Then the crack propagates in radial form the central part towards the circular steel sample periphery, reaching finally the edge of the fracture [74].

Based on the previous behavior, particularly for this steel in the NS4 solution, the hydrogen generated by corrosion process causes cracking by two complementary hydrogen mechanisms: HELP and the hydrogen embrittlement (HE) cracking mechanisms. So, at an early stage the plasticity was increased by hydrogen-dislocation interactions giving rise to a ductile behaviour. In a second stage the continuous entry of hydrogen eventually weakened and embrittled the steel atomic bonds in both rolling directions, leading to cleavage fracture and blisters.

On the other hand, steel microstructure plays an important role for the mehcanical behavior of the steel. The acicular and widmastaten like ferrite present in the steel have a caothic structure which contribute to increment the yield strength and impose good toughness to the steel [76]. Also, these microstructural characterisitics were important to the SCC steel susceptibility in the test solutions employed to asses its cracking behavior.The intrincsic steel high yield strength produces a vast strain field when a high stress level is also applied to the steel samples in both rolling directions in the SSR tests. This strain field seems to facilitate the hydrogen entry to the stressed and highly deformed steel lattice zone, making the steel more susceptible to SCC and hydrogen effects under the test environments assesed.

It is well known that for pure iron a γ-Fe_2O_3 film is the responsible to the steel passivation [77-79], while a Fe_3O_4 layer promotes an oxidation state over the steel surface. Thus, the Fe_3O_4 layer doesn't contribute to disminish the steel anodic dissolution current. In addition, to get a real steel surface passivation state, first a Fe_3O_4 corrosion products deposit has to be formed, which in turn has a low oxidation state and is highly susceptible to chemical dissolution. That's why until the conditions necessary for the formation and relative stabilization of this layer (Fe_3O_4) was present, the γ-Fe_2O_3s protective film could not be formed and the dissolution of the steel will continue as observed in Figure 1a. Particularly in this study, the displacement of the steel E_{corr} to more anodic values could be related to the previous mentioned corrosion mechanism. This statement could be explained on the characteristics of the Widmanstätten or acicular ferrite structure, in which the rapid cooling from high temperatures led to a ferrite chaotic growth with a needle like form pointing to random directions. This growth leads elongated grains nucleating inside prior austenite grains. This type of ferrite microstructure tends to create galvanic couples with inclusions and precipitates rounding them that could slow down the ferrite corrosion rate, limiting the Fe_3O_4 formation and stabilization promoting the continuous corrosion of the steel.

Higher steel grades, with more strength levels, are more susceptible to SCC and hydrogen embrittlement. This statement is in accordance with this study, being that the steel longitudinal rolling direction is more susceptible to cracking and hydrogen embrittlement than the transveral rolling direction, as shown in Figure 2. As illustrated before in table 2 the steel longitudinal rolling direction has the better mechanical properties, thus applying higher stresses to the SSRT specimens, major strain field zones could be produced, promoting the hydrogen entrance and interaction with the steel lattice and related defects.

In the NS4 solution with hydrogen precharge the steel area reduction in the transversal direction is greater than in the lonely NS4 solution without hydrogen precharge. The presence of transgranular cracks was evident in their fracture surfaces without corrosion products inside the cracks. These cracks apparently grow from the SSRT center specimen to their circular border.

In conclusion the microalloyed steel under study mainly presents a ferritic microstructure with various morphologies (i.e aciular, and polygon) and to a lesser extent bainite and martensite in grain boundaries.This microstructure was the product of controlled rolling and accelerated cooling, which gives the microalloyed steel plate different properties with respect to the rolling direction. The controlled rolling of steel in its longitudinal direction presents best mechanical properties, mainly with respect to its yield strenght and Young modulus.

SSRT testing in both steel rolling directions shows that the microalloyed steel is susceptible to interact with the hydrogen, both pre-charged and produced by corrosion in the NS4 solution and the combination of these two conditions (pre-charged + NS4).

The latter condition simulates the hydrogen that can enterto the steel cathodically protected where the protective coating has suffered a rupture.Steel presents stress corrosion cracking (SCC) in NS4 solution, presenting mainly anodic dissolution and hydrogen embrittlement with the possible presence of HELP mechanism. The NS4 solution + hydrogen pre-charge present the same mechanisms but hydrogen embrittlement was increased; this was noted by the minor area reductions in the steel SRRT samples in both rolling directions.

In the particular case of this steel under the solution conditions assessed, the E_{corr} presented was more anodic with regard to that reported in the literature. The results suggest that this behavior is closely related to the coexistence of different ferrite morphologies in the steel mainly related to the acicular or Widmansttäten ferrite with bainite and martensite patches at grain boundaries.

This type of ferrite microstructure tends to create galvanic couples with inclusions and precipitates rounding them that could slow down the ferrite corrosion rate, limiting the Fe_3O_4 formation and stabilization promoting the continuous corrosion of the steel. The grain boundary martensite and bainite present in the steel could enhance this behavior acting as more active anodic zones, restricting more the ferrite corrosion rate.

It has also determined that this microstructure promotes a synergistic relationship between the hydrogen diffusion in the steel and their mechanical properties reduction. This was especially enhanced with respect to their ductility loss indicated by their area reduction parameter, and cleavage characteristics in their fracture surface.

INTRODUCTION REFERENCES

[1] H.E. Townsend, Jr, *MP,* 1972,10,33.
[2] P.J. Kentish, *Brit. Corrosion. J.,* 1985, 20, 139.
[3] B. Delanty and J. O Bernie, *Oil Gas J.,* 1992 June 15, 39.
[4] R.L. Wenk, 5thSymp. On LinePipe Research, Catalog No. L30174, Arlington, VA, *American Gas Association*, 1974, T-1.
[5] J. M. Sutcliffe, R. R. Fessler, W. K. Boyd and R. N. Parkins, *Corrosion,* 1972, 28, 313.
[6] R. N. Parkins and B.S. Greenwell, *Met. Sci.,* 1977, 11, 405.
[7] R.R. Fessier, 6thSymp. On Line Pipe Research, Arlington, VA, *American Gas Association, 1979*, R-1.
[8] J. A. Beavers and R. N. Parkins, 7thSymp on Line Pipe Research, Arlington, VA, *American Gas Association,* 1986, 25-1.
[9] J. A. Beavers, W.E. Berryand, R. N. Parkins, *MP,* 1986, 25 (6), 9.
[10] J.A. Beavers, T.K. Christman, and R.N. Parkins, *MP,* 1988, 27(4), 22.
[11] R.N. Parkins, C*orrosion,* 1990, 48, 178.
[12] R. N. Parkins, A.J. Markworth and J.H. Holbrook, *Corrosion,* 1988, 44, 572.
[13] R.N. Parkins and P.M. Singh, *Corrosion,* 1990, 46, 485.
[14] T.K. Christman, *Corrosion,* 1990, 46, 450.
[15] R. N. Parkins, E. Belhimer and W.K. B_lanchardJr, *Corrosion,* 1993, 49, 951.
[16] J.T. Justice and D.J. Mackenzie, Proceedings of the NG-18/EPGR *SevethBiennal Joint Technical Meeting on Line Pipe Research, Pipeline Research Committee of the American Gas Association,* Paper No. 28, 1988.
[17] B. Delanty and J. O _Beirne, *Oil Gas J.,* 1985, 20, 139.
[18] R.N. Parkins, W.K. _Blanchard Jr, and B.S. Delanty, *Corrosion,* 1994, 50, 394.
[19] Z. Szklarska-Simialowska, Z. Xia and R.B. Rebak, *Corrosion,* 1994, 50, 334.
[20] B.A. Harle and J.A. Beavers, *Corrosion,* 1993, 49, 861.
[21] A. Plumtee and S.B. Lambert, I*nt. Pipeline Conf.*, New York, NY, ASME, 1996, 1, 565.
[22] T.M. Ahmed, S.B. Lambert, R. Sutherby and A. Plumtree, *Corrosion,* 1998, 54, 115.
[23] L.V. Nielsen, "Hydrogen-Related Stress Corrosion Cracking in Line Pipe Steel", *European Federation of Corrosion Publications, No. 11, Papers from Eurocorr 97,* and Eurocorr 98.
[24] R. N. Parkins, AGA NG-18 *Report* 205, 1992, 61.
[25] National Energy Board, MH-2-95 *Hearing Transcript*, 1996, 16 April, 139.
[26] CEPA, *Submission to the National Energy Board, Proceeding* MH-2-95, 1996, 2, 7.
[27] J.A. Beavers and B.A. Harle., in *Proc. 1st Intern. Pipeline Conf.,* IPC/96, ASME, 1996, 573.
[28] M.J. Wilmott, T.R. Jack and G. Van Boven, *Corrosion/96, Paper No. 242,* (NACE, Houston, TX).
[29] CEPA, *Submission to the National Energy Board, Proceeding MH-2-95,* 1996, 1, Issue 2, NEB.
[30] CEPA, *Submission to the National Energy Board, Proceeding MH-2-95,* 2, 8.
[31] A. Plumtree, S.B. Lambert and R. Sutherby, *European Federation of Corrosion Publications,* 1996, 26.

[32] R.N. Parkins, *AGA PRC Report PR-232-9401,* 1994, November, 54..

SCC Factors, Cathodic Protection, Hydrogen EmbrittlementReferences

[33] Danielson, M.J., and R.H. Jones. 2001. Effect of Microstructure and Microchemistry on the SCC Behavior of Archival and Modern Pipeline Steels in a High pH Environment. *NACE International Corrosion 2001*. Paper No. 01211.

[34] Beavers, J.A. and W.V. Harper. 2004. *Stress Corrosion Cracking Prediction Model. NACE International CORROSION 2004*. Paper 04189.

[35] Torres-Islas, A., Gonzalez-Rodriguez, J.G., Uruchurtu, J., Serna, S. *Stress corrosion cracking study of microalloyed pipeline steels in dilute NaHCO solutions Corrosion Science* volume 50, issue 10, year 2008, pp. 2831 – 2839.

[36] M.A., Arafin Szpunar *Effect of bainitic microstructure on the susceptibility of pipeline steels to hydrogen induced cracking Materials Science and Engineering* A (2010) doi: 10.1016/j.msea.2011.03.036.

[37] Lu. B.T., Luo J. L. *Relationship between yield strength and near-neutral pH stress corrosion cracking resistance of pipeline steels : An effect of microstructure.* Corrosion ISSN0010-9312, 2006, vol. 62, no2, pp. 129-140.

[38] J.T. Bulger, B.T. Lu, J.L. Luo Microstructural effect on near-neutral pH stress corrosion cracking resistance of pipeline steels J Mater Sci 41 (2006) pp. 5001-5005

[39] Chen W., Van Boven G., Rogge R. The role of residual stress in neutral pH stress corrosion cracking of pipeline steels – Part II: Crack dormancy. *Acta Materialia* (2007) 43-53.

[40] Baotong Lu, Jing-Li Luo and Douglas G. Ivey. Near-Neutral pH Stress Corrosion Cracking Susceptibility of Plastically Prestrained X70 Steel Weldment. *Metallurgical and Materials Transactions A Volume 41*, Number 10, 2538-2547, DOI:10.1007/s11661-010-0283-6 (2009).

[41] Li M.C. and Cheng Y.F. Corrosion of the stressed pipe steel in carbonate–bicarbonate solution studied by scanning localized electrochemical impedance spectroscopy. *Electrochimica Acta* 53(2008) 2831-2836.

[42] Torres-Islas A., Serna S., and Campillo B.. 2011 *Hydrogen Effect on Recently Developed Microalloyed Pipeline Steel Stress Corrosion Cracking (SCC) under near-Neutral pH NS4 test Solution.* On manuscript phase.

[43] Beavers, J.A. 1999. On the Mechanism of Stress Corrosion Cracking of Natural Gas Pipelines. 1999 *AGA Symposium*.

[44] CEPA. 1997. *The CEPA Report on Circumferential Stress Corrosion Cracking. Submitted to the National Energy Board.* Canadian Energy Pipeline Association. December.

[45] Zheng WY.*Stress corrosion cracking of oil and gas pipelines in near neutral pH environment: review of recent research Materials: Materials Science and Engineering* DOI: 10.1179/174892309X12555944292234 (2008).

[46] Beavers J.A., and Thompson N.G.. Effects of coatings on SCC of pipelines: new developments. *Proceedings of the First International Pipeline Corrosion Conference*

and Exhibition, (held October 17-20, 1994, in Houston, Texas), Paper No. 14. Copies avail. from Gulf Publishing Co., P. O. Box 2608, Houston, Texas 77252.

[47] Thomas R. Jack, Fraser King. *A Permeable Coating Model For Predicting The Environment at The Pipe Surface Under Cp-Compatible Coatings Paper* 04158, NOVA Chemicals Corp.; NOVA Research & Technology Center; Miroslav Kolar, LS Computing; Robert Grant Worthingham, TransCanada Pipelines CORROSION 2004, March 28 - April 1, 2004 , New Orleans, La 2004. NACE International

[48] Niu L. and Cheng Y.F.*Construction and Building MaterialsVolume 22,* Issue 4, April 2008, Pages 417-422.

[49] Wenk, R.L. 1974. Field Investigation of Stress Corrosion Cracking. In *Proceedings of the Fifth Symposium on Line Pipe Research.* PRCI. November.

[50] Fessler, R.R, T. Groeneveld, and A. Elsea. 1973. Stress-Corrosion and Hydrogen-Stress Cracking in Buried Pipelines. *International Conference on Stress Corrsoion Cracking and Hydrogen Embrittlement of Iron Base Alloys.* June. Published in Stress Corrosion Cracking and Hydrogen Embrittlement of Iron Base Alloys. NACE. 1977.

[51] Mercer, W.L. 1979. Stress Corrosion Cracking — Control Through Understanding. In Proceedings from the 6th Symposium on Line Pipe Research. PRCI. NACE. 2003. External Stress Corrosion Cracking of Underground Pipelines. NACE International. Publication 35103, Item No. 24221. October. NEB. 1996. *Stress Corrosion Cracking on Canadian Oil and Gas Pipelines. Report of the Inquiry.* National Energy Board. MH-2-95. December.

[52] Delanty, B.S., and J. O'Beirne. 1991. Low-pH Stress Corrosion Cracking. In *Proceedings from the 6th Symposium on Line Pipe Research.* PRCI.

[53] Delanty, B.S., and J. O'Beirne. 1992. Major Field Study Compares Pipeline SCC with Coatings. *Oil and Gas Journal* 90, 24.

[54] Marr, J.E. 1990. *The Relationship Between External Pipeline Stress Corrosion Cracking and the Environment Within TCPL's Facilities Across Canada.* Unpublished TCPL Research Report. May.

[55] Delanty, B.S., and J.E. Marr. 1992. Stress Corrosion Cracking Severity Rating Model. In *Proceedings from the International Conference on Pipeline Reliability.* CANMET. June.

[56] Beavers, J.A., and K.C. Garrity. 2001. *100 mV Polarization Criterion and External SCC of Underground Pipelines.*NACE International Corrosion 2001. Paper No. 01592.

[57] Beavers, J.A., C.L. Durr, B.S. Delanty, D.M. Owen, and R.L. Sutherby. 2001. *Near-Neutral pH* SCC: *Crack Propagation in Susceptible Soil Environments. NACE International Corrosion2001.* Paper No. 01217.

[58] Benmoussa A., Hadjel M. and Traisnel M. Corrosion behavior of API 5L X-60 pipeline steel exposed to near-neutral pH soil simulating solution. *Materials and corrosion(2006)* 57, No 10. DOI: 10.1002/maco.200503964.

[59] Ping Liang, Xiaogang Li, Cuiwei Dua, Xu Chen. Stress corrosion cracking of X80 pipeline steel in simulated alkaline soil solution. *Materials and Design* 30 (2009) 1712-1717.

[60] Shreir, LL, Jarman, RA and Burstein, GT(editors) *Corrosion,* Vols I and II, Newnes Butterworth, 1994.

[61] Jeffrey Xie, Lin Yang. *Hydrogen effects on high strength pipeline steels.* Paper Number 09120. CORROSION 2009, March 22 - 26, 2009 , Atlanta, GA Copyright 2009. NACE International.

[62] Dong C.F., Liu Z.Y., Li X.G., Cheng Y.F. Effects of hydrogen-charging on the susceptibility of X100 pipeline steel to hydrogen-induced cracking. *Int.Journal of hydrogen energy* 34 (2009)9879-9884.

[63] Julien Capelle, Igor Dmytrakh, Guy Pluvinage. Hydrogen effect on local fracture emenating from notches in pipeline from steel API X52. *Strength of materials, vol.*41, No 5. 2009.

[64] G.A. Zhang, Y.F. Cheng. Micro-electrochemical characterization of corrosion of welded X70 pipeline steel in near-neutral pH solution. *Corrosion science* 51 (2009) 1714-1724.

[65] Capelle J., Dmytrakh I., Pluvinage G.. Comparative assessment of electrochemical hydrogen absorption by pipeline steels with different strength. *Corrosion science* 52 (2010) 1554-1559.

[66] Rong Wang. Effects of hydrogen on the fracture toughness of a X70 pipeline steel. *Corrosion science* 51 (2009) 2803-2810.

[67] Alhussein A., Capelle J., Gilgert J., Dominiak S., Azari Z.. Influence of sandblasting and hydrogen on tensile and fatigue properties of pipeline API 5L X52 steel. *Int. Jaournal of hydrogen energy.* 36 (2011) 2291-2301.

[68] Toribio J. Fracture Mechanics Approach to Stress Corrosion Cracking of Pipeline Steels: When Hydrogen Is the Circumstance. *Integrity of pipelines transporting hydrocarbons, NATO science for peace and security series C: Environmental security,* DOI 10.1007/978-94-007-0588-3_4.

RECENT INVESTIGATIONS OF THE AUTHORS REFERENCES

[69] Y.F. cheng, *J.Mater.Sci.* 42 (2007) 2701.

[70] G.VanBoven, W.Chen, R. Rogge, *Acta Mater. 55*(2007)29.

[71] W. Chen, G.VanBoven, R. Rogge, *Acta Mater.* 55(2007)43.

[72] B.Y. Fang, E.H. Han, J.Q. Wang, W. Ke., *Corros. Eng. Sci. Technol.* 42 (2007)123.

[73] B.Fang, E.H. Han, J.Wang, W.Ke.,*Corrosion* 63(2007)419.

[74] Z.Y.Liu, X.G. Li, C.W. Du, G.L. Zhai, Y.F. Cheng, *Corr.Sci. 50* (2008) 2251.

[75] A.Torres-Islas, J.G. Gonzalez-Rodriguez, J. Uruchurtu, S Serna, *Corr. Sci.* 50 (2008) 2831.

[76] Zhang Liang, Li Xiaogang, Du Cuiwei, *Wang Yizhong Materials and Design* 30 (2009) 2259.

[77] Maoqiu Wang, Eiji Akiyama, KaneakiTsuzaki.,*Corr.Sci.* 49(2007)11.

[78] H.K. Birnbaum "Mechanisms of Hydrogen Related Fracture of Metals" *Hydrogen Effects on Material Behavior.* The Minerals, Metals & Materials Society, 1990.

[79] G.Z. Koval'chuck, V.N. Geichenko, V.N. Yarmosh, L.V. Podobedova.,*Metal Science and Heat Treatment,* 21(1979)2

[80] S. Szklarska-Smialowska, *Passivity of Metals" Electrochemical SocietyCorrosion Monograph Series,* Princeton, NJ, 1978.

[81] J.A. Bardwell, B. MacDougall, and M.J. Graham, *J.Electrochem. Soc.* 135 (1998) 413.
[82] J.L. Ord, D.J. DeSmet, *J. Electrochem. Soc.* 113 (1966) 1258.

In: Pipelines: Design, Applications and Safety
Editors: M. G. Rivero et al. pp. 69-113

ISBN 978-1-62100-178-2

Chapter 3

A SCIENCE-BASED MODEL FOR CRACK GROWTH OF BURIED PIPELINES UNDERGOING HIGH PH SCC

***B. T. Lu*[1]*, F. Song,*[1] *M. Gao*[2] *and M. Elboujdaini*[3]**

[1]Environmental Performance of Materials, Materials Engineering Department, Mechanical Engineering Division, Southwest Research Institute, 6220 Culebra Rd. San Antonio, TX 78238, USA

[2]Blade Engineering Partners, 16235 Park Ten Place, Suite 450, Houston, TX 77084, USA

[3]CANMET Materials Technology Laboratory, Nature Resources Canada, 568 Booth St., Ottawa, ON, K1A 0G1, Canada

ABSTRACT

A predictive crack growth model for high pH stress corrosion cracking of pipelines is postulated based on the fundamental understandings of the film rupture mechanism. It is known that the cracking process is governed by the factors in three catalogs: (1) the mechanical properties and microstructure of material that can affect the crack-tip strain rate; (2) the parameters characterizing the external loads and (3) the environmental variables that can alter the kinetics of anodic dissolution and repassivation. The theoretical approaches and/or experimental methods are described for determining the parameters that dominate the crack growth.

The experimental validation indicates that the model gives a reasonably good prediction to the effects of important factors relating to materials, environment and loading conditions.

Although fatigue damage is negligible in operation of pipelines for gas transportation, the accelerated dissolution at the crack tip due to the cyclic deformation is still required to be considered in the crack velocity estimation. Finally, the procedure for the field application of the new model is outlined.

Keywords: stress corrosion cracking, passive film, plastic strain, steel.

1. INTRODUCTION

A significant concern for the pipeline operating companies is how to guarantee the safety of aged pipelines as many of them have been approaching to their design life. As one of the leading causes for failure of underground pipelines, stress corrosion cracking (SCC) is may occur in the two broad conditions [1]. After the operation for years, the coating on the external pipe surface may be disbonded at some locations and crevices with holidays would form in these areas, so that groundwater would enter the coating-disbonded region. The transgranular SCC can occur as the external pipe surface is exposed to the dilute groundwater with near-neutral pH. Hydrogen embrittlement is commonly regarded to be the dominant mechanism for cracking [1]. Under the combined action of cathodic polarization and evaporation, concentrated carbonate-bicarbonate solutions with high pH would form in the crevices that can induce the intergranular SCC [2,3]. The high pH SCC is believed to be governed by the repeated rupture of passive film at the crack tip [4-8] and will be the focus of this study.

A predictive model of crack growth would be very helpful to set up reasonable schedule for in-line inspection (ILI) after cracks are detected on the pipe surface, to improve the pipeline safety, and to reduce the maintenance cost of aged pipelines. As pointed out by Parkins [1], the SCC development in pipelines can be divided into 5 stages: (1) the establishment of the corroding environment on the external pipe surface in the coating-disbonded region that is suitable for the occurrence of SCC, (2) the crack initiation; (3) the small crack propagation accompanying with crack initiation; (4) the main crack growth and (5) the pipe failure (see Figure 1) [1,6,9].

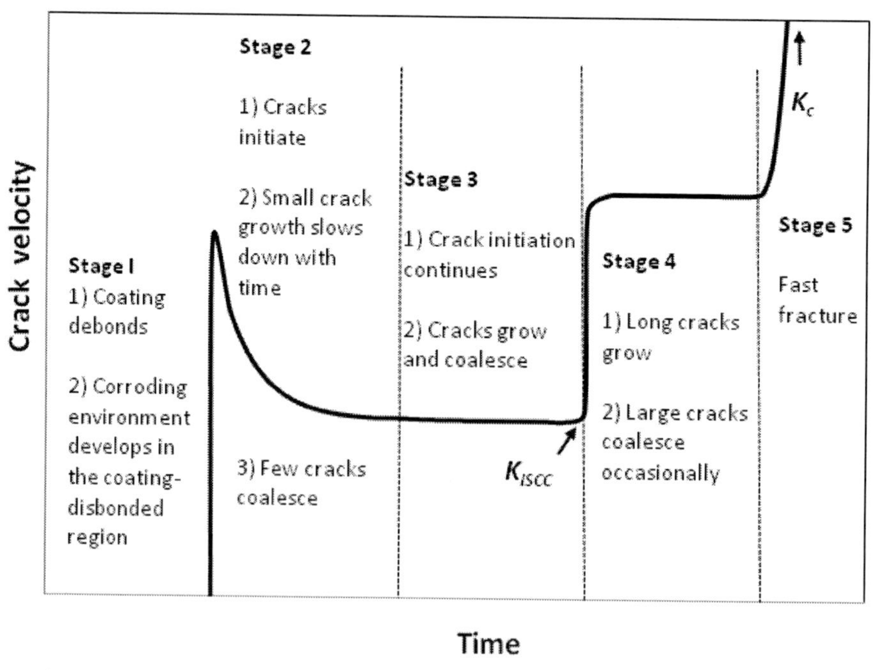

Figure 1. Schematic illustration for the SCC process of underground pipelines [1].

In stage 3, the growth of small cracks decelerates with time and the crack coalescence occurs simultaneously [6,10,11]. The minimum crack size in the depth direction that is detectable by the in-line inspection techniques is typically larger than 1 mm or 10% of pipe wall [1]. In this situation, the stress intensity factor of the cracks on the external surface of high pressure pipe would be close to or exceed K_{ISCC}. Therefore, the objective of this research is to develop a crack growth model to predict the crack velocity in the 4th stage based on our fundamental understanding of the film rupture mechanism.

Efforts were made to build a crack growth model that can quantitatively predict the service life of buried pipelines exposed to concentrated carbonate-bicarbonate solutions [10-13], but few have been directly validated by laboratory investigations. Various techniques have been developed to monitor and measure the crack size in field, such as the in-line inspection, hydrostatic failure test and SCC direction assessment [1].

The crack velocity is the basis for the remaining service life prediction. The average crack velocity of real pipeline is often estimated from a_{max}/t_{age}, where a_{max} is the maximum crack depth detected by the in-line inspection (ILI) and t_{age} is the pipe age [1].

Obviously, it will underestimate the crack growth rate in reality because the pipe age includes the total time spending in stages 1 through 4, namely, the periods required for establishment of local corrosion environment in the coating-disbonded region, crack initiation and small crack propagation. Employing a_{max}/t_{age} would give a prediction on unsafe side.

Therefore, it is necessary to improve the existing crack growth model for the safe operation of underground pipelines. A new crack growth model described herein is mainly based on a recent research of the authors [14,15].

2. Basic Concepts for Modeling

The film rupture or slip-dissolution model has been postulated for more than half century [16-20]. When the environmentally assisted cracking is dominated by the film rupture mechanism, the average crack velocity is formulated by Faraday's law [6,7,19]:

$$\dot{a}_e = \frac{M}{zF\rho}\frac{Q_f}{t_f} = \frac{M}{zF\rho}\frac{\dot{\varepsilon}_{ct}}{\varepsilon_F}\int_0^{\varepsilon_F/\dot{\varepsilon}_{ct}} i_a(t)\mathrm{d}t \tag{1}$$

where the specimen is assumed to have unit thickness; F is Faraday's constant; z is the number of valence electrons of metal, M and ρ are the molar mass and density of steel, respectively; ε_F is the rupture ductility of passive film ε_F and $\dot{\varepsilon}_{ct}$ is the crack tip strain rate. Experimental measurements have indicated that the transient anodic dissolution of carbon steel after the film rupture can be approximately formulated by [20-23]

$$i_a = i_a^*\left(\frac{t}{t_0}\right)^{-n};\ t \geq t_0 \tag{2}$$

When $t < t_0$, the fresh surface created by the rupture of passive film is believed to be film-free. Hence, i_a^* is treated as the anodic current density on the bare metal surface. Normally, $t_0 \ll t_f = \varepsilon_F/\dot{\varepsilon}_{ct}$ and the repassivation exponent n is in a range from 0.5 to 1. The transient anodic current density during the repassivation i_a is well characterized by i_a^* and *n*. After substituting Eq.(2) into Eq.(1), the crack velocity will be further formulated as [19]

$$\dot{a}_e \approx \frac{M}{zF\rho} i_a^* \frac{1}{1-n} \left(\frac{t_0}{\varepsilon_F}\right)^n \dot{\varepsilon}_{ct}^n = \frac{\dot{a}^*}{1-n} \left(\frac{t_0}{\varepsilon_F}\right)^n \dot{\varepsilon}_{ct}^n \quad (3)$$

In Eq.(3), $\dot{a}^* = \frac{M}{zF\rho} i_a^*$ represents the crack velocity is produced by the active dissolution of crack tip while the crack tip is not able to be passivated. Since *F, M, z* and ρ are the material constants, the crack velocity is controlled by the frequency of film rupture events ($= \dot{\varepsilon}_{ct}/\varepsilon_F$) and the average electric charge passing between two successive film rupture events, $Q_f = \int_0^{\varepsilon_F/\dot{\varepsilon}_{ct}} i_a(t)dt$. In accord with Eq.(3), the important parameters to be considered include

- the crack-tip strain rate (CTSR or $\dot{\varepsilon}_{ct}$);
- the fracture ductility of passive film (ε_F);
- the anodic dissolution rate of bare metal surface (i_a^*) and
- the repassivation kinetics (n, t_0).

As schematically illustrated by Figure 2, these parameters are heavily affected various factors relating to material, environmental and mechanical aspects, as well as their interactions.

The transient corrosion rate relies (i_a) heavily on the chemical compositions of both steel and corroding medium at crack tip, as well as the temperature, surface condition and electrochemical potential on the external pipe surface where the SCC occurs [5,8,10,24].

The local solution chemistry at the crack tip is affected by the electrochemical characteristics of steel, permeability of coating, soil type and chemistry, temperature, seasonal changes, mass transfer / electrochemical reactions in the crack and in the coating-disbonded region, as well as the pumping action due to the repeated open-closure of crack under cyclic loading [24,25].

It is still technically difficult to measure the exact values of t_0 and ε_F . The value of t_0 is commonly in the order of 10^{-2} s [13,23]. In accord with the experimental data reported by Ford [19], Harrison [26], Diegle and Vermilyea [27], ε_F is in a range from 5×10^{-4} to 5×10^{-3}. In equation (3), *A* is the function of the rupture ductility of passive film ε_F and the transient anodic dissolution behavior that is assessed by i_a^*, t_0 and *n*.

In the practical applications, the kinetic parameters of repssivation (n, t_0) and the ductility of passive films (ε_F) are often roughly taken as the constants for a given material-environment system [12,13,19]. In this study, the mean value of experimental data, *i.e.*, $\varepsilon_F = 1\times10^{-3}$, is adopted as a reasonable approximation for the film rupture ductility, based on our best knowledge currently available.

The material response to the mechanical force in the cracking process will be characterized by the CTSR, which is a function of mechanical properties of material, the stress intensity factor and its fluctuations during the operation of pipelines [28,29]. The anodic dissolution of crack tip has also an impact. It was experimentally observed the dynamic plastic deformation in the surface layer of a metallic electrode was accelerated when they were exposed to corroding media [30-32].

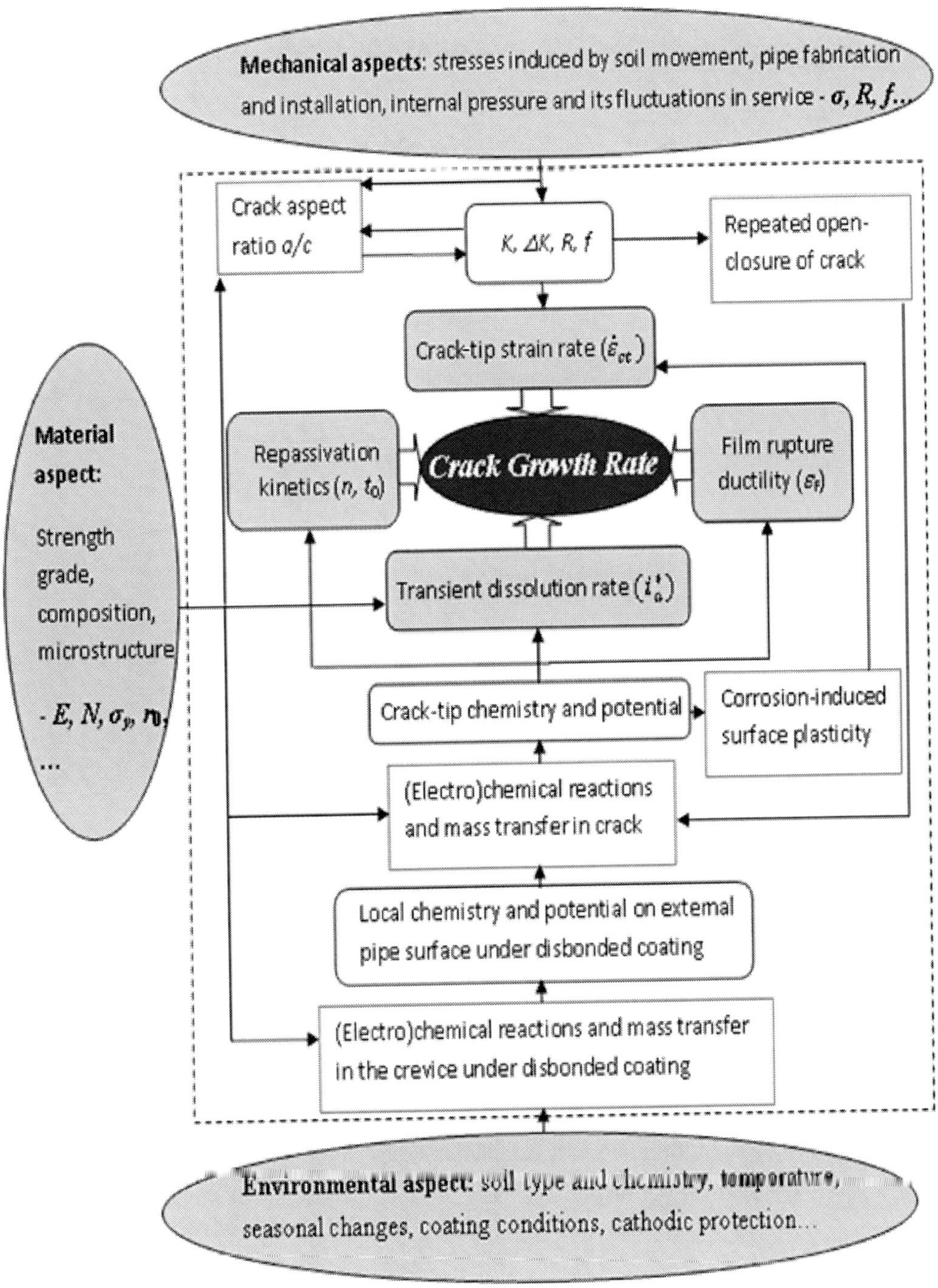

Figure 2. Various factors affecting high pH SCC of pipelines.

The chemo-mechanical effect or corrosion-induced surface plasticity was enhanced with increasing anodic current density [33,34]. Because of difficulty in quantitative assessment, the chemo-mechanical effect will not be considered in the present modeling. As such, the CTSR is regarded as a mechanical quantity.

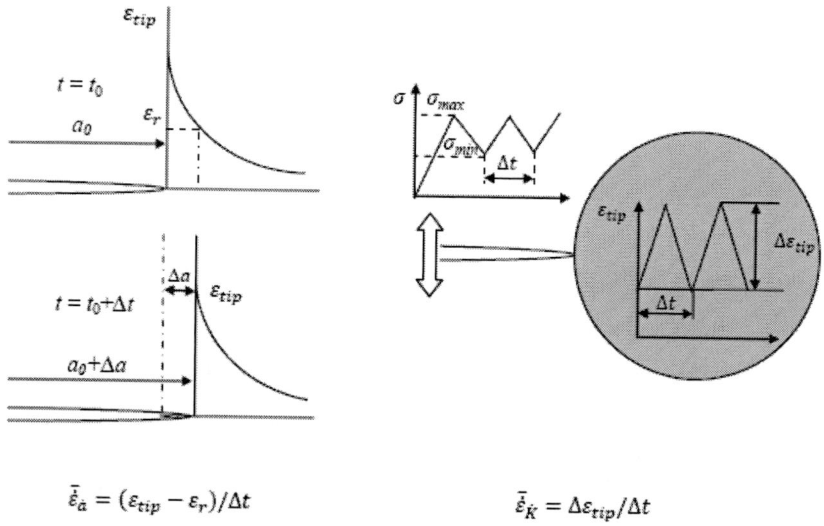

Figure 3. CTSRs produced by the crack tip advance and cyclic loading.

3. Local Strain Rate at Crack Tip

Under the small-scale yield condition, the local stress/strain ahead of the crack tip is K-dominated. If the time-dependent plastic strain (creep) is ignored, the near-tip strain is given by

$$\varepsilon_{ct} = \varepsilon_{ct}(r, K) \tag{4}$$

where r is the distance to the crack tip and K is the stress intensity factor that is a function of external loading, and crack size and geometry. Consequently, the local strain rate at the crack tip will be [35],

$$\dot{\varepsilon}_{ct} = \left(\frac{\partial \varepsilon_{ct}}{\partial K}\right)\dot{K} + \left(\frac{\partial \varepsilon_{ct}}{\partial r}\right)\dot{r} = \left(\frac{\partial \varepsilon_{ct}}{\partial K}\right)\dot{K} - \left(\frac{\partial \varepsilon_{ct}}{\partial r}\right)\dot{a} \quad = \dot{\varepsilon}_{\dot{K}} + \dot{\varepsilon}_{\dot{a}} \tag{5}$$

In Eq.(5) $\dot{r} \equiv -\dot{a}$ by definition and $\dot{a}$ is the instant crack velocity. The physical meanings of $\dot{\varepsilon}_{\dot{K}}$ and $\dot{\varepsilon}_{\dot{a}}$ are schematically illustrated in Figure 3. $\dot{\varepsilon}_{\dot{a}} = -\dot{a}\left(\frac{\partial \varepsilon_{ct}}{\partial r}\right)$ stands for the CTSR component produced by the crack tip advance. $\dot{\varepsilon}_{\dot{K}} = \left(\frac{\partial \varepsilon_{ct}}{\partial K}\right)\dot{K}$ is attributed to the change of stress intensity factor caused by the load fluctuations and/or the increment of crack size.

3.1. Near-Tip Strain Field of a Growing Crack

It is known that the singularity at a non-stationary crack tip is weaker than that at a stationary one [36]. For a strain-hardening material, one of most widely cited solution for the

CTSR for a growing crack is proposed by Shoji [37]. It is based on the formulation of near-tip strain field postulated by Gao and Hwang in 1981[38],

$$\varepsilon_{ct} \propto \left[\ln\left(\frac{R^*}{r}\right)\right]^{\frac{N^*}{N^*-1}} \tag{6}$$

In Eq.(6), N^* is the strain-hardening exponent that is defined by the following equation [38],

$$\frac{\varepsilon}{\varepsilon_y} = \frac{\sigma}{\sigma_y} + \alpha^* \left(\frac{\sigma-\sigma_y}{\sigma_y}\right)^{N^*} \tag{7}$$

where ε_y is the yield strain, σ_y and E are the yield strength and Young's modulus of material, the coefficient α^* is a material constant. 0.2% of plastic strain is often defined as the yield strain. Gao and Hwang [38] did not specify the parameter R^* in Eq.(6).

It is often replaced by the plastic zone ahead of crack tip, R_p. At the plain strain state, $R_p \approx \frac{1}{3\pi}\left(\frac{K}{\sigma_y}\right)^2$ [39]. The typical N^* value for pipeline steel is about 1.6. When steel is exposed to a corroding environment under sustained loading, the environmentally assisted cracking is normally the mechanism of crack growth, *i.e.*, $\dot{a} = \dot{a}_e$. Seeking the time-derivative of Eq.(7), we will have Shoji's equation [37],

$$\dot{\varepsilon}_{ct} = \dot{\varepsilon}_{\dot{K}} + \dot{\varepsilon}_{\dot{a}} = \left(2\frac{\dot{K}}{K} + \frac{\dot{a}_e}{r_0}\right)\frac{N^*}{N^*-1}\frac{\beta\sigma_y}{E}\left\{\ln\left[\frac{1}{3\pi r_0}\left(\frac{K}{\sigma_y}\right)^2\right]\right\}^{\frac{1}{N^*-1}} \tag{8}$$

where the coefficient β is a constant with a typical value 5.08 [37]. The first term at the left side of Eq.(8) represents $\dot{\varepsilon}_{\dot{K}}$ and the second term equals to $\dot{\varepsilon}_{\dot{a}}$. Owing to the singularity of the strain field at crack tip, the local strain and strain rate tend to be infinite as $r \rightarrow 0$. In the practical calculation, the specific length r_0 is introduced into Eq.(8) to replace r. r_0 is commonly treated as a material constant depending on the microstructure of steel and the crack growth mechanism [37].

According to the finite element analysis of Peng et al. [40], r_0 is in the range of 1×10^{-6} m to 5×10^{-6} m for steels. In the present analysis, $r_0 = 1\times10^{-6}$ m is employed.

In 1983, Gao, Zhang and Hwang [41] revised their model to fit the boundary conditions of Model III crack better and they found that the following expression was more suitable to be used to formulate the strain field ahead of the steadily growing crack,

$$\varepsilon_{ct} \propto \left[\ln\left(\frac{R^*}{r}\right)\right]^{\frac{2N}{N-1}} \tag{9}$$

Again, $R^* \approx R_p \approx \frac{1}{3\pi}\left(\frac{K}{\sigma_y}\right)^2$. In Eq.(9), the strain-hardening behavior of material is assumed to obey Ramberg-Osgood's law ($1 < N < \infty$) [42]:

$$\frac{\varepsilon}{\varepsilon_y} = \frac{\sigma}{\sigma_y} + \alpha\left(\frac{\sigma}{\sigma_y}\right)^N \tag{10}$$

The coefficient α is a material constant depending on the definition of the yield strength. The typical N value for pipeline steels is about 6. Seeking the time-derivative of equation (9), we can obtain a different equation for the CTSR ahead of a steadily growing crack [43],

$$\dot{\varepsilon}_{ct} = \dot{\varepsilon}_{\dot{K}} + \dot{\varepsilon}_{\dot{a}} = \left(\frac{\dot{K}}{K} + \frac{\dot{a}_e}{r_0}\right)\frac{2N}{N-1}\frac{\beta\sigma_y}{E}\left\{\ln\left[\frac{1}{3\pi r_0}\left(\frac{K}{\sigma_y}\right)^2\right]\right\}^{\frac{N+1}{N-1}} \tag{11}$$

The solution in the form of Eq.(11) was confirmed by Fan et al. [44] in a plate with a Mode I crack at the plain stress state and by Gao [45] for a Model II crack but the strain-hardening behavior of material is described by Eq.(7). SCC develops normally alone the plain where the maximum tensile stress is applied (a Mode I crack). It is still hard to conclude which constitutive relation is more suitable to the near-tip strain field ahead of a moving crack in pipeline steel.

Actually, the strain-hardening exponent N instead of N^* was often employed, when Shoji's equation (Eq.(8)) was applied to estimate the near-tip strain [35-37,40], although it is not rigorous in theory. Shoji et al. [43] reviewed recently the effects of strain-hardening exponent on the CTSR estimation. They thought the both constitutive models were applicable if a suitable r_0 value was chosen. With the aid of crack-tip strain model formulated by Eq.(9), Hall [35] proposed an alternative CTSR equation as follows.

$$\dot{\varepsilon}_{ct} = \dot{\varepsilon}_{\dot{K}} + \dot{\varepsilon}_{\dot{a}} = \frac{2N}{1+N}\frac{\alpha\sigma_y}{E}\left(\frac{K^2}{\alpha I_n r_0 \sigma_y^2}\right)^{\frac{N}{N+1}}\frac{\dot{K}}{K} + \frac{2\beta N}{N-1}\frac{\sigma_y}{E}\frac{\dot{a}_e}{r_0}\left\{\ln\left[\frac{1}{3\pi r_0}\left(\frac{K}{\sigma_y}\right)^2\right]\right\}^{\frac{N+1}{N-1}} \tag{12}$$

The difference between Eq.(11) and (12) is the first term on the right side: the expression for $\dot{\varepsilon}_{\dot{K}}$. Hall [35] assumed $\dot{\varepsilon}_{\dot{K}} = (\partial\varepsilon_{ct}/\partial t)_{\dot{a}=0}$ and utilized the time-derivative of the near-tip strain field of a stationary crack given by Hutchinson [46], Rice and Rosengren [47] (the HRR field) defined by Eq.(13) to replace the first term in Eq.(11), $\dot{\varepsilon}_{\dot{K}}$.

$$\varepsilon_{ct} = \alpha\frac{\sigma_y}{E}\left(\frac{K^2}{\alpha I_n r_0 \sigma_y^2}\right)^{\frac{N}{1+N}} \tag{13}$$

where α is the material constant in the Ramberg-Osgood law. I_n is a constant depending on the strain-hardening exponent and the stress state of cracked body. When $N \approx 6$, close to the strain-hardening exponent of pipeline steels, $I_n \approx 4.9$ under the plain strain condition [46]. Since $\varepsilon = \delta/r_0$ (δ is the crack tip opening displacement), Hall's equation will become the solution of Rice et al [48] for elastic and perfectly plastic material ($N \to \infty$) at plain strain state.

$$\dot{\delta} = \alpha'\frac{\dot{J}}{\sigma_y} + \beta\dot{a}\left(\frac{\sigma_y}{E}\right)\ln\left(\frac{R_p}{r_0}\right) \tag{14}$$

3.2. CTSR due to Crack-Tip Advance

Under sustained load, the term of $\frac{\dot{K}}{K}$ results from the increase of crack size. In this case, $\dot{\varepsilon}_{\dot{K}} \ll \dot{\varepsilon}_{\dot{a}}$ and $\dot{\varepsilon}_{ct} \approx \dot{\varepsilon}_{\dot{a}}$. In line with Eq.(8) and Eq.(11),

$$\dot{\varepsilon}_{ct} \approx \dot{\varepsilon}_{\dot{a}} = \frac{N^*}{N^*-1}\frac{\dot{a}_e}{r_0}\frac{\beta\sigma_y}{E}\left\{\ln\left[\frac{1}{3\pi r_0}\left(\frac{K}{\sigma_y}\right)^2\right]\right\}^{\frac{1}{N^*-1}} \tag{15}$$

$$\dot{\varepsilon}_{ct} \approx \dot{\varepsilon}_{\dot{a}} = \frac{2N}{N-1}\frac{\dot{a}_e}{r_0}\frac{\beta\sigma_y}{E}\left\{\ln\left[\frac{1}{3\pi r_0}\left(\frac{K}{\sigma_y}\right)^2\right]\right\}^{\frac{N+1}{N-1}} \tag{16}$$

The difference in the two above CTSR equations is the exponents $[(1/(N^*-1)$ vs. $(N+1)/(N-1)]$ and a term in the coefficient $[N^*/(N^*-1)$ vs. $2N/(N-1)]$. Considering $N^* \approx 1.6$ and $N \approx 6$, the ratio of exponents will be 1.67:1.4 and that of coefficient will be 2.67:2.4. It means that the CSTR estimated by these two equations are rather close. Therefore, in the following sections, only Eq.(16) is examined because the Ramberg-Osgood's law is more commonly employed in practice.

3.3. CSTR due to Cyclic Load

Several methods were proposed to estimate the CTSR under cyclic load but few have been directly confirmed by experiments. Owing to the heavily plastic deformation in the plastic zone ahead of crack tip and cyclic hardening/softening, the correlation between the CTSR and $\dot{K}$ (or K) is likely to be nonlinear. Under the small-scale yield condition, the formulation of local strain rate at crack tip caused by the cyclic load is often correlated the alterative crack tip opening displacement (CTOD) range $\Delta\delta$ [28],

$$\Delta\varepsilon_{ct} = \frac{\Delta\delta}{r_0} = d_n\frac{\Delta\delta}{E\sigma_y r_0} = \frac{d_n}{r_0}\frac{\Delta K^2}{E\sigma_y} \tag{17}$$

where d_n is a constant depending on the stress state and the strain-hardening exponent [49], $\Delta K(= K_{max} - K_{min})$ is the cyclic stress intensity factor range, where K_{max} and K_{min} are the maximum and minimum stress intensity factors during cyclic loading, respectively. The relation as Eq.(17) has been experimentally confirmed by the in-situ stereo-image technique [22,50]. Therefore, the average CTSR of a stationary crack in a stress cycle can be approximated by

$$\dot{\varepsilon}_{\dot{K}} = \frac{\Delta\varepsilon_{ct}}{\Delta t} = f d_n\frac{\Delta K^2}{E\sigma_y r_0} = \varepsilon_0 f(1-R_\sigma)^2 K_{max}^2 \tag{18}$$

the stress ratio $R = \frac{K_{min}}{K_{max}}$, $\Delta t(= f^{-1})$ is the time duration of a stress cycle; for pipeline steel at the plain strain state, $d_n \approx 0.3$ [49]; ε_0 is regarded as a material constant depending on the environment of testing [50] because hydrogen dissolved in material [25,51,52] and/or the adsorption of surface-active species on surface [30,53] may alter the mechanical properties of material. For instance, the CTODs measured from cracks in hydrogen-charged steels are smaller than those measured from the non-charged ones [51,52], owing to the embrittlement of material. The typical values of yield strength and Young's modulus of pipeline steel are 480 MPa and 2×10^6 MPa, respectively. In line with Eq.(18),

$$\varepsilon_0 = \frac{d_n}{E\sigma_y r_0} = \frac{0.3}{2\times10^6\times480\times1\times10^{-6}} = 3.13\times10^{-4}\ \text{MPa}^{-2}\text{m}^{-1}$$

It is close the ε_0 value ($\approx 1.56\times10^{-4}$ MPa^{-2}m^{-1}) reported by Hudak et al.[50], who measured the crack-tip strain of type 304 stainless steel using the stereo-imaging technique. This fact suggests that $r_0 = 1\times10^{-6}$ m is likely a reasonably good approximation in the CTSR estimation at the plain stress state.

3.4. Threshold of Crack Growth

Based on the concept of the film rupture model, the local strain accumulated will eventually reach the critical value ε_F to cause the rupture of passive film at the crack tip no matter how small the CSTR is. If $K_{ISCC} = 0$, a crack would not stop propagating so long as $K > 0$. Numerous evidences have indicated the existence of a threshold stress intensity K_{ISCC} below which the crack arrests under either sustained or cyclic load. As such, modifications are required for the CTSR modeling to account on the existence of K_{ISCC}.

Under cyclic loading, the crack tip cannot open until the stress intensity factor exceeds a critical value [52,54]. The crack closure results from the reversed plastic deformation in the plastic zone ahead of the crack tip, the surface roughness in crack walls and the build-up of oxides/corrosion products within the crack enclave [54].

The crack closure effect is significant for a long crack and it declines and eventually disappears as the crack size is reduced. This is why the short cracks can propagate faster [54,55]. The details of crack closure mechanisms are out of scope of this study. The concept of crack closure will be employed to introduce the threshold in crack growth, considering the fact that the stress fluctuations cannot affect the CTSR as long as the crack surfaces keep in touch. As such, the CTSR is likely a function of effective CTOD [50,54,56]. According to McEvily [56], the effective CTOD of a fatigue crack is expressed as follows,

$$\Delta\delta_{eff} = \Delta\delta - \Delta\delta_{th} \tag{19}$$

The fatigue crack will cease propagation as $\Delta\delta_{eff}(\propto \Delta K^2 - \Delta K_{th}^2)$ approaches to zero. ΔK_{th} is the threshold of crack growth. Obviously, $\Delta\delta_{th} = d_n\Delta K_{th}^2/E\sigma_y$ [56]. $\dot{\varepsilon}_{ct} = 0$ whenever $\Delta\delta \le \Delta\delta_{th}$. No matter the cyclic load is present to not, a stress corrosion crack ceases propagating as $K \le K_{ISCC}$ [11,57]. Thus, it is reasonable to assume,

$$\Delta\delta_{th} = \frac{(1-R)^2K_{ISCC}^2}{E\sigma_y} = r_0\varepsilon_0(1-R_\sigma)^2K_{ISCC}^2 \tag{20}$$

The CSTR created by the crack tip advance must also be zero as long as $K_{max} \leq K_{ISCC}$. After replacing K^2 in Eqs.(16) and K_{max}^2 in Eq.(18) with $K_{max}^2 - K_{ISCC}^2$, the CTSR will be re-formulated into,

$$\dot{\varepsilon}_{ct} = \dot{\varepsilon}_{\dot{a}} + \dot{\varepsilon}_{\dot{K}} \approx \frac{\dot{a}_e}{r_0}\frac{2N}{N-1}\frac{\beta\sigma_y}{E}\left[\ln\left(\frac{K_{max}^2-K_{ISCC}^2}{3\pi r_0\sigma_y^2}\right)\right]^{\frac{N+1}{N-1}} + f\varepsilon_0(1-R_\sigma)^2(K_{max}^2 - K_{ISCC}^2) \quad (21)$$

The values of parameters in Eq.(21) are given in Table 1.

4. Local Environment at Crack Tip

The local chemistry at the crack tip depends on the solution chemistry at the crack mouth (or the bulk solution chemistry used in the SCC tests), the kinetics of electrochemical reactions and mass transfer in the crack. The crack-tip solution chemistry is difficult to be determined using experimental methods.

To assess the significance of local chemistry changes in the crack during the crack propagation under sustained loading, the coupling process of mass transfer and electrochemical reactions are numerically simulated for a crack exposed to 0.5 M Na_2CO_3 + 1 M $NaHCO_3$ solution at 75 °C using the program that was originally developed by Song [13], where the anodic kinetics was characterized by an anodic polarization curve of steel measured in the test solution on interest. In the present study, the following modifications were made in the numerical simulation.

- The surface except at the crack tip was assumed to be under the passive condition and the passive current density was assumed to 10^{-6} A/m^2.
- Following Parkins [7,9], the dependence of anodic current density of the bare metal surface on the potential were characterized using the potentiodymanic curve that was measured with the potential-scanning rate 1 V/min.
- Crack width δ_r was formulated by [24,25]

$$\delta_r = \delta + \theta r \quad (22)$$

where δ was the crack tip opening displacement; r was the distance from the crack tip; θ was the crack opening angle which could be calculated in line with the equations given by [20,58].

A typical the polarization curve applied in the crack velocity estimation is illustrated in Figure 4 [14], which was measured from pipeline steel in 0.5 M Na_2CO_3+ 1 M $NaHCO_3$ aqueous solution at 75 °C with potential-scanning rate of 1 V/min [57]. The anodic current density reaches its maximum at a potential around -0.725 V_{CSE} (the potential vs. Cu/$CuSO_4$ electrode), where pipeline steel displays its maximum SCC susceptibility, as demonstrated by the slow strain rate tensile (SSRT) tests [4,59].

If the polarization curves reported by different researchers are compared, it will be found that the shapes of polarization curves are similar and the peak anodic current density appears at a potential very close to -0.725 V_{CSE}. However, the peak anodic current densities are in a range from 140 A/m^2 to 700 A/m^2 [7,12,13,57].

According to the experimental measurements [13], $n \approx 0.667$ is a good approximation for pipeline steel in aqueous solution of 0.5 M Na_2CO_3 + 1 M $NaHCO_3$ at 75 °C. In the present study $n = 0.667$ will be utilized for the crack velocity estimations except mentioned. The parameters utilized in the simulation are listed in Table 1.

Table 1. Parameters used in calculation

n	Repassivation kinetic exponent	0.667
ε_F	Rupture ductility of passive film	0.001
t_0(s)	Incubation of repassivation	0.01
r_0 (m)	Specific length for CTSR calculation	1×10^{-6}
E (GPa)	Young's modulus	200
σ_y (MPa)	Yield strength	480
N	Strain hardening exponent defined by Ramberg-Osgood law	6
ν	Poisson ratio	0.3
σ (MPa)	Stress	341
ε_0 ($MPa^{-2}m^{-1}$)	Cyclic CTSR coefficient	3.13×10^{-4}
β	Rice's coefficient	5.08
M (kg/mol)	Atomic mass of iron	55.845×10^{-3}
F (C/mol)	Faraday's constant	96485
Q (kJ/mol)	Thermal activation energy of crack growth	42
z	Number of electrons exchanged in corrosion	2
ρ (kg/m^3)	Density of iron	7.847×10^{6}
d_n	Coefficient for CTOD calculation	0.3
w (m)	Pipe wall thickness	9.42×10^{-3}
D (m)	Pipe diameter	1.0668
a_0 (m)	Initial crack depth	1×10^{-3}
a_f (m)	Final crack depth	4.71×10^{-3}

The other technical details about the numeric simulation have been described elsewhere [13]. The results in Figure 5 indicate that, when the initial value of i_a^* is lower than 400 A/m^2 (corresponding to a plateau crack velocity below 1×10^{-9} m/s), the potential drop within the crack with a depth of several mm is only a few mV, resulting in a very small driving force for the electric immigration in the crack. This is why the changes of ion concentration (Figure 6) and pH (Figure 7) in the solution at the crack tip are insignificant. Consequently, the variation of anodic current density on the fresh crack-tip surface (i_a^*) is also negligible (Figure 8).

It was assumed in the numeric simulation that the chemical composition of bulk solution (at the crack mouth) was assumed to be 0.5 M Na_2CO_3 + 1 M $NaHCO_3$ solution, the temperature was 75 °C and the applied potential at the crack mouth was -0.725 V_{CSE}.

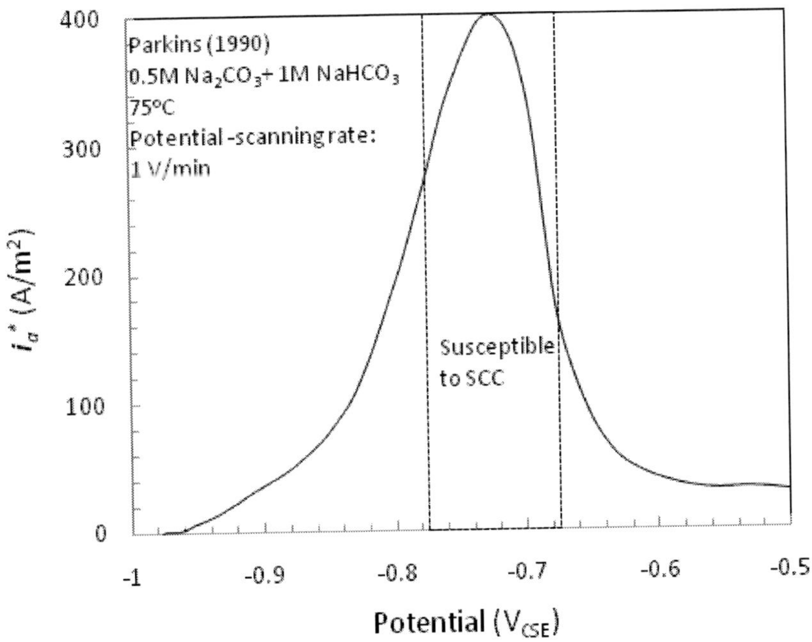

Figure 4. Potentiodynamic curve used to simulate the anodic dissolution kinetics of bare surface of pipeline steel [7].

During the practical operations, the pipelines are exposed to the cathodic protection of certain extent and the service temperature is often lower than 75°C. Actually, the combined actions of cathodic polarization and evaporation are necessary for the formation of concentrated carbonate/bicarbonate solutions with high pH in the coating-disbonded regions [2,3]. The water samples collected from field indicated that the local chemistries were concentrated as the model one (0.5 M Na_2CO_3 + 1 M $NaHCO_3$) [6]. The cathodic polarization and less concentrated solution will result in lower crack velocities [4-6,8]. The average crack velocities reported from the field observations were frequently lower than 10^{-11} m/s [1,6]. For this reason, the potential drop and the mass transfer in crack will not be considered in the modeling. The following simplifications will be adopted in the crack velocity prediction.

- $E_{crack\ tip} = E_{crack\ mounth}$.
- The chemical composition of solution at the crack tip is the same as that of bulk solution.
- The anodic current density on the bare metal surface at crack tip can be estimated from the polarization curve measured by the fast potential scanning (1 V/min) when the potential at the crack mouth is known.

However, the local environments over the external pipe surface, *i.e.*, the potentials and solution chemistries at the crack mouths, are affected by the mass transfer in the crevice under the disbonded coating, cathodic polarization, seasonal changes and evaporation [61-64].

The solution compositions on the real pipe surface demonstrated in Table 2 are different from those used in the laboratory investigations [2-4,6,8]. The impacts of solution chemistry will be further discussed in the following sections.

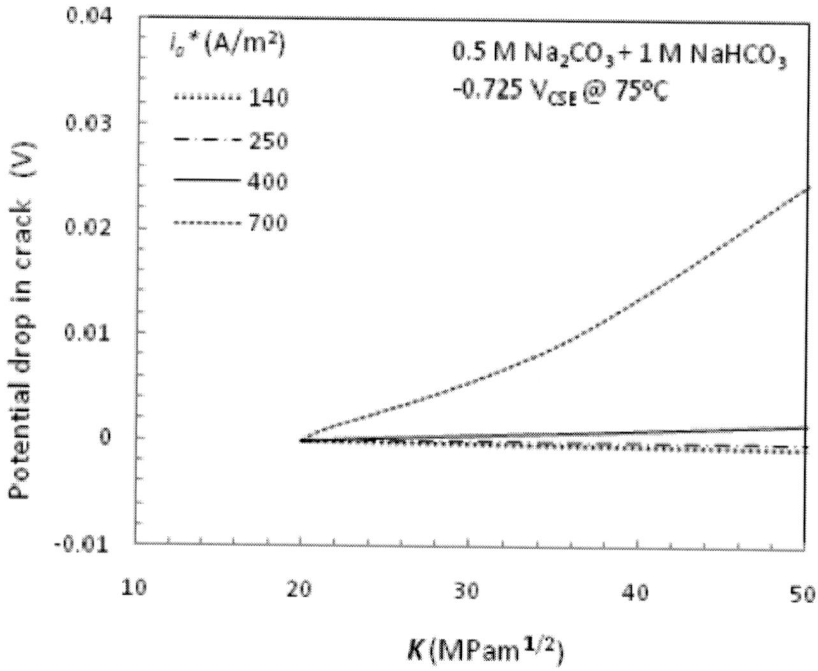

Figure 5. The potential drop in crack during crack propagation.

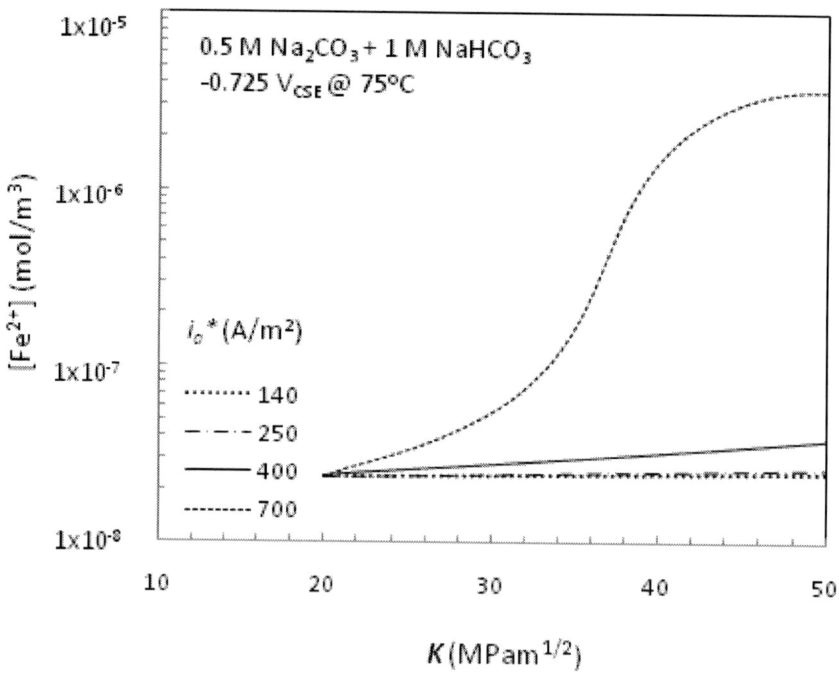

Figure 6. The change of concentration of Fe^{2+} at the crack tip during crack propagation.

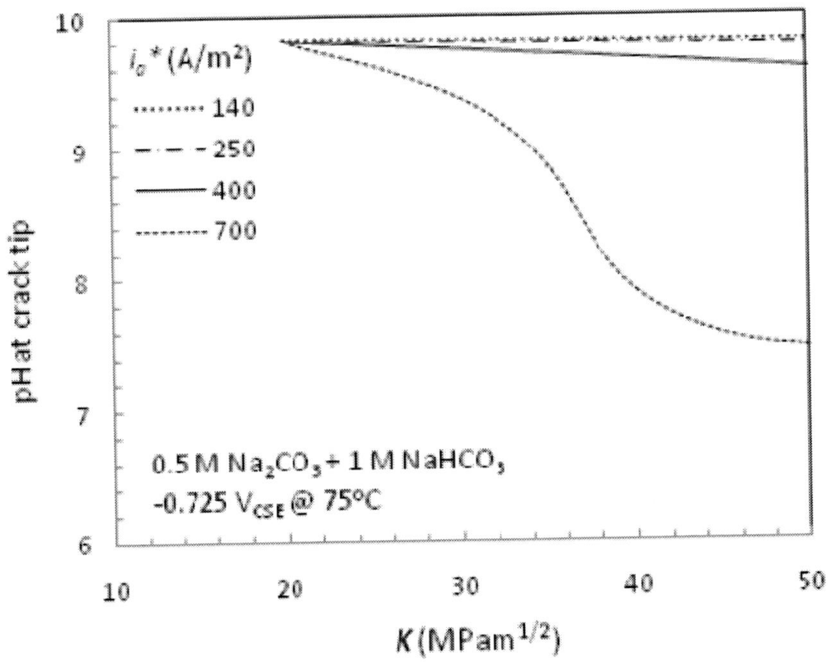

Figure 7. The change of pH at the crack tip during crack propagation.

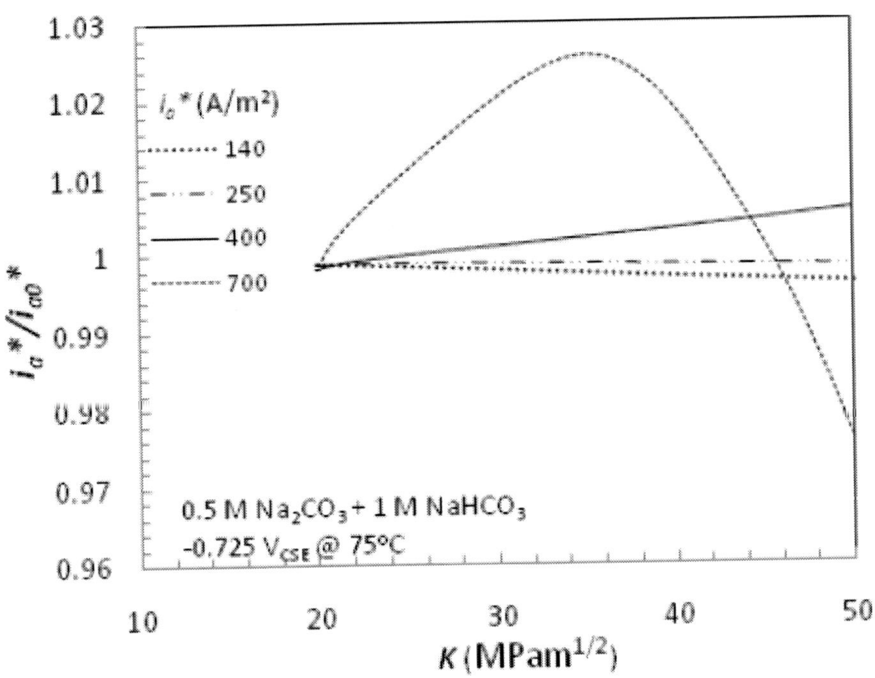

Figure 8. The relative change of anodic current density of the bare crack-tip surface during crack propagation (i_a^* is normalized by its initial value).

Table 2. Ion concentrations of the solutions in coating-disbonded region adjacent to stress corrosion cracks# [6]

State	pH	$[CO_3^{2-}]$ (M)	$[HCO_3^-]$ (M)	$[Cl^-]$ (M)	$[NO_3^-]$ (M)
Alabama	9.7	0.083	0.042	-	-
Arizona	12.3	0.17	N/A	0.0014	0.00056
Mississippi	10	0.47	0.042	0.017	0.00038
Mississippi	10	0.3	0.066	0.017	<0.0008
Mississippi	9.6	0.083	0.049	N/A	-
North Carolina	10.5	0.12	0.033	-	N/A

Na^+ is predominant cation in each case.

5. Formulation of Crack Gowth Model

Inserting Eq.(21) into Eq.(3), we have,

$$\dot{a}_e = A\left\{\frac{\dot{a}_e}{r_0}\frac{2N}{N-1}\frac{\beta\sigma_y}{E}\left[\ln\left(\frac{K_{max}^2-K_{ISCC}^2}{3\pi r_0\sigma_y^2}\right)\right]^{\frac{N+1}{N-1}} + \xi(K_{max}^2 - K_{ISCC}^2)\right\}^n \quad (23)$$

This is an expression of crack velocity dominated by the film rupture mechanism, where

$$A = \frac{\dot{a}^*}{1-n}\left(\frac{t_0}{\varepsilon_F}\right)^n = \frac{i_a^*}{1-n}\frac{M}{zF\rho}\left(\frac{t_0}{\varepsilon_F}\right)^n \quad (23a)$$

$$\xi = f\varepsilon_0(1-R_\sigma)^2 \quad (23b)$$

A is the important parameter characterizing the material-environment interactions under the action of applied stresses and ξ is the unique parameter to characterize the cyclic loading effect on the CTSR. The second term in the right side of Eq.(23) represents the contribution of cyclic loading-enhanced SCC.

6. Prediction of Crack Velocity

6.1. *K* - and *f* - Dependences

The predicted curves for crack velocity under sustained load vs. *K* are shown in Figure 9. The experimental data reported by Parkins [57] are also presented for comparison. The results indicate that, when the parameter values listed in Table 1 are applied, Eq.(23) gives a fairly good description for the dependence of crack velocity on the stress intensity factor (*K*). The predicted crack velocity depends on the value of i_a^*. It seems that the crack growth model will provide a reasonably good prediction to the experimental data after the i_a^* value is properly selected. In Figure 9, the estimated crack velocities under the sustained load with and without considering the potential drop/mass transfer in the crack are compared. It seems to be a reasonably good approximation to ignore the effects of potential drop and mass transfer in crack when pipeline steel is exposed to the concentrated carbonate-bicarbonate solution, when the crack velocity is not higher than 1×10^{-9} m/s.

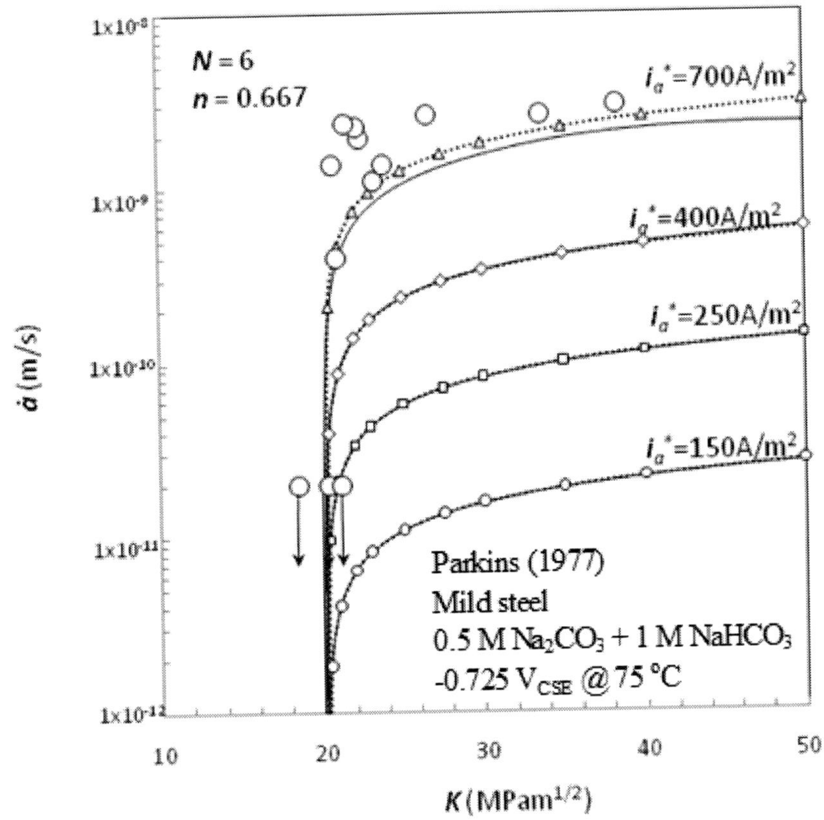

Figure 9. Dependence of crack growth rates on the anodic current density of bare crack-tip surface. Solid lines: the predictions accounting on the mass transfer effect in the crack. Dotted line + small open points: the predictions without considering the mass transfer effect. Large open circle points: the experimental data reported by Parkins [57].

In Figure 10, the predicted curves obtained from Eq.(23) are compared with the test data obtained under the cyclic loading conditions [60]. A good agreement is achieved if the peak anodic current density of the polarization curve at -0.725 V_{CSE} is assumed to be 300 A/m^2, which falls well in the range of experimentally measure values.

In the crack velocity estimation, K_{ISCC} = 25 MPam$^{1/2}$ reported by the authors is adopted. As indicated by Figures 10 and 11, the cyclic load can almost double the crack velocity, when stress ratio is 0.82 and loading frequency is 4.63×10^{-3}Hz (corresponding to $\xi \approx 5\times10^{-8}$ MPa^{-1}m^{-1}).

The impacts of cyclic load produced by the internal pipe pressure fluctuations on the crack propagation of pipeline steel in the high pH chemistry are related to the following three aspects,

- enhancing exchange of the liquid in the crack with bulk solution;
- affecting the crack tip strain rate via $\dot{\varepsilon}_{\dot{K}}$;
- producing fatigue damage.

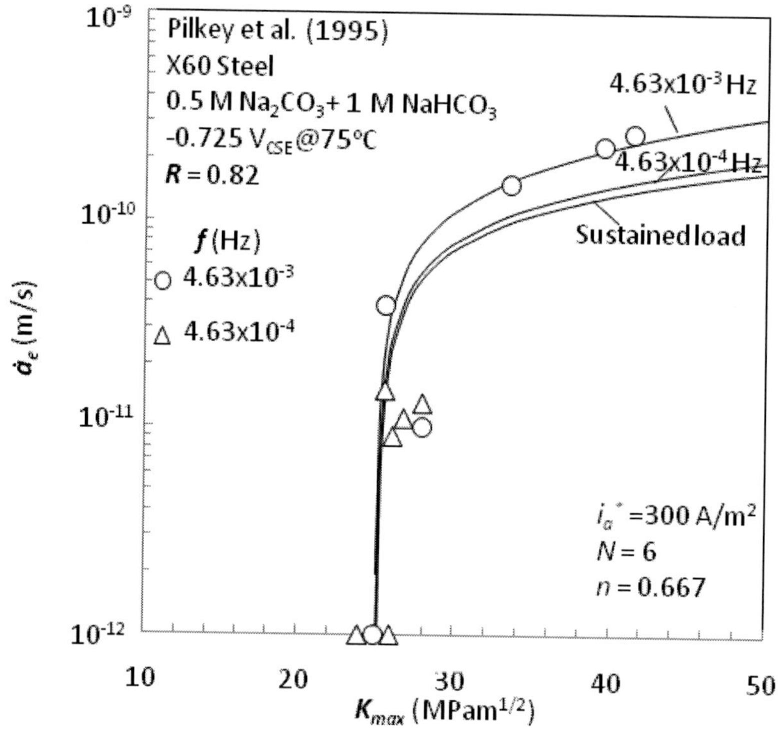

Figure 10. Comparison of the experimental and predicted crack velocities (the test data are quoted from Ref. [60]).

The pumping action produced by the repeated opening and closing under cyclic load will enhance the exchange of the solutions in and out of the crack, and further reduce the difference between the local chemistries in crack and bulk solution [24,25]. The pumping effect would be insignificant because the mass transfer in crack plays a minor role in the cracking process.

The effect of cyclic loads on crack tip dissolution can be assessed using the CTSR that is a linear function of ξ when K_{max} is fixed. An increase of loading frequency and/or a decrease of stress ratio will lead to an increase of CTSR, resulting in the cyclic-load-enhanced SCC.

Fatigue may be a factor to be considered in life prediction of the pipeline for liquid transportation. It will be discussed in section 6.5.

6.2. Effects of Potential and Anodic Current Density

As illustrated in Figures 9 and 10 suggests, the crack growth model developed in Section 4 can provide a good prediction to the dependence of crack growth on the mechanical parameters ($\overline{K}, f$ and R_σ). However, the local chemistry and electrochemical potential on pipe surface under the disbonded coating also have an impact on the crack propagation because they alter the transient anodic dissolution kinetics [3-8, 61-64].

As indicated by a typical dependence of crack velocity on the electrochemical potential ($\dot{a}$: the solid line in Figure 11), pipeline steel displays the maximum crack velocity at the potential where the polarization curve shows an anodic current peak (the fine dotted line on top of Figure 11) and the crack velocity declines as i_a^* is reduced. This phenomenon has been observed experimentally [4,7,8,59,64].

To demonstrate the impacts of different CTSR components, an example for the crack velocity under the sustained load ($\dot{a}_s$: the dashed-dotted line) and the crack velocity component caused purely by the cyclic loading-enhanced SCC ($\dot{a}_c$: the dashed line) are compared in Figure 11. $\dot{a}_s$ and $\dot{a}_c$ are respectively given by the following equations,

$$\dot{a}_s = A\dot{\varepsilon}_{\dot{a}}^n = \left(\frac{\dot{a}^*}{1-n}\right)^{\frac{1}{1-n}} \left(\frac{t_0}{\varepsilon_F}\right)^{\frac{n}{1-n}} \left(\frac{2N}{N-1}\frac{\beta\sigma_y}{r_0 E}\right)^{\frac{n}{1-n}} \left[\ln\left(\frac{K_{max}^2 - K_{ISCC}^2}{3\pi r_0 \sigma_y^2}\right)\right]^{\frac{N+1}{N-1}\frac{n}{1-n}} \quad (24)$$

$$\dot{a}_c = A\dot{\varepsilon}_K^n = \frac{\dot{a}^*}{1-n}\left(\frac{t_0}{\varepsilon_F}\right)^n [\xi(K_{max}^2 - K_{ISCC}^2)]^n \quad (25)$$

The dependence of crack velocity on the kinetics of anodic dissolution is more clearly shown in Figure 12, where the crack velocities are plotted against i_a^* in the log-log scale. The CTSR produced by cyclic load is a function of ξ independent of the crack growth rate, so that $\dot{a}_c$ is a linear function of i_a^*, as indicated by a group of parallel lines with a slope of 1. Because the CTSR under sustained load is proportional to the crack velocity, $\dot{a}_s$ is a power function of i_a^* with an exponent of $1/(1-n)$, as demonstrated by the dotted-dashed line in Figure 12. The fact of $n < 1$ implies that the crack velocity under sustained load is more sensitive to i_a^*, which is, in turn, a function of electrochemical potential as described by the polarization curve.

As shown by Figure 11, the significant crack growth under sustained load is only observed within a narrow range of potential (about 0.1 V), which is smaller than the potential range for the crack propagation under cyclic load. This agrees with the fact that corrosion fatigue can occur at the potentials where SCC under sustained load does not.

As shown in Figure 12, the impact of cyclic load is enhanced with decreasing aggressiveness of corroding medium that is characterized by i_a^*, indicating the synergistic effects of mechanical and environmental factors in the cracking process. There is a critical value of ξ.

When $\xi > \xi_c$, the effect of the corroding medium investigated will tend to be saturated because the crack tip is unable to be repassivated. According to the data in Figure 12, the critical value $\xi_c \approx 5 \times 10^{-6}$ $MPa^{-2}m^{-1}s$.

This prediction agrees with the experimental observations reported by Wei et al. [65] who measured the corrosion fatigue crack velocity of X70 steel in quiescent solution of 0.5 M Na_2CO_3 + 1 M $NaHCO_3$ at 70 °C when $\Delta K = 40$ MPa and $R_\sigma = 0.1$ within the range of cyclic loading frequency from 0.03 Hz to 10 Hz (the corresponding value of $\xi > 3\times10^{-6}$ $MPa^{-2}m^{-1}s$) and found crack velocity was independent of loading frequency in the stagnant solution.

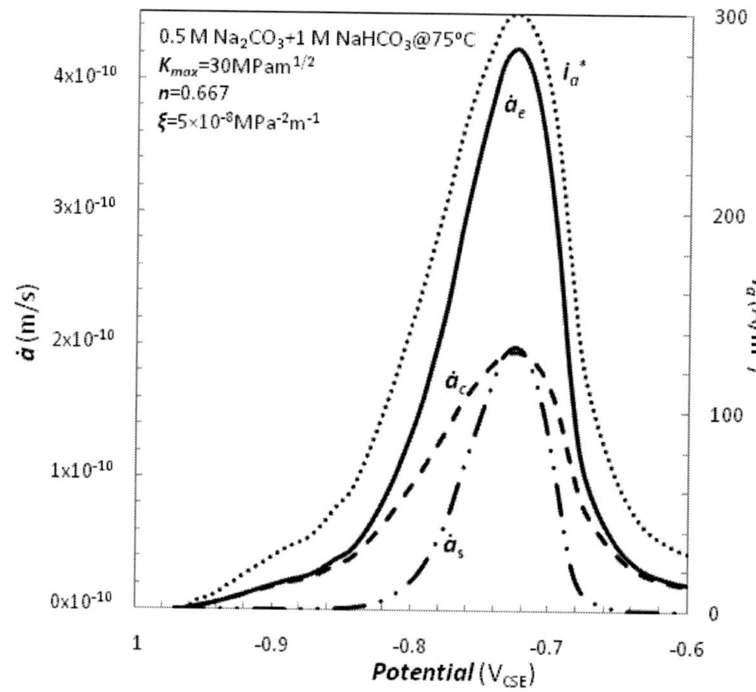

Figure 11. Effect of potential on crack growth rate ($R_\sigma = 0.82; f = 4.63 \times 10^3$Hz).

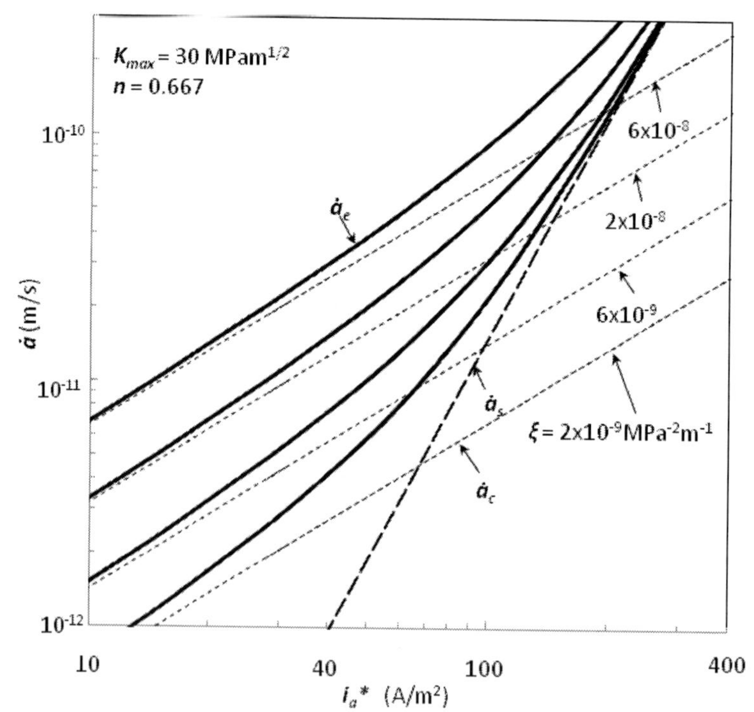

Figure 12. Effect of i_a^* on crack velocity.

6.3. Impact of Repassivation Kinetics

The repassivation exponent n is the parameter characterizing the repassivation capacity. A larger n value corresponds to a strong passivating capacity. When i_a^* and $t_f = \varepsilon_F/\dot{\varepsilon}_{ct}$ are held unchanged, the crack velocity decreases with an increase of n, as demonstrated by some examples of crack velocity under sustained load in Figure 13. As predicted by Eq.(25), the crack velocity component due to the cyclic-load-enhanced SCC is proportional to ξ^n, if the other parameters are held unchanged. It suggests that the crack velocity is more sensitive to the change of cyclic loading condition as the repassivating capacity is increased.

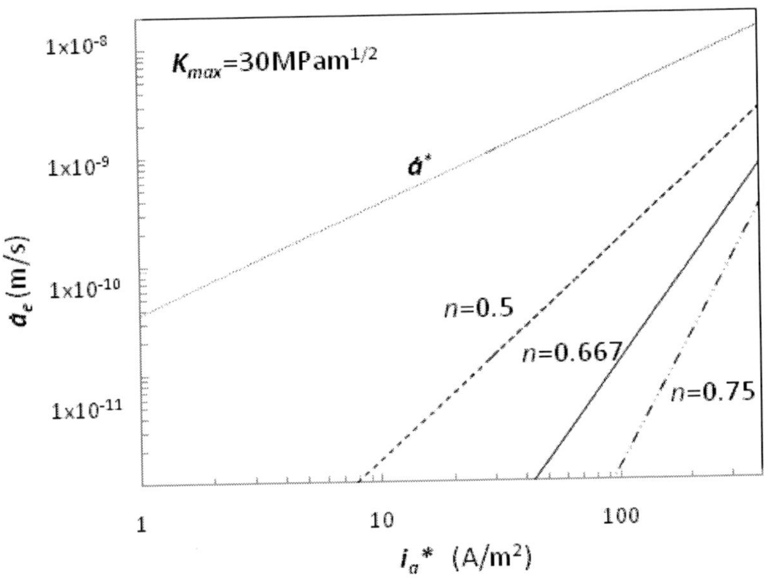

Figure 13. Effect of repassivation kinetics exponent on crack velocity under sustained loading.

The repassivation kinetics is affected by the environmental factors, such as water chemistry, temperature and potential [13,19], as well as the compositions and microstructures of steels. However, the experimental data of repassivation exponent is very scattered [13] so that it is still hard to evaluate quantitatively the effects of environmental variables on the repassivation kinetic exponent. In the present study, n = 0.667 is used as an approximation when the experimental data in various solution chemistries are lack.

7. ENGINEERING APPLICATIONS

7.1. Estimation of Stress Intensity Factor

Under the action of internal pipe pressure, the cracks on the external pipe surface will develop in the longitudinal direction. Before the crack penetrates the pipe wall, it is normally regarded as a partially penetrated half-elliptical surface crack. The stress intensity factor of the crack is given by [58]

$$K = Y\sigma\sqrt{\frac{\pi a}{E_c}} \tag{26}$$

The crack shape factor is square of the completed elliptical integral of the second kind and approximately given as a function of the aspect ratio of crack a/c [58],

$$E_c = 1 + 1.464\left(\frac{a}{c}\right)^{1.65} \tag{27a}$$

where a and c are the depth and surface length of crack, respectively. The geometric factor of crack is a function of crack and pipe geometry [66,67],

$$Y = Y\left(\frac{a}{w}, \frac{a}{c}, \frac{w}{D}, \varphi\right) \tag{27b}$$

where D and w are the pipe diameter and wall thickness, φ is the angle between the pipe surface and an arrow from the middle of surface length of crack to crack tip and $\varphi = 90^0$ corresponds to the deepest point of crack. Some field observations have indicated that the aspect ratios of surface crack on the external pipe surface are very small (< 0.06).

In this case, the effect of aspect ratio is negligible. For a thin wall pipe with large diameter ($w/D \leq 1/50$) with longitudinal cracks, they can be approximated as a single edge notch tensile panel with width w, the stress intensity factor is formulated as follows [58],

$$K = Y\left(\frac{a}{w}\right)\sigma\sqrt{w} \tag{28}$$

The non-dimensional coefficient Y is given by

$$Y\left(\frac{a}{w}\right) = \left(0.752 + 0.202\frac{a}{w} + 0.37\left(1 - \sin\left(\frac{\pi a}{2w}\right)\right)^3\right)\sqrt{2\tan\left(\frac{\pi a}{2W}\right)}/\cos\left(\frac{\pi a}{2w}\right) \tag{28a}$$

In some cases, the aspect ratios of short cracks are not very small [68]. The more complex expressions for stress intensity factor calculation may be required to improve the accuracy of prediction [66,67]. The geometric restrict at the deepest tip point of a surface crack increases with the aspect ratio, when the crack depth and applied stress are held unchanged, and it leads to a decrease of stress intensity factor at the deepest point. In reality, numerous cracks initiate on the external pipe surface under disbonded coating. Many of them will be dormant and some develop into main cracks [6]. The cracks may coalesce during propagating [11]. The coalescence of cracks will lead to a decrease in the aspect ratio of crack because the surface length of crack increases while the maximum depth of crack is almost unaffected [11]. Eq.(28) assumes the aspect ratio to be zero, so that it would overestimate the stress intensity factor for cracks with non-zero aspect ratios.

Sometimes, the bending of pipeline induced by the soil movement will result in the tensile/compressive stresses in the longitudinal direction. Finite element analysis is often required to estimate the stress intensity factor to account on the impact of soil movement. It is out of scope of this research. Some useful solutions can be found in handbooks such as [58].

7.2. Effect of Crack Size

It has been reported that the threshold of crack growth measured from the full-sized pipe, when exposed to the near-neutral pH groundwater, is only about ½ of those determined from the pre cracked compact tension specimens [14,29]. The discrepancy has been attributed to (1) the tensile residual stresses and (2) small crack size in the real pipe [69]. Beavers et al. [70] measured that the residual stresses on the external pipe surface after long term service utilizing small hole-drilling technique. They found the tensile residual stresses in the area where SCC was detected were higher than those in non-SCC area. The maximum tensile stresses might exceed 200 MPa. The tensile residual stresses will raise K_{max} and result in a higher crack growth rate. It is worthy to note that the residual stresses will relax and re-distribute with increase of crack size [71]. As such, the impact of residual stress will be more pronounced when crack size is small.

Small cracks can propagate at K-levels well below K_{ISCC} determined from specimens with a long crack but most of them will arrest after propagating a short distance. Parkins et al.[10,11] reported the initiation and growth behavior of small cracks on pipe specimen surface in 0.5 M Na_2CO_3 + 1 M $NaHCO_3$ solution at 75 °C under cyclic loads with low frequencies (2.8×10^{-6} Hz to 2.8×10^{-3} Hz).

They found the crack number increased, the crack growth rate declined with time and the coalescence of small cracks was an important mechanism for the crack growth. These findings suggest that the cracks will propagate in a different manner when the crack size is smaller than a critical value which depends on the material-environment system [14,72]. The crack growth data of full-sized pipe exposed to concentrated carbonate–bicarbonate solution are still unavailable.

The only clue available is the experimental data in Figure 10. The experiments were conducted with the notched and pre-cracked specimens with the shape of a quart of ring cut from an aged pipe (16 years old, 508 mm diameter, and 6.4 mm wall thickness). The total depth of notch + crack in the experiment was less than 2.5 mm [60], which is quite close to the crack size range in the full-sized pipe experiments [29].

Same specimens and experimental setup were employed to investigate the crack growth in the near-neutral pH groundwater [72] and the threshold experimental observed was about ½ of those measured from the pre-cracked compact-tension ones [29].

In contrast, the experimental data of high pH SCC in Figure 6 display a threshold about 25 MPam$^{1/2}$, higher than the threshold determined under sustained load (about 20 MPam$^{1/2}$, as shown in Figure 9).

This fact also implies that the critical crack size leading to the crack growth behavior transition may be smaller for the high pH SCC, compared with the critical crack size to induce same transition in the near-neutral pH soil environment.

The difference of K_{ISCC} in Figures 9 and 10 may be related to different steels used in the tests. To account on the crack size effect, a threshold K_{ISCC} = 11.3 MPam$^{1/2}$, as a first order of approximation, will be employed in the crack velocity estimation. This value is about ½ of average K_{ISCC} experimentally measured that are shown in Figure 9 and 10.

7.3. An Example of Remaining Lifetime Prediction

In practice, a constant crack velocity 0.3 m/y (9.5×10^{-12} m/s) is assumed in the life prediction for pipeline undergoing high pH SCC, if the field data are lack [1]. As we know, the crack velocity is a constant and it a constant and it affected by the parameters in the following three groups.

- The mechanical properties and microstructural characteristics of steels that affect the CTSR, including Young's modulus E, the yield strength σ_y, the strain-hardening exponent N, the specific length r_0 and the coefficient β and ε_0.
- The parameters to characterize the external loads, such as the maximum stress σ_{max}, the stress intensity factor K, the cyclic loading frequency f and the stress ratio R.
- The service environments (soil chemistry, local solution chemistry in the area in the coating-disbonded region, temperature etc.[2-4,61,62,73]) that is characterized by the kinetic parameters of transient anodic dissolution, i_a^* and n, as well as the potential range in which pipeline steel is susceptible to SCC [5,7,8,59].

The variables in the first category are mainly the material properties. Some of them are essentially constants (e.g., E and β) or ready to be measured by the standardized experiments (the yield strength and strain-hardening exponent). The specific length r_0 is regarded as a material constant depending on microstructure of steel and the crack growth mechanism [35,37,40], in spite of uncertainties about its exact value. It is worthy to point out that the plastic deformation produced by the field bending or soil movement can alter the yield strength and strain-hardening exponent, reduce ductility of steel and enhance the SCC susceptibility of pipeline steels [74,75]

The second class variables depend mainly on the substance in pipe to be transported (liquid or gas). There are some statistical results on the loading history of various pipelines [76,77]. Moreover, the soil movement is another source of external loads [1]. It was found that the transient increase of CTSR caused by the occasional soil movement could activate the dormant cracks and promote crack growth [29].

The last kind of variables relate to the material-environment interactions. The rupture ductility of passive film on pipeline steel surface in concentrated carbonate-bicarbonate solutions is still not available. $\varepsilon_F = 0.001$ can be used a rough estimation in line with the measurements in the other systems. In principle, i_a^*, n, and t_0 are measurable using the strained electrode or the scratch electrode technique [13,17], but the experimental data are often highly scattered. $n = 0.667$ and $t_0 = 0.01$ s are regarded as good approximations. The poteniodynamic technique is the suitable method to measure i_a^* [7]. Utilizing i_a^* measured from the material of pipeline in the solution which has a chemical composition from the field survey as those shown in Table 2 can reduce the uncertainties due to the material and local environments. When the local solution chemistry is unknown, the data obtained in 0.5 M Na_2CO_3 + 1 M $NaHCO_3$ solution can be employed in the initial estimation.

The example to be considered is a pipeline of X65 steel for gas transportation reported in literature, where high pH SCC was identified in the part of pipe close to compressor station after 30 years service [12]. The pipe diameter is 1066.8 mm and pipe wall was 9.42 mm. The maximum hoop stress produce by the internal pressure was 341 MPa, about 75% of SMYS

(Specific Minimum Yield Strength) of material. The temperature at the discharge of compressor station was 41.6 °C. Cracks produced by the high pH SCC were observed after a hydrostatic failure test [12]. Because the field data for crack evolution are absent, we are unable to validate the predicted results directly but the operating parameters of the pipeline will be employed in the calculation.

The procedure of life prediction is schematically illustrated in Figure14. The details will be described in this and following subsections. Generally, the hoop stress (σ) caused by the maximum operating pressure of a pipeline is relatively stable, as indicated by high stress ratio (> 0.8) and low loading frequency (in the order of 10^{-4} Hz) so that it can be approximated as an operating constant. As such, the cracking process is similar to a fatigue test conducted under a constant σ_{max} (or constant mean stress) with high stress ratio. In the following analysis, a constant maximum hoop stress is adopted. Because the real-time loading spectrum for the pipeline to be examined is absent, the loading spectra determined from the load-historical records of pipelines operating under similar conditions can be employed. The original loading spectrum reported was treated with the rain flow method and the information of load sequence has been lost. It still records (1) the operating pipe pressure from which the maximum hoop stress is determined; (2) the frequencies of both stress ratios and loading frequencies appearing in a block of loading spectrum and (3) the information about the combinations of loading frequencies and stress ratios. Figure 15 shows the typical combinations of loading frequencies and stress ratios in a loading spectrum of gas line [70]. These data will be utilized in the calculation. This load block is assumed to be applied on the pipeline repeatedly and it can be regarded as the unit of crack growth life. We define

$$\bar{\xi} = \varepsilon_0 N_L^{-1} \sum_{j=1}^{N_L} f_j (1-R_{\sigma,j})^2 \tag{29}$$

$$\bar{K_i} = N_L^{-1} \sum_{j=1}^{N_L} K_{max,ij} \tag{30}$$

where N_L and $\bar{\xi}$ are the number of loading cycles and the average ξ in one load block, respectively; $\bar{K_i}$ is the average K_{max} value in the ith load block; $K_{max,ij}$ is the maximum stress intensity factor produced by the jth cyclic load in the ith load block. $\bar{\xi} \approx 4.2 \times 10^{-9}$ MPa^{-2}m^{-1} for the loading block shown in Figure 14 and the time required to complete the loading block is

$$\tau = \sum_{i=1}^{N_L} \Delta t_j \approx 2.1 \text{days}. \tag{31}$$

where $\Delta t_j (= f_j^{-1})$ is the time duration of the jth cyclic load in the load block. The increment of crack size produced in the ith load block Δa_i is given by

$$\Delta a_i = \sum_{j=1}^{N_L} \Delta a_{ij} \tag{32}$$

In Eq.(32), Δa_{ij} is the increment of crack size produced by the jth cyclic load in the ith load block. Since $\bar{K_i}$ increases with the number of loading blocks experienced, the maximum crack increment created in single loading block $\Delta a_{\max} = \Delta a_{N_B}$. N_B is the number of loading blocks that would be experienced in the remaining service life. The calculation indicates $\Delta a_{\max} <$ 0.08 mm. As such,

$$\bar{K}_i \approx \sigma_{max}\sqrt{w}Y\left(\frac{\bar{a}_i}{w}\right) \approx \sigma_{max}\sqrt{w}Y\left(\frac{a_i}{w}\right), \quad (33)$$

where $\bar{a}_i$ and a_i are the average and initial size of crack in load block i ($i = 1,2, \ldots N_B$). When the aspect ratios of cracks are very small (< 0.1), Eq.(28) can be approximately utilized in calculating stress intensity factor.

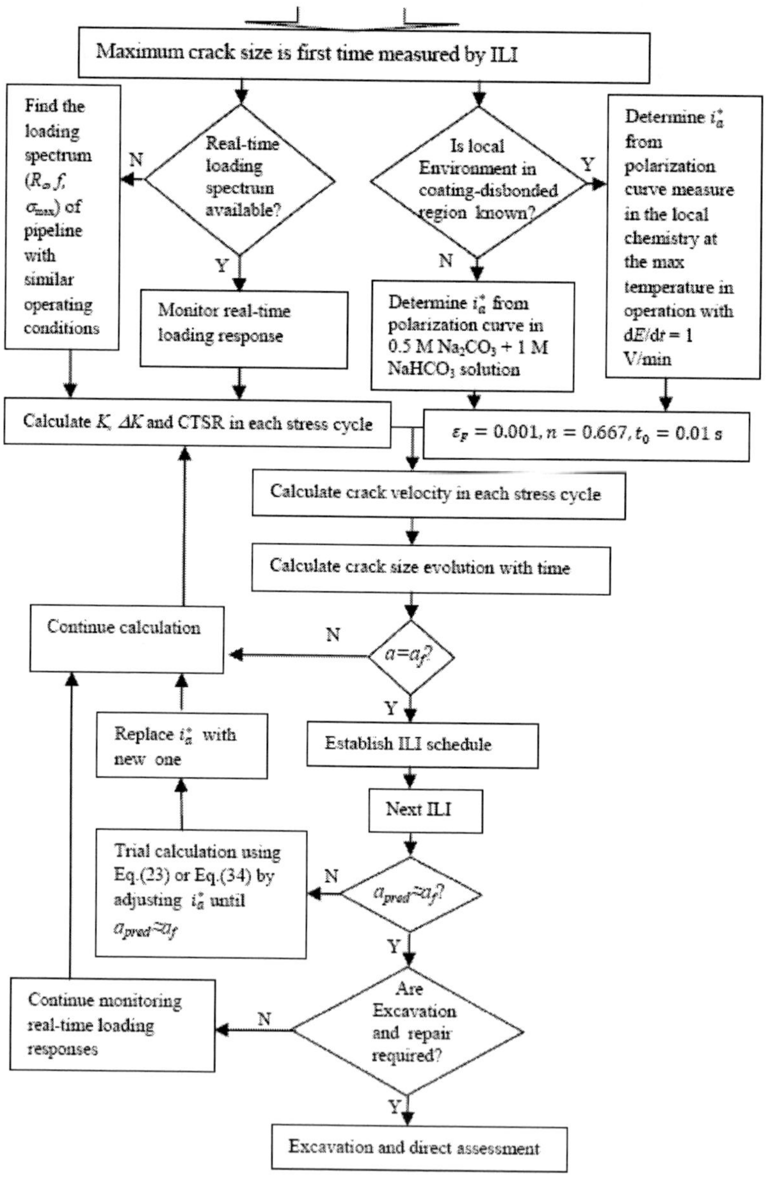

Figure 14. Procedure for life prediction of pipelines.

It will be seen in section 6.5 that the fatigue damage produced in the operation of pipelines for gas transportation is negligible, *i.e.*, $\dot{a} \approx \dot{a}_e$. The average crack velocity in the ith loading block is approximately formulated as follows,

$$\bar{\dot{a}}_{e,i} = \frac{\sum_{j=1}^{N_L} \Delta u_j}{\sum_{j=1}^{N_L} \Delta t_j} \approx \frac{i_a^*}{1-n} \frac{M}{zF\rho} \left(\frac{t_0}{\varepsilon_F}\right)^n \left\{ \frac{\dot{a}_e}{r_0} \frac{2N}{N-1} \frac{\beta \sigma_y}{E} \left[\ln \left(\frac{\bar{K}_i^2 - K_{ISCC}^2}{3\pi r_0 \sigma_y^2} \right) \right]^{\frac{N+1}{N-1}} + \bar{\xi} (\bar{K}_i^2 - K_{ISCC}^2) \right\}^n \quad (34)$$

The operation temperature of pipeline is an important factor to be considered in the crack velocity estimation. The temperature-dependence of anodic current densities over the bare metal surface can be formulated using Arrhenius law,

$$i_a^* = i_{a,ref}^* \exp\left[\frac{Q}{R}\left(\frac{1}{T_{ref}} - \frac{1}{T}\right)\right] \quad (35)$$

where R is the gas constant, $i_{a,ref}^*$ is i_a^* at T_{ref}. The activation energy Q is about 42 kJ/mol, for the anodic dissolution in the material-environment system investigated [6]. In this study, T_{ref} = 75 °C. If the local solution chemistry in the coating-disbonded region is assumed to be 0.5 M Na_2CO_3 + 1 M $NaHCO_3$, $i_{a,ref}^* \approx 140 \sim 700$ A/m^2. When T = 41.6 °C, $i_a^* \approx$ $30 \sim 140$ A/m^2. Eq.(34) can be used to assess the effect of i_a^* on the average crack velocity under the loading spectrum. The correlations of $\bar{\dot{a}}$ vs. $\bar{K}$ vary with i_a^* are presented in Figure 16 by a group of parallel curves, where $\dot{a}$ = 0.3 m/y is plotted for comparison.

Assume that a_0 is the maximum crack depth detected by the ILI technique and that a_f is the maximum crack depth when the following ILI is to be conducted. The time interval between these two ILIs t_R will be

$$t_R = N_B \tau = \int_{a_0}^{a_f} \frac{da}{\bar{\dot{a}}_e} \quad (36)$$

The increment of crack size in one loading block is given by

$$\Delta a_i = \tau \bar{\dot{a}}_{e,i} \quad (37)$$

After experiencing N_p loading blocks, the crack size will be

$$a_{pred} = a_0 + \sum_{i=1}^{N_p} \Delta a_i = a_0 + \tau \sum_{i=1}^{N_p} \bar{\dot{a}}_{e,i}. \quad (38)$$

when $N_p = N_B$, $a_{pred} = a_f$

$$a_f = a_0 + \tau \sum_{i=1}^{N_B} \bar{\dot{a}}_{e,i} \quad (39)$$

The hoop stress of high pressure pipeline may be up to 72% of yield strength of pipeline steel. When crack depth exceeds 50% of pipe wall, the plastic zone ahead of crack tip may penetrate the pipe wall, so that the stress intensity factor K would no long be a valid mechanical parameter to characterize the near-tip strain field. In the following calculation, we let $a_f = 0.5w$.

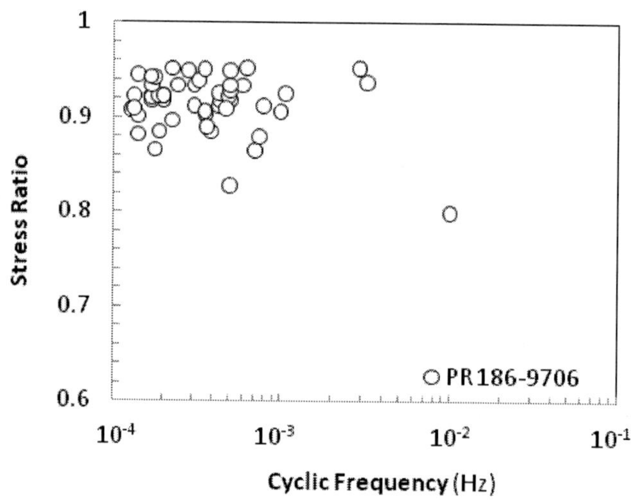

Figure 15. A typical load spectrum of pipeline for gas transportation [70].

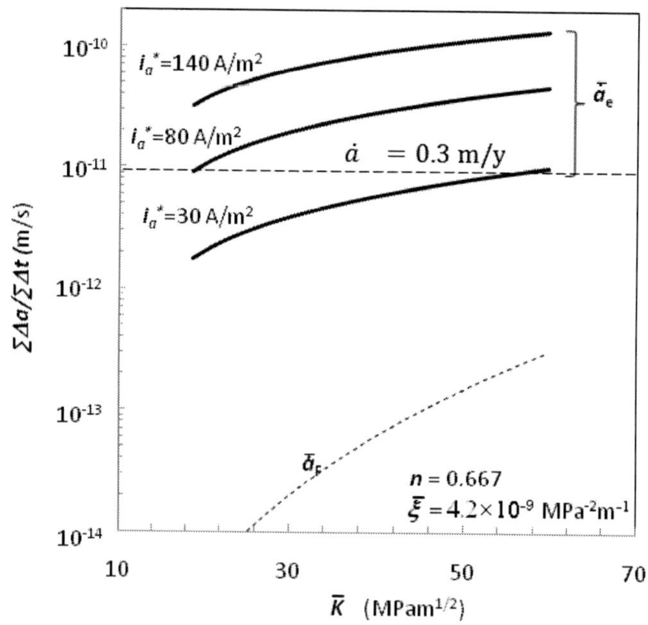

Figure 16. Predicted curves for average crack growth rate vs. stress intensity factor ($\bar{R}_\sigma = 0.9$ was assumed in calculating $\bar{a}_F$).

The crack growth lifetime predicted using Eq.(34) through Eq.(39) is a function of i_a^*, as indicated by Figure 17. The crack growth lifetime is about 1.8, 5.7 and 25.2 years when i_a^* is equal to 140, 80 and 30 A/m^2, respectively. If we assume $\dot{a} \equiv 0.3$ m/y, $t_R \approx 13.3$ years, which is independent of solution chemistry and loading history. In accord with the prediction using the model developed in this study (Figure 17), if the and remaining lifetime $t_R \approx 13.3$ years, the equivalence anodic current density would be around 46 A/m^2.

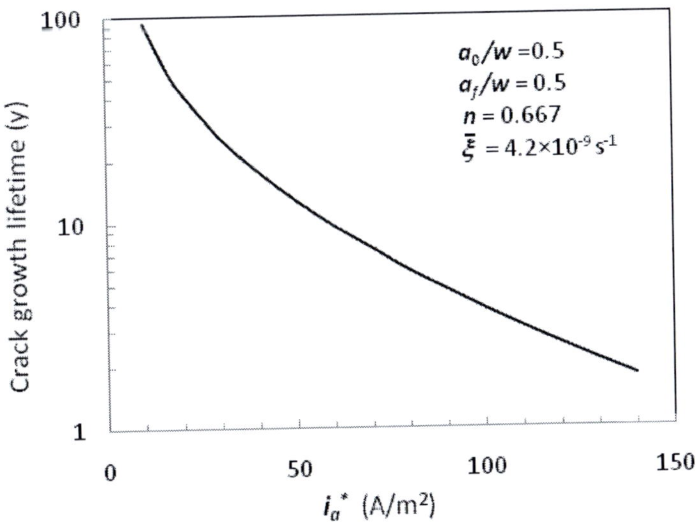

Figure 17. Effect of the anodic current density on bare crack-tip surface on the crack growth lifetime of the gas line having the loading spectrum shown in Figure 15.

Actually, the crack velocity in reality is not a constant independent of environment and mechanism driving force. As predicted, the crack velocity will increases with crack depth and anodic current density immediately after the rupture of passive film (i_a^*) for a pipeline having a fixed loading spectrum, as demonstrated in Figures 16 and 18. After i_a^* and loading spectrum is known, the remaining lifetime of the pipeline can expressed as the initial crack depth detected by the ILI, as typically shown in Figure 19, where the predicted lifetime obtained by the constant crack velocity 0.3 m/y is also presented for comparison. From Figure 18 and Figure 19, the crack velocity and the remaining lifetime of pipeline can be estimated after the crack size is determined using ILI.

If the real-time load spectra of the pipeline to be inspected are known, the crack size increment in each loading cycle will be given by

$$\Delta a_k = f_k^{-1}\dot{a}_{e,k} = \dot{a}_{e,k}\Delta t_k; \quad k = 1,2,3,\ldots N_f \tag{40}$$

where $\dot{a}_{e,k}$ and $f_k^{-1}(=\Delta t_k)$ are the average crack growth rate and the time duration of the kth loading cycle; N_f is the remaining crack growth life. $\dot{a}_{e,k}$ is calculated using Eq.(23) after let $R = R_o$ and $K_{max} = K_{max,k}\left[= K_{max}\left(\frac{a_k}{w}\right)\right]$. Note that N_f is given in cycle while N_B is represented in number of loading blocks. The crack depth a_f can be estimated after the average crack velocity in each loading cycle is calculated using Eq.(23).

$$a_f = a_0 + \sum_{k=1}^{N_f} f_k^{-1}\dot{a}_{e,k} \tag{41}$$

N_f will be determined as $a = a_f$. The correlation between the remaining service lifetime and N_f is

$$t_R = \sum_{k=1}^{N_f} f_k^{-1} \tag{42}$$

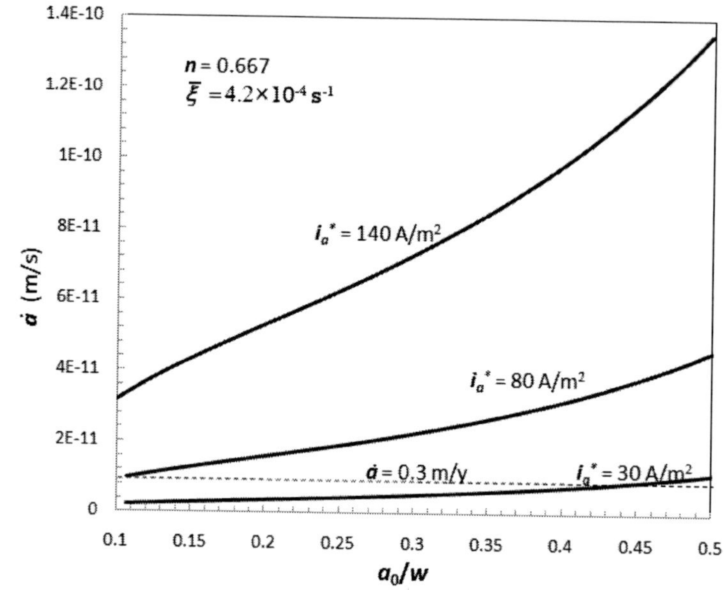

Figure 18. The correlation of crack depth and crack velocity of the gas line having the loading spectrum in Figure 15.

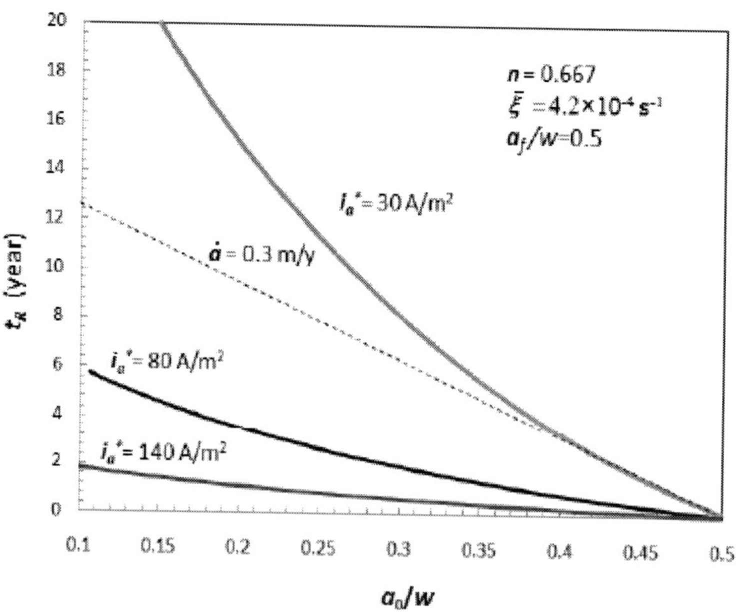

Figure 19. The correlation of initial crack depth and remaining lifetime of the gas line having the loading spectrum in Figure15.

7.4. Local Environment in Coating-Disbonded Region

$\dot{a}$ = 0.3 m/y can be regarded as a rough estimation of average crack velocity that was obtained from the field experience. The results in Figs.16 though 17 indicate that the predicted remaining lifetime for the pipeline with a crack with initial depth 0.1 w (5.72 years) based on the mean value of i_a^* ($\approx$ 80 A/m^2) is comparable with but shorter than the prediction (13.3 years) determined from $\dot{a}$ = 0.3 m/y. The life predictions shown in these figures are based on the following assumptions:

- the local chemistry in the coating-disbonded region is 0.5 M Na_2CO_3 + 1 M $NaHCO_3$ aqueous solution;
- the temperature is held constant at 41.6 °C;
- the pipe surface is immersed in the solution all the time and
- the potential is held unchanged where the peak i_a^* appears.

Experiments [2,61,62] and numeric simulation [3,63] have indicated that the local solution chemistry on the external pipe surface relies on various factors including the soil type, soil chemistry, terrain and drainage, climate, temperature and electrochemical potential distribution over the pipe surface, as well as the mass transfer in coating-disbonded region. In reality, the local solution chemistry in the areas where SCC takes place is often different from and less concentrated than the test solution employed in the laboratory investigations, as indicated by the water samples listed in Table 2 that were collected from different sites in field [6]. The local potential at the crack mouth may deviate from the potential where the peak i_a^* appears. Therefore, the crack velocity in the real pipelines is likely lower than the maximum i_a^* in 0.5 M Na_2CO_3 + 1 M $NaHCO_3$ aqueous solution at same temperature. To reduce uncertainties in the life estimation, the effects of environmental variables need to be further examined.

For the material-environment system of interest in this work, the most important species affecting the corrosion behavior of pipelines are H^+, CO_3^{2-}, HCO_3^- and Cl^-. The anodic dissolution rate of pipeline steel in the active potential range is not very sensitive to the concentration of CO_3^{2-}, HCO_3^- and pH, as well as the potential scanning rate. In the potential range for the active-to-passive transition, the solution chemistry and temperature have a significant effect. The reduced concentration of CO_3^{2-} and HCO_3^-, lower pH and temperature will result in a higher peak current potential. Davies and Burstein [78] pointed out that, under certain conditions, the presence of HCO_3^- could accelerate the anodic dissolution of iron in the both active and passive regions, owing to the formation of the soluble complex anion $Fe(CO_3)_2^{2-}$. Same phenomenon was also found in carbon steel/concentrated bicarbonate solution system [5,79,80]. Using SSRT technique, Parkins [81] examined the cracking velocity of C-Mn steel in a 0.5 M $NaHCO_3$-0.25 M Na_2CO_3 solution with the additions of sodium chloride, over a range of 0 to 200 g/l, sodium sulfate, in a range of 0 to 50 g/l, and sodium nitrate, with concentrations of 10 to 100 g/l, respectively. He found that crack velocity decreased with increasing chloride concentration and the impacts of the sulfate and nitrate were insignificant. A numeric simulation indicated that the cathodic polarization might repulse the chloride ions out of the disbonded coating areas [62]. Parkins and Zhou [2,3] evaluated that influence of pH in 1 M $NaHCO_3$, obtained by varying the concentrations of

CO_2 and HCO_3^-, on the potential range and severity of cracking. They observed that the pH range for cracking extended from 6.7 to 11 at 25 °C and the potential range for cracking shifted to the negative direction with an increase of pH. The crack velocity increased with decreasing pH. Their findings are quite similar to those observed at 75 °C. However, the crack velocity decreases with increasing pH. The potential range for cracking at lower temperatures extended from 8.5 to 10.5 and narrowed with increasing pH [81].

Experimental evidence has indicated that the IGSCC susceptibility of pipeline steels declines as the concentration of carbonate/bicarbonate is reduced [63,71]. Electrochemical measurements and SSRT tests have indicated that the crack velocities in the less concentrated solutions are lower than those observed in 0.5 M Na_2CO_3 + 1 M $NaHCO_3$ solution [4-8]. This is agreement with the observations from anodic polarization measurements [3,5,78-80]. The calculations in Figure 9 indicate that the crack velocity may change more than 30 times within the range of peak anodic current densities (i_a^*) measured in 0.5 M Na_2CO_3 + 1 M $NaHCO_3$ solution at 75 °C. To reduce uncertainty caused by the pipeline steel and the local solution chemistry in the areas where IGSCC occurs, a practical approach is to estimate anodic current density on the bare metal surface from the polarization curve of the pipeline steel that is measured at the highest operating temperature using a potential-scanning rate 1 V/min in the solution with chemical composition determined from the field survey.

Most IGSCC failures have been observed on pipelines coated with coal tar. This may be related to the fact that this type of coating was frequently applied on the pipelines installed in 1960's to 1970's [6,12]. The failures have also been found from asphalt coated pipes and they are rare in the tape coated ones [1,6]. In practices, underground pipelines are operated under cathodic protection to certain extent. In light of the experimental research [2,62] and numeric simulation [3,63], cathodic polarization and evaporation due to the seasonal changes and/or heat from the compressor stations are necessary for the formation of concentrated carbonate/bicarbonate chemistry in the coating-disbonded region. As such, the presence of permeable, degraded, and/or disbonded coating may promote the occurrence of intergranular SCC [4,6,10]. When the cathodic potential is presented at the outside of disbonded coating with holidays, the potential at the crack mouths within the crevice will shift to the negative direction. The extent of potential shift and the anodic current density depend on the solution chemistry, crevice geometry, seasonal changes, terrain and drainage [4-6,60-63].

When pipeline steel is exposed to a solution with a fixed concentration of CO_3^{2-}-HCO_3^-, the maximum crack velocity will be observed at the potential where the peak anodic current appears. The cathodic polarization over the pipe surface will result in a lower crack velocity when the solution chemistry is unchanged, as typically demonstrated in Figure 11. The water samples shown in Table 2 were collected during the SCC direct assessment, when the cathodic polarization was likely to have been stopped. Experimental observations [2,62] and numeric simulations [3,63] showed that the instant solution chemistries changed with the potential fluctuations that were used to simulate the seasonal changes. The solution pH is raised when steel is cathodically polarized and is decreased when cathodic current cannot reach the pipe surface [2,3,62,63]. The SSRT test results indicated that the intergranular SCC susceptibility increased when the pH of carbonate-bicarbonate solution was reduced and the concentration HCO_3^- was increased [3,4,8].

Further, the intergranular SCC is commonly observed in pipes close to compressor station where the temperature is relatively higher. Evaporation due to seasonal changes and/or heat from compressor station may lead to the potential fluctuation and the dry-wet cycles in areas where the SCC occurs. The potential window for the cracking becomes smaller as the temperature is lowed [4,6]. This would make the crack velocity more sensitive to the fluctuations of potential and result in the uneven crack growth or the crack dormancy. Therefore, the calculations based on the peak anodic current density measured in the simulated local chemistry at the highest service temperature is likely to overestimate on the crack velocity of pipelines during practical operation.

If the predicted crack size a_f is found to be significantly larger than a_f^* that measured in the successive ILI after the cracks were detected, the trial calculations can be utilized to re-estimate the average value of i_a^* in the particle operation of pipeline, so that the uncertainty produced by the changes of local environments will be reduced in the successive life prediction. The calculations are done by adjusting i_a^* value in Eq.(23) or Eq.(34), until $a_f \approx a_f^*$. The i_a^* value so-determined can be further used in the remaining service life prediction. If the real-time loading spectrum is available, Eq.(23) will be employed in crack velocity estimation. If the loading block used is obtained from another pipeline operated under similar conditions, Eq.(34) will be applied.

7.5. Fatigue and Cyclic-Load-Enhanced SCC

As long as ΔK exceeds the fatigue threshold ΔK_{th}, a crack can propagate under action of cyclic load in a non-corrosive environment. In accordance with ASME boiler and pressure vessel code, Section XI [82], when fatigue and SCC mechanisms coexist, the crack growth rate is estimated using the linear superposition model proposed by Wei [83]. It can be formulated as follows

$$\dot{a} = \dot{a}_F + \dot{a}_e \tag{43}$$

In Eq.(43), $\dot{a}$ is the crack velocity produced by corrosion fatigue; $\dot{a}_e$ is formulated by Eq.(23), and

$$\dot{a}_F = f\left(\frac{da}{dN}\right)_F \tag{44}$$

where $\left(\frac{da}{dN}\right)_F$ is the fatigue crack growth rate in air that is approximately formulated by Paris' law

$$\left(\frac{da}{dN}\right)_F = C\Delta K^m \text{ (m/cycle)} \tag{45}$$

For pipeline steels, $m = 3.97$ and $C = 2.03 \times 10^{-14} + 5.4 \times 10^{13} R_\sigma$ (MPa$^{-3.97}$m$^{-0.985}$) to account on the effect of stress ratio [60]. In the ASME code [82], the impact of loading frequency (f) on $\left(\frac{da}{dN}\right)_F$ is ignored. The relative contributions of fatigue and SCC rely on the anodic dissolution rate of crack tip and loading conditions. The loading spectrum in Figure 15 indicates an average loading frequency around 5×10^{-4} Hz and a mean stress ratio about 0.9.

The maximum crack velocity produced purely by fatigue in K_{max} range of interest (15 to 60 MPam$^{1/2}$), according to Eq.(45), will be in a range from 3×10^{-14} m/s to 3×10^{-12} m/s.

As depicted in Figure 16, it is less than 2% of the estimated crack velocities produced by the SCC, Consequently, it is a good approximation to ignore the influence of fatigue in the life prediction of pipelines for gas transportation. However, for the pipelines transporting liquid, the minimum stress ratio is lower than 0.6 [84].

In certain cases, the contribution of fatigue mechanism may be comparable with that of SCC, depending on the loading spectra of pipeline and the local environments at the external pipe surface below the disbonded coating.

When the contribution of mechanical fatigue needs to be considered, Eq.(41) is still applicable for the crack depth estimation during the operation of pipeline but $\dot{a}_{e,k}$ should be replaced by $\dot{a}_k(= \dot{a}_{F,k} + \dot{a}_{e,k})$.

$$a_f = a_0 + \sum_{k=1}^{N_f} \left[\left(\frac{da}{dN}\right)_{F,k} + f_k^{-1}\dot{a}_{e,k}\right] \tag{46}$$

where $\left(\frac{da}{dN}\right)_{F,k}$ is the fatigue crack growth rate in the kth loading cycle.

When the loading blocks obtained from the pipelines operated under similar conditions are applied in the life estimation, $\bar{\dot{a}}_{e,i}$ in Eq.(39) will be replaced by $\bar{\dot{a}}_i$ $(= \bar{\dot{a}}_{F,i} + \bar{\dot{a}}_{e,i})$,

$$a_f = a_0 + \tau \sum_{i=1}^{N_B} [\bar{\dot{a}}_{F,i} + \bar{\dot{a}}_{e,i}] \tag{47}$$

The average crack velocity produced by fatigue in the jth loading block is given by

$$\bar{\dot{a}}_{F,i} = N_L^{-1} \sum_1^{N_L} \dot{a}_{F,ij} = N_L^{-1} \sum_{j=1}^{N_L} f_j \left(\frac{da}{dN}\right)_{F,ij} \tag{48}$$

where $\left(\frac{da}{dN}\right)_{F,ij}$ is the fatigue crack growth rate in the jth loading cycle of the ith loading block. $\left(\frac{da}{dN}\right)_{F,ij}$ is a function of $\Delta K_{i,j}$ and R_j, as given by Eq.(45). $\Delta K_{i,j}$ the cyclic range of stress intensity factor of the ith cycle in the jth loading block. R_j is the stress ratio of ith cycle in the loading block.

Although the direct contribution of fatigue damage can be ignored in the service life estimation of pipeline for gas transportation, the cyclic loading can still affect the SCC development.

To demonstrate this effect, the ratios of crack velocities $\bar{\dot{a}}_s/\bar{\dot{a}}$ are plotted against the stress intensity factor in Figure 20. When $i_a^* = 80$ A/m^2, $\dot{a}_e \approx 1\text{~}3 \times 10^{-11}$ or 0.3~ 1 mm/y (Figure 16), $t_R \approx 5.7$ years (Figure 17). In accord with Figure 20, $\frac{\bar{\dot{a}}_s}{\bar{\dot{a}}_e} \approx 0.5$ or $\bar{\dot{a}}_e \approx 2\bar{\dot{a}}_s$. It means that the internal pipe pressure fluctuations can almost let the average crack velocity double.

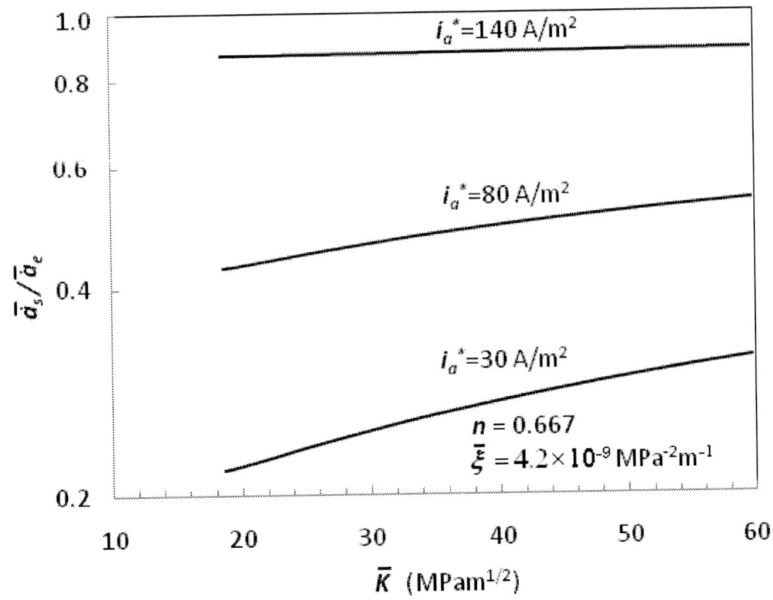

Figure 20. Effect of i_a^* on the ratio of $\bar{\dot{a}}_s/\bar{\dot{a}}_e$.

As pointed out in the discussion in section 5, the effect of cyclic loading declines with the corrosivity of media that is characterized by i_a^*. When i_a^* is reduced to 30 A/m^2 ($\dot{a}_e \approx 2\text{~}8 \times 10^{-12}$ or 0.06 ~ 0.23 mm/y ; $t_R \approx 25.2$ years), the ratio of $\frac{\bar{\dot{a}}_s}{\bar{\dot{a}}_e}$ would be further decreased to 0.22 ~ 0.33, indicating cyclic load would dominate the crack growth. If the local environment is highly aggressive ($i_a^* \geq 140$A/m^2), the effect of cyclic load ceases ($\bar{\dot{a}}_s/\bar{\dot{a}}_e \approx 1$). In this case, the crack velocities are in a range of 4 ~ 12×10^{-11} m/s or 1.2 ~ 4.8 mm/y (Figure 16) and the remaining lifetime for the gas line (t_R) would be less than 2 years (Figure 17). where $\left(\frac{da}{dN}\right)_{F,ij}$ is the fatigue crack growth rate in the jth loading cycle of the ith loading block. $\left(\frac{da}{dN}\right)_{F,ij}$ is a function of $\Delta K_{i,j}$ and R_j, as given by Eq.(45). $\Delta K_{i,j}$ the cyclic range of stress intensity factor of the ith cycle in the jth loading block. R_j is the stress ratio of ith cycle in the loading block.

CONCLUSIONS

- A science-based crack growth model is postulated for pipelines exposed to the concentrated carbonate-bicarbonate solution. It gives reasonable predictions for the crack velocities measured in laboratory. This model is developed on the basis of the following assumptions.
- The crack growth is dominated by the repeated film rupture at the crack tip.
- The crack tip strain rate due to crack advance can be formulated by the model described in section 3.1.

- The crack tip strain rate induced by the cyclic load is a result of the alternative crack tip opening displacement.
- The anodic dissolution rate over the fresh metal surface produced by the rupture of passive film can be characterized using the polarization curve of pipeline steel measured in the bulk solution with a potential scanning velocity of 1 V/min.
- The repassivation kinetics is formulated by the power law.
- In accordance with the proposed model, pipeline steel displays the maximum crack velocity at the potential where the peak anodic current appears. The numerical simulations indicate that, when the pipeline is exposed to concentrated Na_2CO_3 - $NaHCO_3$ solutions and the crack velocity is lower than 1×10^{-9} m/s, the effects of mass transfer in crack on the local electrolyte evolution is negligible.
- The preliminary analysis indicates that the new model can be used to assess the effects of various factors relating to material, mechanical and environmental aspects and their interactions on the crack growing during high pH SCC of pipelines.
- After cracks are detected using ILI techniques and the maximum depth of the crack is known, the remaining service life of pipeline undergoing the high pH SCC on the safe side can be estimated using the following procedures.
- When the real-time loading spectrum is available but the local solution chemistry in the coating-disbonded region is unknown, the peak anodic current density i_a^* determined from the polarization curve measured in 0.5 M Na_2CO_3 + 1 M $NaHCO_3$ can be approximately used for the initial crack velocity estimation. Eq.(35) will be used to estimate i_a^* at the operating temperature if it is not 75 °C.
- In the life prediction of pipelines for gas transportation, the fatigue damage can be neglected and the crack velocity can be estimated using Eq.(23). The increment of crack depth and the remaining service life of pipeline can be approximately predicted by means of Eqs.(40 - 42)
- If the local solution chemistry in the coating-disbonded region is known, the polarization curve of the pipeline steel can be measured in this solution at the highest operating temperature using the potential-scanning rate of 1 V/min and the peak anodic current density determined from this polarization curve can be employed in the life estimation to reduce the uncertainty produced by the local solution chemistry.
- If the crack size detected by the successive ILI is significantly smaller than the predicted value, the crack size measured in field can be used for a trial calculation to reduce the uncertainty caused by the seasonal changes and local potential over the pipe surface under the disbonded coating. The trial calculation can be done by adjusting the i_a^* value in Eq.(23) until the predicted crack size equals to the measured one. In accord with i_a^* so-determined and the crack size measured by the ILI, the remaining service lifetime of the pipeline will be estimated by re-using Eqs.(40 - 42).
- If the real-time loading spectrum is unavailable, the loading spectrum of pipelines operated under similar condition can be employed. Eqs.(34 – 39) will be used in the calculation .
- In the life prediction of pipeline for gas transportation, the impact of fatigue damage can be ignored but the accelerated dissolution at the crack tip due to the cyclic deformation should be considered.

- At the crack tip in pipeline used liquid transportation, the local strain rate is governed by the cyclic loading. In the life prediction, the comparison should be made between the crack velocities caused by SCC and fatigue to see if the contribution of fatigue crack growth is necessary to be considered in the life prediction.

ACKNOWLEDGEMENTS

A part of this research was supported by the Department of Transportation (DOT) under Contract DTPH 56-08-T-000001. Vincent Holohan and James Merritt of PHMSA/DOT provided management oversight for the project. This work was co-sponsored by CANMET Materials Laboratory of the Natural Resources of Canada (Winston Revie), Williams Pipeline Company (Sergio Limon), and TransCanada Pipeline Company (Richard Kania and Robert Worthingham).

REFERENCES

[1] National Energy Board of Canada, 1995.Public Inquiry Concerning Stress Corrosion Cracking on Canadian oil and Gas Pipelines, *Report of the Inquiry,* MH-2-95, 1995.
[2] Charles, E. A.; Parkins, R. N. *Corrosion*, 1995, vol.51, 519-527.
[3] Song, F. M. Corrosion, 2010, vol.66:095004-1-14.
[4] Parkins, R. N.; Zhou S. *Corros. Sci.,* 1997, vol. 39, 159-173.
[5] Parkins, R. N.; Zhou S., *Corros. Sci.*, 1997. vol.39, 175-191.
[6] Parkins, R. N. In Research Topic Symposia of NACE, eds. Andersen P. L.; Parkins, R. N., Corrosion'96, NACE International, Houston, TX, 1996, 1-49.
[7] Parkins, R. N. *Corrosion*, 1990, vol.46, 178-189.
[8] Parkins R. N. *Corrosion,* 1996, vol.52, 363-374.
[9] Parkins, R. N. *Corrosion,* 1987, vol.43, 130-139.
[10] Parkins, R. N.; Belhimer, E.; Blanchard, W. K. *Corrosion,* 1993, vol.49, 951-953.
[11] Parkins, R. N.; Singh, P. M., *Corrosion,* 1990, vol.46, 486-499.
[12] Been, J.; King, F.; Fenyvesi, L; Sutherby, R. In *Proc. 5th Biennial Inter. Pipeline Conf., ASME*, Calgary, Alberta, Canada, 2004, Vol.1, pp.83-100.
[13] Song, F. M. Corros. Sci., 2009, vol.51, 2657-2674.
[14] Lu, B. T.; Song, F. M. Gao,; M., Elboujdaini, M. *Corros. Sci.*, 2010, vol 52, 4668-4072.
[15] Lu, B. T.; Song, F. M. Gao, M.; Elboujdaini, M. *Eng. Fract. Mech.,* 2011, doi:10.1016/ j.engfracmech.2011.01.004
[16] Vermilyea, D.A. *J. Electrochemical Soc.*, 1972, vol.119, 405-407.
[17] Staehle, R. W., In Stress Corrosion Cracking and Hydrogen Embrittlement of Iron-Based Alloys, ed. Staehle, R. W.; Hochmann, J.; McCright, R. D; Slates, J. E. *NACE*, Houston, TX, 1977, 180-205.
[18] Scully, J. C. *Corros. Sci.*, 1980, vol. 20, 997-1016.
[19] Ford, F. P. *Corrosion,* 1996, vol.52, 1996, 375-395.
[20] Uligh, H. H. *Fracture*, 1972, vol.3, 645-677.
[21] Keddam, M., Silva, J. V. D. *Corros. Sci.*, 1980, vol. 20 167-175.

[22] Ford, R. P. *J. ASME,* 1988, vol.110, 113-128.
[23] Staehler, R. W. *Corros. Sci.*, 2007, vol.49, 7-19.
[24] Turnbull, A., *Rev. Coat. Corros*., 1982, vol.5, 43-111.
[25] Turnbull, A. *Corrosion,* 2001, vol.57, 175-189.
[26] Harrison, P. L. *Corros. Sci.,* 1967, vol.7, 789-804
[27] Diegle, R. B.; A. Vermilyea, D. A. *Corrosion,* 1976, vol.32, 411-414.
[28] Lidbury, D. P. G. In Embrittlement by the Localized Crack Environment, ed. Gangloff, R. P. *Proc. Conf. AIME, Metallurgical Soc of AIME;* Warrendale, PA, USA, 1984, pp.140-172.
[29] Zheng, W.; Tyson, B.; Shen, G.; Revie, R. W.; Williams; G.; Bibby, D.; Effects of Operating Practices on Crack Dormancy and Growth, Report for Gas Research Institute, GPI-05/8668, CANMET, MTL-2005-10, 2005.
[30] Revie, R. W.; Uligh, H. H. *Acta Metall.,* 1974, vol.23, 619-627.
[31] Gu, B; Chu, W. Y.; Qiao, L. J.; Hsiao, C. M. *Corros. Sci*., 1994, vol. 36, 437-445.
[32] Jones, D. A. *Corrosion*, 1996, vol.52, 356-367.
[33] Gutman, E. M. *Soviet Powder Metal. Met. Ceram*., 1970, vol. 9, 328-334.
[34] Lu, B. T.; Luo, J. L. J. *Phys. Chem*. B, 2006, vol.110, 4217-4231.
[35] Hall, M. M. Jr. *Corros. Sci.,*2008, vol.50, 2902-2905.
[36] Gerberich, W. W.; Davidson, D. L.; Kaczoroski, M. *J. Mech. Phys. Solids.*, 1990, vol.38, 87-113.
[37] Shoji, T. In Predictive Capacities in Environmentally-Assisted Cracking, ed. Ravi, R. *Proc. ASME-PVP Symposium, ASME,* New York, NY, USA, 1985, 127-142.
[38] Gao, Y. C.; Hwang, K. C. In Advances in Fracture Mechanics, ed. Francois, D.; *5th Intern. Conf. on Fracture*, 1981, Vol.2, pp.669-682.
[39] Irwin, G. R. In *Proc. 7th Sagamore Ordnance Materials Conf.*, Vol.IV, New York, Syracuse University, 1960, pp.63-78.
[40] Peng, Q. J, Kwonm J.; Shoji, T. J. *Nuclear Materials*, 2004, vol.324, 52-61.
[41] Gao, Y. C.; Zhang, X. T., Hwang, K. C. *Int. J. Fract*., 1983, vol.21, 301-317.
[42] Ramberg, W.; Osgood, W. D. Description of Stress-Strain Curves by Three Parameters, NACA TN 902, Washington, DC, USA, 1943.
[43] Shoji, T.; Lu, Z. P.; Murakami, H. *Corros. Sci*., 2010, vol.52, 769-779.
[44] Fan, Y. T.; Sutton, M. A.; Zhang, L. X. *Int. J. Fract*., 1997, vol.86, 327-341.
[45] Gao, Y. C. *Theor. Appl. Fract. Mech*., 1996, vol.25, 179-186.
[46] Hutchinson, J. W., J. *Mech. Phys. Solids*, 1968, vol.16, 13-31.
[47] Rice, J. R.; Rosengren, C. F.; *J. Mech. Phys. Solids,* 1968, vol.16, 1-12.
[48] Rice, J. R.; Dugan, W. J.; Sham, T. L., in *Fracture Mechanics*, 12th Conference of ASTM, ASTM 700, ASTM, Philadelphia, Pa, USA, 1980, pp.189-231.
[49] Shih, C. F. *J. Mech. Phys. Solids*, 1981, vol.29, 305-326.
[50] Hudak, Jr. S. J.; Davidson, D. L.; Page, R. A. In *Embrittlement by the Localized Crack Environment,* ed. Gangloff, R. P. Proc. Conf. AIME, Metallurgical Soc of AIME; Warrendale, PA, USA, 1984, pp.172-196.
[51] Dietzel, W. *Mater. Sci.,* 2004, *vol.*40, 749-755.
[52] Kondo, Y.; Kubota, M.; Mizobe, K. Trans. *Jpn Soc. Mech. Eng*. A, 2010, vol.A76, 594-601.

[53] Likhtman, V.; Rebinder, P; G. Karpenko, G. Effect of a Surface-Active Medium on the Deformation of Metals, Academy of Science Publishing House, Moscow, English Edition Published by H. M. M. O. (1958).
[54] Ritchie, R. O. *Intern, J. Fract.*, 1982, vol.100, 55-83.
[55] Gangloff, R. P. In *Environment-Induced Cracking of Metals, eds. Gangloff,* R. P. and Ives, M. B., NACE Houston, ADA228834, 1990, 55-109.
[56] McEvily, A. J. In *Fatigue Mechanisms: Advances in Quantitative Measurement of Physical Damage*, ed. Lankford, L; Davidson, D. L.; Morris, W. L.; Wei, R. P. ASTM STP 811 , ASTM, Philadelphia, Pa, USA, 1984, pp. 283-312.
[57] Parkins, R. N.; Greenwell, B. S. *Met. Sci.*, 1977, vol.11, 405-413.
[58] Tata, T.; Paris, P.; Irwin, G.; The Stress Analysis of Cracks Handbook, 2nd Edition, *ASME*, New York, NY, USA, 1985, 2-10-2.12.
[59] Sutcliffe, J. M.; Fessler, R. R.; Boyd, W. K.; Parkins, R. N., *Corrosion,* 1972, vol.28, 313-320.
[60] Pilkey, A. K.; Lambert, S. B.; Plumtree, A. *Corrosion,*1995, vol.51, 91-96.
[61] Yan, M. S.; Wang, J. Q.; Han, E. H.; Ke, W. *Corros. Sci.*, 2008, vol.50, 1331-1399.
[62] Fessler, R. R.; Markworth, A. J.; Parkins, R. N. *Corrosion,* 1983, vol.39, 20-25.
[63] Song, F.; Sridhar, N. *Corros. Sci.*, 2008, vol.50, 70-83.
[64] Beavers, J.; Durr, C. L; Delanky, B. S. In *2nd Intern Pipeline Conf. ASME,* IPC1998, Calgary, AB, Canada, 1998, pp.423-437.
[65] Wei, R. P.; Song, C. *Eng. Fract. Mech.*, 1992, vol.41, 463-473.
[66] Raju, I. S.; Newman, Jr. J. C. *ASME J.* Pres. Ves. Pip., 1982, vol.104, 293-298.
[67] Wang, X.; Lambert, S. B. *Eng. Fract. Mech.*, 1995, vol.51, 517-532.
[68] Chen, W. X.; King, F.; Vokes, E. *Corrosion*, 2002, vol.58, 267-275.
[69] Chen, W. X.; Kaniam, R.; Worthingham, R.; Kariyawasam, S. In *Proc. Biennial Intern. Pipeline Conf., ASME*, IPC2008, Calgary, AB, Canada, 2008, pp. 1-10.
[70] Beavers, J. A.; Johnson, J. T.; Sutherby, R. L.; In *Proc.3rd Intern. Pipeline Conf..*, Vol. 2, October1–5, Calgary, AB, Canada; 2000. pp. 979–988.
[71] Liljedahl, C. M. D.; Zanellato, O.; Fitzpatrick, M. E.; Lin, J.; Edwards, L. Intern. *J. Fatigue*, 2010, vol.32, 735-743.
[72] Zhang, X. Y.; Lambert, S. B.; Sutherby, R.; Plumtree, A. *Corrosion,* 1999, vol.55, 297-305.
[73] Beavers, J. A.; Worthingham, R. G., In *4th Proc. Biennial Intern. Pipeline Conf.*, (IPC'02) ASME, IPC2002, Calgary, AB, Canada, 2002, pp. 1671-1678.
[74] Lu, D. T.; Luo, J. L.; Ivey, D. *Metall. Mater. Trans. A*, 2010, vol.41A, 2438-2547.
[75] Yunovich, M.; Xia, Z.; Szklarska-Smialowska, Z. *Corrosion*, 1998, vol.54, 154-161.
[76] Been, J.; Fessler, R. R.; Keane, S.; Kresic, W. In *6th Biennial Inter. Pipeline Conf. ASME*, Calgary, Alberta, Canada, 2007, Vol.2, pp.579-588.
[77] Beavers, J. A.; Maier, C. J.; Jaske, C. E. In *Corrosion'2007*, NACE International, Houston, TX, USA, 2007, paper # 128.
[78] Davies, D. H.; Burstein, G. T. *Corrosion*, 1990, vol.36, 416-423.
[79] Dong, D. J.; Nishimura, T.; Kodoma, T. In Scientific Basis for Nuclear Waste Management XXV; ed. McGrail, B. P., *Mater. Res. Soc. Symp. Proc.* 713, 2002, pp.105-112.

[80] Torres-Islas, A.; Gonzalez-Rodriguez, J. G. *Intern. J. Electrochem.,* 2009, vol.4, 640-653.

[81] Parkins, R. N., In 5th *Symp. Line Pipe Researchm Pipeline Research of AGA,* America Gas Association, Inc. 1515 Wilson Boulevard, Arlington VA 23209, Catalog No. L30174, Paper U, 1974.

[82] ASME, Rules for In-Service Inspection of Nuclear Power Plant Components, in *ASME Boiler and Pressure Vessel Code*, Section XI, Div. 1 (ASME, New York, NY, 1994).

[83] Wei, R. P. Fatigue Fract. *Eng. Mater. Struct*., 2002, vol.25, 845-854.

[84] Van Boven, G.; Sutherby, R.; King, F. In 4th *Proc. Biennial Intern. Pipeline Conf.,* (IPC'02) ASME, IPC2002, Calgary, AB, Canada, 2002, 1687-1698.

[85] Harris, G.H.; Lorenz, A. New coatings for the corrosion protection of steel pipelines and pilings in severely aggressive environments. *Corrosion Science*, 1993 35, 1417-1423.

[86] Guidetti, G.P.; Rigosi, G.L.; Marzola, R. The use of polypropylene in pipeline coatings. *Progress in Organic Coatings*, 1996 27, 79-85.

[87] Kamimura, T.; Kishikawa, H. Mechanism of cathodic disbonding of three polyethylene coated steel pipes, *Corrosion*, 1998 54, 979-987.

[88] Leng, A.; Streckel, H.; Stratmann, M. The delamination of polymeric coatings from steel. Part 2: first stage of delamination, effect of type and concentration of cations on delamination, chemical analysis of the interface. *Corrosion Science*, 1998 41, 579-597.

[89] Leng, A.; Streckel, H.; Hofmann, K.; Stratmann, M. The delamination of polymeric coatings from steel. Part 3: effect of the oxygen partial pressure on the delamination reaction and current distribution at the metal/polymer interface. *Corrosion Science*, 1998 41, 599-620.

[90] Melot, D.; Paugam G.; Roche, M. Disbondments of pipelines coatings and their consequences on corrosion risks. *17th International conference on Pipeline Protection, Edinburgh: BHR Group*, 2007.

[91] Guermazia, N.; Elleuchb, K.; Ayedia, H.F.; Kapsac, P. Aging effect on thermal, mechanical and tribological behaviour of polymeric coatings used for pipeline application, *Journal of materials processing technology*, 2008 203, 404-410.

[92] Kehr, J.A.; Enos, D.G. FBE, a foundation for pipeline corrosion coatings. *Corrosion*, 2000. 00757(NACE international Conference).

[93] Coeuille, F. Revêtements polymères de canalisation de fluide: caractérisation et évolution de l'adhésion en milieu agressif. *Ph.D. Dissertation*, Ecole des Mines de Paris, France, 2002.

[94] Perera, D.Y. On adhesion and stress in organic coatings. *Progress in Organic Coatings*, 1996 28, 21-23.

[95] Abdelkader, A.; White, J., Comparison of internal stresses in coatings cured on rigid substrates and on unrestrained thin substrates. *Progress in organic coatings*, 2002 44, 121-129.

[96] Yan, G.; White, J.R. Residual Stress Development in a Bi-Layer Coating. *Polymer Engineering and Science*, 1999 39, 1856-1865.

[97] Schmidt, R.G.; Bell, J.P. Epoxy adhesion to metals. *Advances in Polymer Science*, 1986 75, 33-71.

[98] Kamarchik, P. Crosslink density and the development of internal stress in organic coatings. Macromolecular. Symposia, 2001 175, 411-419.

[99] Ramani, K.; Zhao, W. The evolution of residual stresses in thermoplastic bonding to metals. *International Journal of Adhesion and Adhesives*, 1997 17, 353-357.

[100] Wang, H.; Siow, K.S. Measurement of Tg in Epoxy Resins by DSC-Effects of Residual Stress. *Polymer Engineering and Science*, 1999 39, 422-429.

[101] Perera, D.Y.; Oosterbroek, M. Hygrothermal stress evolution during weathering in organic coatings. *Journal of coatings technology*, 1994 66, 83-88.

[102] Legghe, E. Etude des paramètres et des mécanismes conditionnant l'adhérence entre un primaire époxy et un substrat acier: Cas des revêtements tricouche de pipelines. *Ph.D. Dissertation*, Université de Toulon et du Var, 2009.

[103] Timoshenko, S.P.; Goodier, J.N. *Theory of elasticity*, 1st edition, Mc Graw-Hill Book Company, p 399.

[104] Chu, J.L.; Sanboh, L. Thermal stresses in a long coaxial cylinder of two materials: Zero axial force. *Mechanics of Materials*, 1995 20, 105-123.

[105] Chang, B.; Sue, H.-j.; Wong, D.; Kehr, A.; Pham, H.; Siegmund, A.; Snider, W.; Jiang, H.; Browning, B.; Mallozzi, M.; Guo, S. Integrity of 3LPE Pipeline Coatings - Residual Stresses and Adhesion Degradation. *7th international Pipeline Conference*, Calgary, 2008.

[106] Chang, B.T.A.; Jiang, H.; Sue, H.J.; Guo, S.; StJean, G.; Pham, H.; Wong, D.; Kehr, A.; Lo K.H. Disbondment mechanism of 3LPE pipeline coatings. *17th International conference on Pipeline Protection*, Edinburgh, 2007.

[107] Corcoran, E. M. Determining Stresses in organic coatings using plate beam deflection. *Journal of Paint technology*, 1969 41, 635-640.

[108] Perera, D.Y.; Eynde, D.V. Considerations on a cantilever (Beam) Method for measuring the internal stress in organic coatings. *Journal of Coatings Technology*, 1981 53, 39-41.

[109] Negele, O.; Funke, W. Internal stress and wet adhesion of organic coatings. *Progress in organic coatings*, 1996 28, 285-289.

[110] Tsui, Y.C.; Howard, S.J.; Clyne, T.W. The effect of residual stresses on the debonding of coatings-II. An experimental study of a thermally sprayed system. *Acta Metallurgica et Materialia*, 1994 42, 2837-2844.

[111] Howard, S.J.; Tsui, Y.C.; Clyne, T.W. The effect of residual stresses on the debonding of coatings-I. A model for delamination at a bimaterial interface. *Acta Metallurgica et Materialia*, 1994 42, 2823-2836.

[112] Saarnak, A.; Nilsson, E.; Kornum, L.O. Usefulness of the measurement of internal stresses in paint films. *J. Oil. Col. Chem. Assoc.*, 1976 59, 427-432.

[113] Legghe, E.; Joliff, Y.; Belec, L.; Aragon, E. Computational analysis of a three layer pipeline coating: Internal stresses generated during the manufacturing process. *Computational Materials Science*, 2011 50, 1533-1542.

[114] Legghe, E.; Joliff, Y.; Belec, L.; Aragon, E.; Margaillan, A. Computational analysis of internal stresses generated during the manufacturing process of a monolayer or three-layer pipeline coating. *Computational Materials science*, 2010 48, 360-365.

[115] Cardarelli, F. *Materials Handbook - A Concise Desktop Reference*, 2nd Edition, Springer-Verlag London Limited, 2008.

[116] Astruc, A. Microstructure et perméabilité de revêtements anticorrosion : influence des contraintes hygrothermiques et rôle des interphases. *Ph.D. Dissertation*, Université du Sud-Toulon-Var, France, 2007.

[117] Nitta, K.H.; Ishiburo, T. Ultimate Tensile Behavior of Linear Polyethylene Solids. *Journal of Polymer Science: Part B - Polymer Physics*, 2002 40, 2018-2026.

[118] ABAQUS. *Analysis User's Manual, Contact formulation for ABAQUS/Standard contact pairs*, chapter 18, section 18.7, 2007.

[119] ABAQUS. *Analysis Manual, Viscoelasticity*, chapter 4, section 4.8, 2007.

[120] Arzoumanidis, G. A.; Liechti, K.M. Linear Viscoelastic Property Measurement and Its Significance for Some Nonlinear Viscoelasticity Models. *Mechanics of Time-Dependent Materials*, 2003 7, 209-250.

[121] Absi, J.; Glandus, J.C.; Improved method for severe thermal shocks testing of ceramics by water quenching. *Journal of the European Ceramic Society*, 2004 24, 2835-283.

[122] Legghe, E.; Aragon, E.; Bélec, L.; Margaillan, A.; Melot, D. Correlation between water diffusion and adhesion loss: Study of an epoxy primer on steel. *Progress in organic coatings*, 2009 66, 276-280.

[123] ISO 1663 – 1999. Rigid cellular plastics - Determination of water vapour transmission properties.

[124] Kumins, C.A. Transport through polymer films. *Journal of Polymer Science Part C: Polymer Symposia*, 1965 10, 1-9.

[125] Klute, C.H.; Franklin, P.J. The permeation of water vapor through polyethylene. *Journal of Polymer Science*, 1958 32, 161-176.

[126] Park, G.S. Transport principles: solution, diffusion and permeation in polymer membranes. *Congres - North atlantic treaty organization. Advanced study institute* 1986 181, 57-107.

[127] Peppas, N. A.; Meadows, D. L. Macromolecular structure and solute diffusion on membranes: An overview of recent theories. *Journal of Membrane Science*, 1983 16, 361-377.

[128] Frisch, H.L. Sorption and transport in glassy polymers – A review. Polymer Engineering & Science, 1980 20, 2-13.

[129] Vieth, W.R. *Diffusion in and through polymers*. Munich: Hanser Publishers, 1991.

[130] Crank, J. *The Mathematics of Diffusion*. 2nd Ed. New York: Oxford University Press, 1975.

[131] Hamilton, R.L. Water vapor permeability of polyethylene and other plastic materials, *The Bell System Technical Journal*, 1967 46, 391-415.

[132] Carter, H.G.; Kibler, K.G. Langmuir-type model for anomalous moisture diffusion in composite resins. *Journal of Composite Materials*, 1978 12, 118-131.

[133] Popineau, S.; Rondeau-Mouro, C.; Sulpice-Gaillet, C.; Shanahan, M.E.R. Free/bound water absorption in an epoxy adhesive. *Polymer*, 2005 46, 10733-10740.

[134] Weitsman, Y. A continuum diffusion model for viscoelastic materials. *J. Physical Chemistry*, 1990 94, 961-968.

[135] De Kee, D.; Liu, Q.; Hinestroza, J. Viscoelastic (Non-Fickian) Diffusion. *The Canadian Journal of Chemical Engineering*, 2005 83, 913-929.

[136] Apicella, A.; Nicolais, L.; Astarita, G.; Drioli, E. Effect of thermal history on water sorption, elastic properties and the glass transition of epoxy resins. *Polymer*, 1979 20, 1143-1148.

[137] Nicaise, E.; Gandini, A.; Cheradame, H. Behaviour of Composites Used in the Confinement of Nuclear Wastes 2 - Water Absorption in Particle-filled Epoxide Networks. British Polymer Journal, 1986 18, 251-255.

[138] Sauvant-Moynot V., Kittel J., Melot D., and Roche M. Three layer polyolefin coatings: how the FBE primer properties govern the long term adhesion. in 17th *International conference on Pipeline Protection*. 2007. Edinburgh: BHR Group.

[139] Wong, T.C.; Broutman, L.J.. Moisture Diffusion in Epoxy Resins - Part I -Non-Fickian sorption processes. *Polymer Engineering and Science*, 1985 25(9): 521-528.

[140] Nogueira, P.; Ramirez, C.; Torres, A.; Abad, M.J.; Cano, J.; Lopez, J.; Lopez-Bueno, I.; Barral, L. Effect of Water Sorption on the Structure and Mechanical Properties of an Epoxy Resin System. *Journal of Applied Polymer Science*, 2001 80, 71-80.

[141] Jeffrey, K.; Pethrick, R.A. Influence of chemical structure on free volume in epoxy resins: A positron anninilation study. *European Polymer Journal*, 1994 30, 153-158.

[142] Soles, C.L.; Chang, F.T.; Gidley, D.W.; Yee, A.F. Contributions of the nanovoid structure to the kinetics of moisture transport in epoxy resins. *Journal of Polymer Science Part B: Polymer Physics*, 2000 38, 776-791.

[143] VanLandingham, M.R.; Eduljee, R.F.; Gillespie, J.W.Jr. Moisture diffusion in epoxy systems. *Journal of Applied Polymer Science*, 1999 71, 787-798.

[144] Morel, E.; Bellenger, V.; Verdu, J. Structure-water absorption relationships of amine-cured epoxy resins. *Polymer*, 1985 26, 1719-1724.

[145] De Neve B.; Delamar M.; Nguyen T.T.; Shanahan M.E.R. Failure mode and ageing of steel epoxy joints. *Applied Surface Science*, 1998 134, 202-212.

[146] De Neve, B.; Shanahan, M. Physical and chemical effects in an epoxy resin exposed to water vapour. *Journal of Adhesion*, 1995 49, 165-176.

[147] Xiao, G.Z.; Shanahan, M.E.R. Swelling of DGEBA/DDA epoxy resin during hygrothermal ageing. *Polymer*, 1998 39, 3253-3260.

[148] Carfagna, C.; Apicella, A.; Nicolais, L. The effect of the prepolymer composition of amino-hardened epoxy resins on the water sorption behavior and plasticization. *Journal of Applied Polymer Science*, 1981 27, 105-112.

[149] Kalenda, P.; Kalendova, A. Improved Chemical resistance of Epoxy resin-based coating composition. *Dyes and Pigments*, 1995 27, 305-312.

[150] Lasoski, Jr.S.W.; Cobbs Jr.W.H. Moisture permeability of polymers. I. Role of crystallinity and orientation. *Journal of Polymer Science*, 1959 36, 21-33.

[151] Klute, C.H. Diffusion of small molecules in semicrystalline polymers: Water in polyethylene. *Journal of Applied Polymer Science*, 1959 1, 340-350.

[152] Alter, H. A critical investigation of polyethylene gas permeability. *Journal of Polymer Science*, 1962 57, 925-935.

[153] Brandrup, J.; Immergut, E.H. *Polymer handbook*, 3rd edition, John Wiley & Sons, New York, 1989.

[154] Kamal, M.R.; Jinnah I.A. Permeability of Oxygen and Water Vapor Through Polyethylene/ Polyamide Films. *Polymer Engineering and Science*, 1984 24, 1337-1347.

[155] Rouw, A.C. Model epoxy powder coatings and their adhesion to steel. *Progress in Organic Coatings*, 1998 34, 181-192.

[156] Li, L.; ShuYong Zhang; Chen, Y.; Liu, M.; Ding, Y.; Luo, X.; Pu, Z.; Zhou, W.; Li, S. Water Transportation in Epoxy Resin. *Chemical Materials*, 2005 17, 839-845.

[157] Al-Harthi, M.; Loughlin, K.; Kahraman, R. Moisture diffusion into epoxy adhesive: testing and modeling. *Adsorption*, 2007 13, 115-120.

[158] McCall, D.W.; Doughlass, D.C.; Blyler Jr.L.L.; Johnson, G.E.; Jelinski, L.W.; Bair, H.E. Solubility and diffusion of water in low-density polyethylene. *Macromolecules*, 1984 17, 1644-1649.

[159] Zhou, J.; Lucas, J.P. Hygrothermal effects of epoxy resin. Part I: the nature of water in epoxy. *Polymer*, 1999 40, 5505-5512.

[160] Popineau, S.; Shanahan, M.E.R. Simple model to estimate adhesion of structural bonding during humid ageing. *International Journal of Adhesion & Adhesives*, 2006 26, 363-370.

[161] Bajat, J.; Dedi, O. Adhesion and corrosion resistance of epoxy primers used in the automotive industry. *Journal of Adhesion Science and Technology*, 2007 21, 819-831.

[162] Funke, W. Thin-layer technology in organic coatings. *Progress in organic coatings*, 1996 28, 3-7.

[163] Allen, K.W. A Review of Contemporary Views of Theories of Adhesion. *The Journal of Adhesion*, 1987 21, 261-277.

[164] Munger, C.G. Surface, adhesion, coatings. *Materials Performance*, 1983 22, 33-37.

[165] Schmidt, R.G.; Bell; J.P. Investigation of Steel/Epoxy Adhesion Durability Using Polymeric Coupling Agents. II. Factors Affecting Adhesion Durability. *The Journal of Adhesion*, 1988 25, 85-107.

[166] Nguyen, T.; Martin, J.W. Modes & Mechanisms of degradation of epoxy coated reinforcing steel in a marine environment. *Durability of building materials and components*, 1996 1, 491-502.

[167] Venables, J.D. Adhesion and durability of metal polymer bonds. *Adhesion*, 1983 7, 87-93.

[168] McKnight, M.; Seiler, J.F.; Nguyen, T.; Rossiter, W.J. Measuring Peel Adhesion of Coatings. *Journal of Protective Coatings and Linings*, 1995 12, 82-89.

[169] Shaw, G.S.; Rogers, C.E.; Payer, J.H. The effect of immersion on the breaking force and failure locus in an epoxy/mild steel system. The Journal of adhesion, 1992 38, 255-268.

[170] Walker, P. The effect of silanes on paint adhesion and sites of failure. *Surface and Interface Analysis*, 1991 17, 465-470.

[171] Keisler, C.; Lataillade, J. L. The effect of substrate roughness characteristics on wettability and on the mechanical properties of adhesive j oints loaded at high strain rates. *Journal of Adhesion Science and Technology*, 1995 9, 395-411.

[172] Sancaktar, E.; Gomatam, R. A study on the effects of surface roughness on the strength of single lap joints. *Journal of Adhesion Science and Technology*, 2001 15, 97-117.

[173] Zhang, S.; Panat, R.; Hsia, K.J. Influence of surface morphology on the adhesion strength of epoxy–aluminum interfaces. *Journal of Adhesion Science and Technology*, 2003 17, 1685-1711.

[174] Meine, K.; Klob, K.; Schneider, T.; Spaltmann, D. The influence of surface roughness on the adhesion force. Surface and Interface Analysis, 2004 36, 694-697.

[175] Namkanisorn, A.; Chaudhury, M.K. An Arrhenius method to study the effect of surface roughness in polymer. *Metal adhesion*,1998, 415-417.

[176] Sargent, J.P. Adherent surface morphology and its influence on the peel strength of adhesive joints bonded with modified phenolic and epoxy structural adhesives. *International journal of adhesion and adhesives*, 1994 14, 21-30.

[177] Zhai, L.; Ling, G.; Li, J.; Wang, Y. The effect of nanoparticles on the adhesion of epoxy adhesive. *Materials Letters*, 2006 60, 3031-3033.

[178] Packham, D.E. The mechanical theory of adhesion. Changing perceptions 1925-1991. *The Journal of Adhesion*, 1992 39, 137-144.

[179] Kinloch, A.J. The science of adhesion. *Journal of materials science*, 1980 15, 2141-2166.

[180] Fourche, G. An overview of the basic aspects of polymer adhesion - Part II: Application to surface treatments. *Polymer Engineering and Science*, 1995 35, 968-975.

[181] Cayless, R.A. Interfacial chemistry and adhesion: The role of surface analysis in the design of strong stable interfaces for improved adhesion and durability. *Surface and Interface Analysis*, 1991 17, 430-438.

[182] Bierwagen, G.P. Film formation and mudcracking in latex coating . *Journal of coatings technology*. 1979 51, 117-129.

[183] Piens, M.; Deurwaerder, H.D. Effect of coating stress on adherence and on corrosion prevention. *Progress in Organic Coatings*, 2001 43, 18-24.

[184] Roche, A.A.; Bierwagen, G.P.; Dolde, P.; Bouzziri, M. *Measurement of the practical adhesion of paint coatings to metallic sheets by the pull-off and three-point flexure tests*. Adhesion measurement of films and coatings, K.L. Mittal , 1995, 299-321.

[185] BS EN ISO 4624: 2003 - Paints and varnishes. Pull-off test for adhesion

[186] Roche, A.A.; Dole, P.; Bouzziri, M. Measurement of the practical adhesion of paint coatings to metallic sheets by the pull-off and three-point flexure tests. *Journal of Adhesion Science and Technology*, 1994 8, 587-609.

[187] Turunen, M.P.K.; Marjamaki, P.; Paajanen, M.; Lahtinen, J.; Kivilahti, J.K. Pull-off test in the assessment of adhesion at printed wiring board metallisation/epoxy interface. Microelectronics Reliability, 2004 44, 993-1007.

[188] Fahmy, A.A.; Hurt, J.C. Stress Dependence of Water Diffusion in Epoxy Resin. Polymer Composites, 1980 1, 77-80.

[189] Yaniv, G.; Ishai, O. Coupling Between Stresses and Moisture Diffusion in Polymeric Adhesives. *Polymer Engineering and Science*, 1987 27, 731-739.

[190] Shanati, S.; Ellis, N.S.; J. Randall, T; Marshall, J.M. Coupled diffusion and stress by the finite element method. *Applied Mathematical Modelling*, 1995 19, 87-94.

[191] Abdel Wahaba, M.M.; Crocombea, A.D.; Beeversb, A.; Ebtehaj, K. Coupled stress-diffusion analysis for durability study in adhesively bonded joints. *International Journal of Adhesion & Adhesives*, 2002 22, 61-73.

[192] Leger, R.; Roy, A.; Grandidier, J.C. Non-classical water diffusion in an industrial adhesive. *International Journal of Adhesion & Adhesives*, 2010 30, 744-753.

In: Pipelines: Design, Applications and Safety
Editors: M. G. Rivero et al. pp. 115-153

ISBN 978-1-62100-178-2

Chapter 4

INTERNAL STRESSES IN PIPELINE COATING: MANUFACTURING PROCESS AND LIFETIME

E. Aragon, L. Belec and Y. Joliff *

MAPIEM, EA 4323, Institut des Sciences de l'Ingénieur de Toulon et du Var, Avenue Georges Pompidou, BP56, 83162 La Valette-du-Var cedex, France

ABSTRACT

Pipelines are used worldwide for the transportation of oil or gas and must be protected against corrosion over long periods of time to avoid any production failure. An anti-corrosion coating system is generally used in addition to cathodic protection. This coating must fulfil specific conditions such as good mechanical strength and good ageing resistance in corrosive soils or water for instance. Two types of coatings are currently used, a monolayer system or a three layer systems, which have both their advantages and drawbacks.

Spontaneous disbonding of the three layer coating was thus observed at the ends of pipelines sections during storage, just after coating application. Moreover, for three-layer systems presenting no apparent defaults, pipelines coating failures have been observed at the epoxy/steel interface after a service period shorter than the expected lifetime. These phenomena were never reported for monolayer systems made with the same epoxy primer, which highlights the influence of internal stresses generated in the thick coating during the process on interfacial strength. The simulation of the process by finite element analysis with adequate material behaviour laws give internal stresses levels in good agreement with experimental measurements and in situ observations.

Moreover, most loss of adhesion of the three-layer systems have been observed when the operating temperature was about 50-60°C in wet environments, which suggest wet disbonding. Diffusion phenomena through the different polymer layers must then be taken into account. Depending on primer nature, Fickian or Langmuir water diffusion kinetics show that water molecules diffuse through the thick coatings up to the epoxy-steel interface very rapidly compared to pipelines' service time. The coating failure can then be attributed to epoxy physical or chemical degradation or to interfacial bonds hydrolysis. The diffusion parameters and then, the failure mode in humid environment,

* Corresponding author: yoann.joliff@univ-tln.fr.

strongly depend on temperature, primer nature and fillers proportion and nature. A good correlation can be found for thin coatings between primer disbonding and its saturation time estimated from diffusion data. For thick coatings, wet disbonding occurs before the primer saturation time, which is linked to the presence of internal stresses stored during the process.

INTRODUCTION

Pipelines are used worldwide for the transportation of oil or gas and must be protected against corrosion over long periods of time to avoid any production failure. A passive anti-corrosion coating system is generally used in addition to cathodic protection [1,2,3,4,5] [This coating must fulfil specific conditions such as good mechanical strength, good ageing resistance in corrosive soils for instance [6,7].

Two main types of coatings are nowadays used to protect pipelines from external events and therefore increasing their service life. The first one, a monolayer system, consists in coating the steel cylinder with about 600 μm thick unique epoxy layer (FBE). This type of pipeline coating is mainly dedicated to American facilities. The second type of pipeline coating, used for the European market, is a multi-layer system that associates three layers (epoxy / polyethylene adhesive / polyethylene topcoat) of different thicknesses at the surface of steel cylinder. In operation, pipelines will be submitted to numerous stresses such as impacts which can damage the coating, in particular for monolayer FBE systems. The Polyethylene topcoat can thus act as a mechanical protection.

PART A: MANUFACTURING PROCESS

1. Three-layer Coating Description

The coating system describe in this part is applied for steel pipeline external protection. This coating what is none as a “passive protection” is associated to an “active” anti-corrosion protection.

The coating is composed by three layers (Figure 1):

1. An epoxy layer between 100 to 200 μm, is directly applied on the steel cylinder. Generally, it is a powder epoxy (Fusion Bonded Epoxy – FBE) which is applied by an electrostatic dusting on the hot substrate (200-230°C). This layer plays the part of joint between the steel and the others layers of the coating assembly due to the excellent properties (mechanical and chemical) of the epoxy-steel linkage.
2. An adhesive layer between 150 to 300 μm which allows to bond the external layer of the coating. It is a polyolefin (polyethylene or polypropylene) graft with a more polar group (maleic anhydride) or a copolymer (ethylene / maleic anhydride). This adhesive applied at high temperature allows a good adhesion by permeation of the polyethylene and the epoxy (polar groups).
3. A polyethylene or a polypropylene (depending on applications) between 2 to 5 mm which plays the part of diffusion barrier (water, oxygen…) due to high thickness.

Moreover this layer has good resistance to UV, corrosive elements, thermal aging, and impacts. The last one limits the coating damage during buried pipeline installation.

The three-layer coating is frequently used in Europe for over ten years. Moreover, before 1980, the coating pipeline was a two-layer composed by an adhesive primer and a polyethylene for the external protection. However, it has a lower adherence at the steel/coating interface. Since, a three-layer have been added for increase the interface adhesion about the steel cylinder. This pipeline allows to carry oils and gas with temperature above 60°C.

Epoxy layer is justified by an excellent adhesion with steel, good chemical resistance and lower oxygen permeation. But it is brittle therefore increase care cost and cathodic protection current. All that explain why it is used as primer and it is protected by a polyolefin layer (mechanical protection). For reduce the coating failure up to steel pipe (coating peeling area), three-layer coating are well suitable.

This coating has good performance where mechanical stresses are significant. But there is two disadvantages: the shield effect to cathodic protection in case of peeling and the compatibility with joint coating between pipeline assemblies.

Indeed, joints applied at the end of pipe must be coated with similar material in term of protection properties. Many solutions can be used but they are usually difficult to realize on the industrial field or they have low-quality in hot and humid environment.

Moreover, corrosion protection is due to the coating and the cathodic protection. The coating protects the great surface of steel and the cathodic protection area where coating have been damaged up to steel.

In case of monolayer system, the cathodic current can go across the coating due to porosities and lower thickness for stop corrosion under peeling coating.

In case of three-layer coating, the polyolefin thickness is more significant and insulant, so steel surface is not protected by the cathodic current as long as there is not coating default. Corrosion must develop under the peeling coating [8].

2. Three-layer Coating Pipeline Process

To make sure to have good coating adherence at the initial stage and a good ageing resistance, some rules must be respect during the coating application. The three layers of coating are successively applied at the steel pipe surface via a semi-continuous process with servaral stage (Figure 2 [9]).

The preparation steel surface is composed by a cleaning with a washing for suppress grease and salt. Next a grit blasting is realized for suppress solid residues (scale, oxide...) and for create a roughness advantageous for the adherence.

Finally a dust extraction is done for suppress the grit blasting residues. The roughness values are between 60 to 90 µm (Rz).

Steel cylinder is heated up to 200°C by induction. The epoxy layer is applied by spraying a powder at the surface of the heated cylinder. The adhesive and polyethylene layers are then extruded on the latter. The system then undergoes a rapid cooling (shower water quenching) in order to solidify the final lining before storing at ambient air.

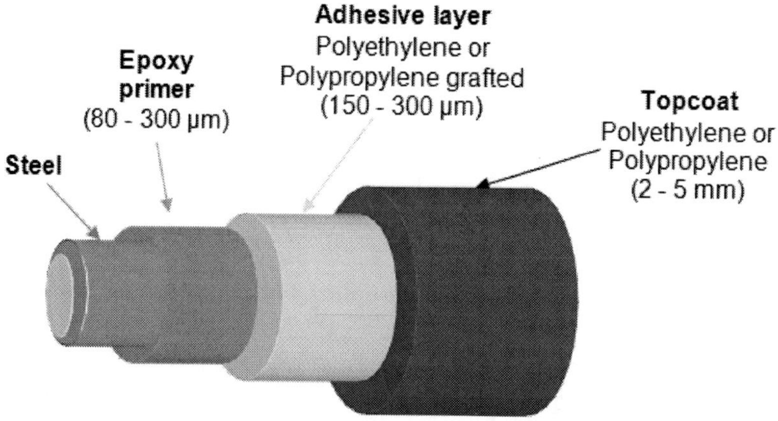

Figure 1. three-layer pipeline coating.

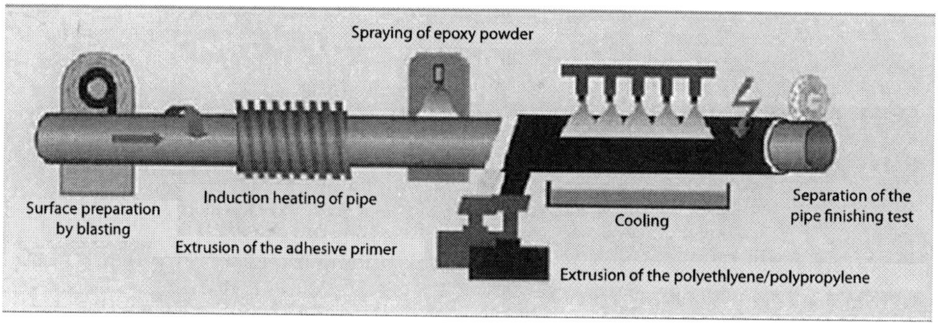

Figure 2. Application process of three-layer pipeline coatings [9].

3. Internal Stresses in Coating

The presence of internal stresses in organic coating is a current phenomenon. Indeed, the adhesion to a substrate with different properties perturbs the molecular reorganization especially during expansion or shrinkage. Theses evolutions can have many origins as thermal variations, relative humidity, crystallisation or crosslinking phenomenon, solvent evaporation [10,11,12,13,14]

Ramani et al. have studied internal stresses evolution during the cooling stage of steel-thermoplastic systems (Steel-PEEK and Steel-PEI). They have observed than a high cooling rate increase the internal stresses in the glass transition area whereas it decreases in the recrystallisation area (lower crystallinity) [15].

Wang et al. have observed that glass transition temperatures of epoxy resins decrease when they were crosslinked on aluminium substrate due to stress producing by epoxy/aluminium adhesion [16].

Another type of stresses can be developing in the coating with humidity. It consists in hygroscopic stresses due to a differential dilatation between the substrate and the coating. At the beginning of the immersion compressive stresses are developing. In the course of time, these stresses tend to decrease by relaxation due to the plasticizing [17].

3.1. Stresses Generated during the Process: Origins

A shrinkage of several centimetres is often observed on pipeline ends after cooling process. Indeed, during the cooling stage between 200°C and room temperature, the thermal contraction of the steel is much lower than the polyethylene or epoxy contraction, which generates internal stresses in polymer layers [10,11,12,13,14]. This phenomenon is accentuated by the polyethylene partial crystallization during the process. Both reasons lead to a strong shrinkage of the thick polyethylene topcoat compared to the steel. A spontaneous disbonding of the lining was thus observed at the ends of pipelines sections during storage at initial state between the epoxy layer and the adhesive layer. Moreover, for three-layer systems presenting no apparent defaults, complete disbonding of the lining has been observed at the epoxy / steel interface after a service period shorter than the expected lifetime of the pipeline. This phenomenon was never reported for monolayer systems. One can then argue that internal stresses generated during the processing of pipelines are higher than the mechanical strength of the interfaces. Internal stresses can also be generated by the crosslink reaction of epoxy on the steel. In monolayer systems, these stresses do not lead to spontaneous disbonding but modify the monolayer system resistance to humid ageing [18].

3.2. Internal Stresses Estimations: Analytical, Experimental and Numerical Approaches

Internal stress can be determined by various ways.

3.2.1. Analytical Approach

An analytical solution has been proposed by Timoshenko et al. for describe thermal residual stress in a long circular cylinder [19].The temperature is taken to be symmetrical about axis and independent of the axial coordinate z. The axial displacement is supposing null along cylinder. The solutions for the radial displacement (u) and thermal stresses (σ_r, σ_θ, σ_z) are :

$$u = \frac{1+\nu}{1-\nu}.\alpha.\frac{1}{r}\int_a^r Trdr + C_1 r + \frac{C_2}{r}$$

$$\sigma_r = -\frac{\alpha E}{1-\nu}.\frac{1}{r^2}\int_a^r Trdr + \frac{E}{1+\nu}\left(\frac{C_1}{1-2\nu} + \frac{C_2}{r^2}\right)$$

$$\sigma_\theta = \frac{\alpha E}{1-\nu}.\frac{1}{r^2}\int_a^r Trdr - \frac{\alpha ET}{1-\nu} + \frac{E}{1+\nu}\left(\frac{C_1}{1-2\nu} + \frac{C_2}{r^2}\right)$$

$$\sigma_z = -\frac{\alpha ET}{1-\nu} + \frac{2\nu EC_1}{(1+\nu)(1-2\nu)}$$

where σ_r is the stress in radial direction, σ_θ is the stress in hoop direction, σ_z is the stress in axial direction, C_1 and C_2 are two constants.

Analytical model can be extend to cylinder with concentric circular hole with a the radius of the hole and b the outer radius of the cylinder.

Chu et al. have proposed an exact analytical solution for thermal stresses in a long cylinder of two materials [20]

Chang has proposed an analytical model which takes in count the three layers of the coating (Figure 3) [21,22]. Model has been defined from thermal stress analysis of a long cylinder [19]. The maximum stresses in FBE can be described by:

$$\sigma_\theta^c = \frac{\alpha_c E_c \Delta T}{1 - \nu_c} + \frac{E_c}{(1+\nu_c)(1-2\nu_c)} H_c + \frac{E_c}{1+\nu_c} \cdot \frac{K_c}{b^2}$$
$$\sigma_r^c = \frac{E_c}{(1+\nu_c)(1-2\nu_c)} H_c - \frac{E_c}{1+\nu_c} \cdot \frac{K_c}{b^2}$$
$$\sigma_z^c = -\frac{\alpha_c E_c \Delta T}{1 - \nu_c} + \frac{2\nu_c E_c}{(1+\nu_c)(1-2\nu_c)} H_c$$

where ΔT is the difference between Room Temperature and Tg, H and K are two constants.

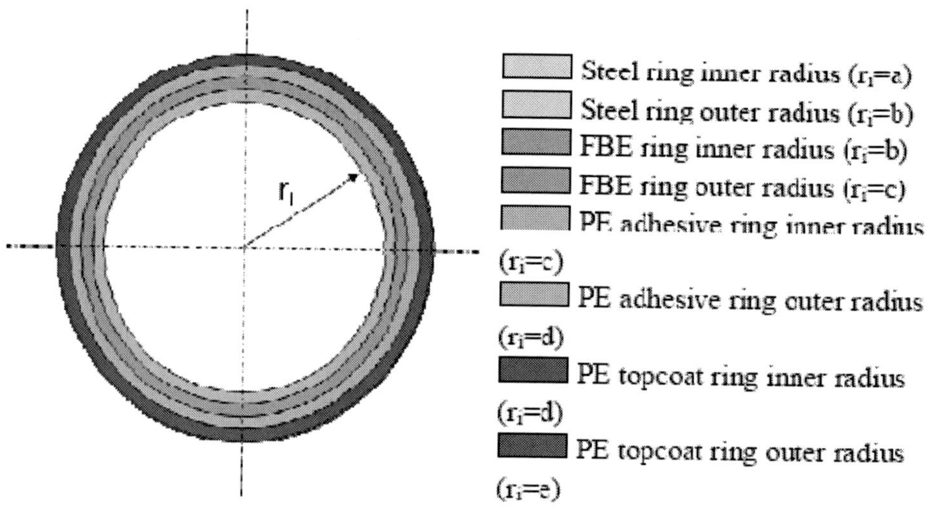

Figure 3. Thermal stresses analysis of a three layers coated pipe.

The analytical stresses analysis based on the plane strain condition is only applicable to the regions away from pipe ends. The maximum stresses have been located at the interface steel/FBE. Analytical results have shown than a significant hoop and radial stresses around 6-7 MPa were present in the interface. On the other hand, the radial stresses can be considerate null (lower than 0.2 MPa). But it could not take in count the edge effects present in these assemblies. Moreover, many assumptions used in the models have not been validated as the axial displacement which is supposed equal to zero or for all the layers of pipe is supposed the same ΔT. All physical properties have been considered constant with temperature.

So it is difficult to use it for quantified the stresses in the pipeline coating.

3.2.2. Experimental Measurements

Some experimental methods are available to measure internal stresses, but they can be quite difficult to apply depending on the samples studied. Deflection test is a current method, developed by Corcoran [22]. It consists in applying a thin layer of the coating on a flexible substrate which will bend under internal stresses. The curvature measurement allows estimating these stresses [23,]:

$$\sigma = \frac{dE_s t^3}{3l^2 c(t+c)(1-\nu_s)} + \frac{dE_c(t+c)}{l^2(1-\nu_c)}$$

where d is the curvature, E_s is the substrate elastic modulus, E_c is the coating elastic modulus, t is the substrate thickness, c is the coating thickness,ν_s is the substrate Poisson coefficient, ,ν_c is the coating Poisson coefficient.

This method, which exists under various forms with different equations, is currently used for internal stress determination [11,12] [25,26,27,28]. In this case, it is impossible to realize a three layers sample representative of the real system (co-extrusion and shower quenching) and then, to reproduce the real internal stress state. So, the Corcoran deflection test cannot be used.

3.2.3. Finite Element Method

Another way to deal with internal stress is finite elements modelling. This method can solve very complex problems even with several layers as long as the required entry parameters are known.

Chang et al. [21,22] have also studied the stresses generated in three-layer pipeline. So, they obtained the same values as for monolayer systems from analytical calculations, and values ranging from 10 to 40 MPa from finite elements modelling. Therefore, the reliability of both the analytical and numerical models can be discussed, since longitudinal shear stresses are not considered for the first one and viscoelastic dissipation is neglected for the second (probably leading to an overestimation of the stresses).

Legghe et al. [29] have tried to calculate internal stresses in more realistic system. The approach has been to write a finite element model based on simple constitutive laws able to quantify internal stress.

For this purpose, several models have been developed and compared: a linear elastic model with all parameters constant (similar to Chang et al.), a linear elastic model with some parameters function of temperature and a viscoelastic model.

Table 1. Origin of parameters used in finite element models

	Epoxy	Polyethylene	Steel
Thermal conductivity (λ)	literature		
Density (ρ)	literature		
Poisson coefficient (ν)	literature		
Specific heat (Cp)	Measured by MDSC*		literature
Thermal expansion coefficient (α)	literature	Evaluated by DMA**	literature
Young modulus (E)	Evaluated by DMA		literature
Shear creep test data	Evaluated by DMA		-
WLF coefficients	literature		-

* MDSC : Modulated Differential Scanning Calorimetry.

** DMA : Dynamic Mechanical Analysis.

4. Finite Element Analysis of Process Pipe Coating

Thermoplastic materials are not supposed to store stresses above their crystallization temperature as they behave as viscous liquids in this domain [15]. Chang et al. [21,22] have realized a finite elements modeling study on epoxy monolayer and on three-layers coatings, using very simple hypothesis. Physical properties were all considered as independent of temperature and the model was based on linear elastic laws. They obtained different stress values for monolayer (10 MPa) and three layers (40 MPa). Recently, a similar study have been realized on both systems shown an increase of about 70% of the internal stresses in the three-layer system in comparison to the monolayer system [30]. Without any viscoelastic dissipation, the stresses are hugely overestimated compared to experimental adherence values (below 30 MPa) between the different layers [18]. Indeed, single lap shear tests on epoxy-steel assemblies give interfacial shear strength around 28 MPa. Moreover, during pipelines storage, industrial observations show debonding at adhesive/epoxy interface. This interfacial strength is consequently lower than 28 MPa. A viscoelastic model must be considered to quantify internal stresses.

4.1. Experimental Parameters Determination

In order to use finite element modelling, various parameters are needed for each materials considered. These parameters can be estimated from literature, but the information collected is often limited, particularly in terms of temperature dependence of data. Furthermore, the materials never correspond exactly to the products. So, it has been measured most of these parameters when it was possible. The origin of all data used is resumed in Table 1.

4.1.1. Parameters Fixed with Temperature

The values of mechanical and thermal properties that are presented in Table 2 have been derived from literature [31]. The evolution of these parameters with temperature can be considered as limited between 20 to 130°C. Therefore, constant values have been used.

4.1.2. Parameters vs Temperature

The specific heat (Cp) has been measured at 2K/min after calibration on a sapphire reference on a Modulated Differential Scanning Calorimeter, Q100 from TA Instruments (Figure 4). Young's modulus (E), thermal expansion coefficients (α) and shear data have been obtained from thermomechanical measurements in dynamic mode (modulus) or static mode (expansion coefficient and shear creep test at 0.1 MPa) on a DMA 2980 from TA Instruments (Figure 5, Figure 6 and Figure 7). A shear sandwich fixture clamp has been used on plane-parallel samples (5x5x1 mm for FBE and 6x6x3 mm for PE). The creep tests have been performed at 100°C and 50°C for FBE and PE respectively and shear strains have been measured versus time. The creep data can be used at any temperature (not too far from test temperature) applying a shift factor in time scale. The shift factor evolution is well described by the Williams-Landel-Ferry (WLF) law (Table 3) which can be written as:

$$-\log A_T = h(T) = \frac{C_1^g(T - T_g)}{C_2^g + (T - T_g)}$$

with C_1^g and C_2^g are universal constants and T_g is the glass transition temperature.

Table 2. Constant parameters used in finite element models

	E (GPa)	ν	α (K^{-1})	ρ ($kg.m^{-3}$)	λ ($W.m^{-1}.K^{-1}$)	Cp ($J.Kg^{-1}.K^{-1}$)
Steel	208	0.29	16.10^{-6}	7 800	22	850
PE	-	0.42	-	940	0.4	-
FBE	-	0.33	-	1 400	0.3	-

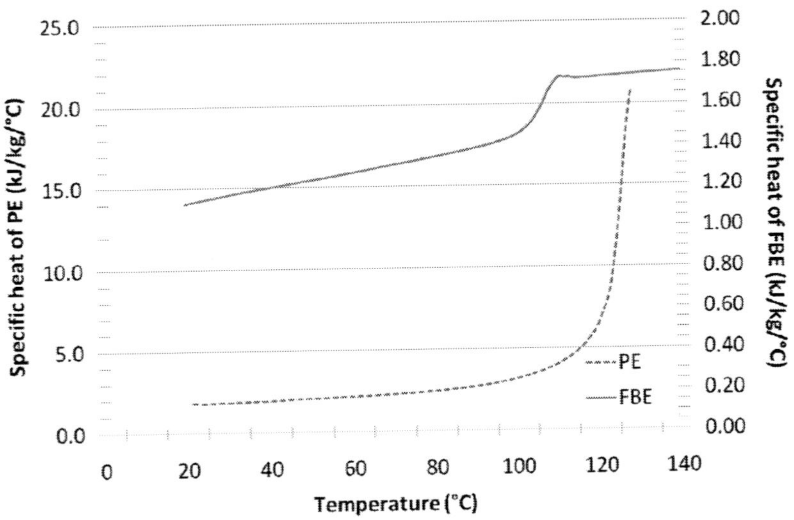

Figure 4. Specific heat as a function of temperature for polyethylene and epoxy (MDSC at 2K/min).

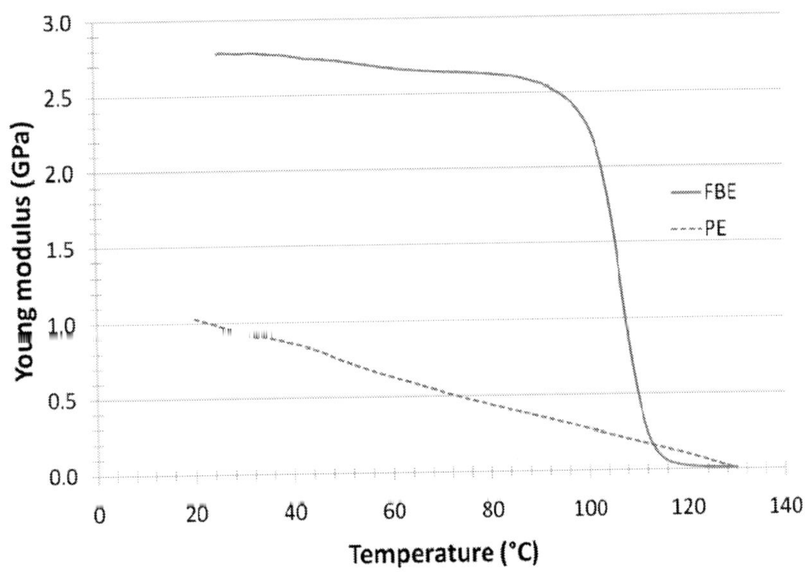

Figure 5. Young modulus as a function of temperature for polyethylene and epoxy (DMA at 1Hz in shear mode in linear viscoelastic range).

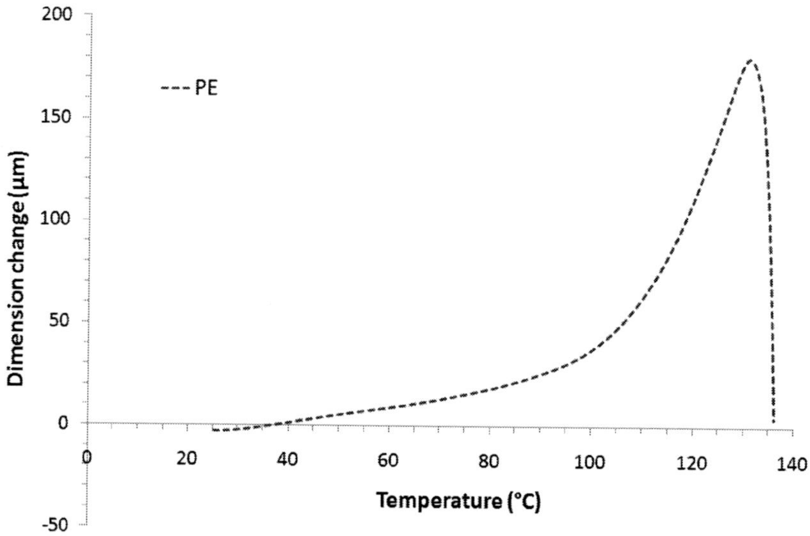

Figure 6. Thermal expansion as a function of temperature for polyethylene.

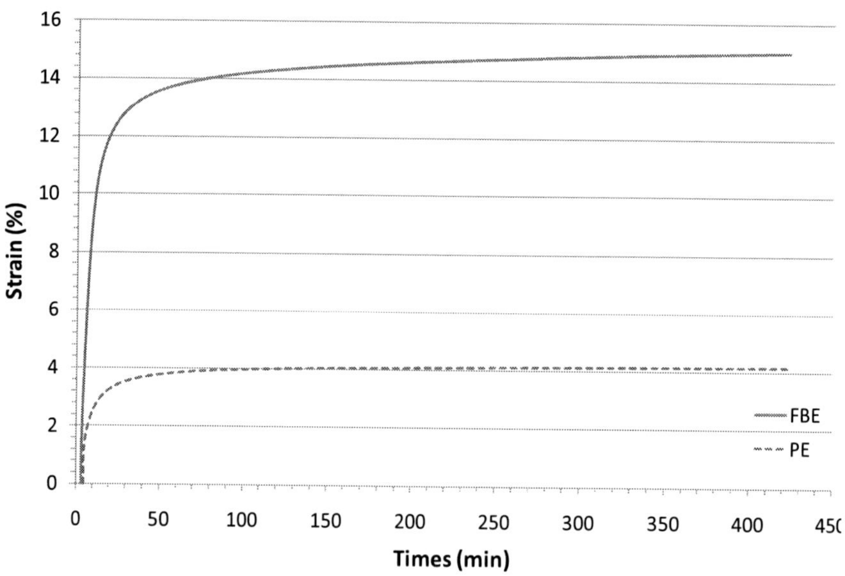

Figure 7. Shear creep test for epoxy at 100°C and polyethylene at 50°C (DMA at constant stress of 0.1 MPa).

Table 3. Coefficients of the Williams-Landel-Ferry law (WLF)

viscoelasticity	WLF		
	T (°C)	C_1	C_2 (K)
Epoxy	100	12	50
PE	25	16.4	112

Astruc [32] has proposed, for similar epoxy (glass transition temperature equal to 100°C), values of C_1 and C_2 which has been used in this chapter (temperature range from 60 to 130°C). The application of this law for polyethylene is more questionable as the range of temperature considered is not around the Tg value for this material. However, coefficients for this law between 30 and 80°C have already been reported in the literature [33].

It has been used to describe a viscoelastic law on polyethylene to take into account viscous dissipation. Even if it only gives an order of magnitude, it can consider that it is sufficient as these values are supposed to have a limited effect on the results of the calculation.

4.2. Viscoelastic Behaviour: Constitutive Laws [34,35]

The real behaviour laws of the epoxy and polyethylene in the temperature range considered are viscoelastic. This behaviour, written in the Abaqus solver, has been described in this part.

4.2.1. Time Domain Viscoelasticity

The time-domain viscoelastic model describes isotropic rate-dependent material behaviour for materials in which dissipative losses primarily caused by viscous (internal damping) effects must be modeled in the time domain [36] It assumes that the shear (deviatoric) and volumetric behaviors are independent in multiaxial stress states.

Time domain viscoelasticity is available in Abaqus for small-strain applications where the rate-independent elastic response can be defined with a linear elastic material model.

4.2.2. Defining the Shear Behaviour

Consider a shear test at small strain in which a time varying shear strain, γ(t), is applied to the material. The response is the shear stress τ(t). The viscoelastic material model defines as:

$$\tau(t) = \int_0^t G_R\,(t-s)\dot{\gamma}(s)ds$$

with $G_R(t)$ is the time-dependent "shear relaxation modulus" that characterizes the material's response. This constitutive behavior can be illustrated by considering a relaxation test in which a strain γ is suddenly applied to a specimen and then held constant for a long time. The beginning of the experiment, when the strain is suddenly applied, is taken as zero time, so that:

$$\tau(t) = \int_0^t G_R\,(t-s)\dot{\gamma}(s)ds = G_R(t)\gamma \qquad \text{(since } \dot{\gamma} = 0 \text{ for } t > 0\text{)}$$

where γ is the fixed strain. The viscoelastic material model is "long-term elastic" in the sense that, after having been subjected to a constant strain for a very long time, the response settles down to a constant stress; i.e., $G_R(t) \rightarrow G_\infty$ as $t \rightarrow \infty$.

The shear relaxation modulus can be written in dimensionless form:

$$g_R(t) = \frac{G_R(t)}{G_0}$$

where $G_0 = G_R(0)$ is the instantaneous shear modulus, so that the expression for the stress takes the form:

$$\tau(t) = G_0 \int_0^t g_R(t-s)\dot{\gamma}(s)ds$$

The dimensionless relaxation function has the limiting values $G_R(0) = 1$ and $g_R(\infty) = G_\infty/G_0$.

4.2.3. Numerical Implementation: Prony Series

Abaqus assumes that the viscoelastic material is defined by a Prony series expansion of the dimensionless relaxation modulus:

$$g_R(t) = 1 - \sum_{i=1}^{N} \bar{g}_i^p \left(1 - e^{-t/\tau_i^G}\right)$$

With N, τ_i^G and $\bar{g}_i^p$ are material constants. Substitution in the expression for the shear stress yields.

$$\tau(t) = G_0\left(\gamma - \sum_{i=1}^{N} \gamma_i\right) \text{ with } \gamma_i = \frac{\bar{g}_i^p}{\tau_i^p} \int_0^t e^{-s/\tau_i^G} \gamma(t-s)ds$$

The γ_i are interpreted as state variables that control the stress relaxation, and γ^{cr} is the "creep" strain i.e. the difference between the total mechanical strain and the instantaneous elastic strain (the stress divided by the instantaneous elastic modulus).

$$\gamma^{cr} = \sum_{i=1}^{N} \gamma_i$$

When creep test data are specified, Abaqus will calculate the terms in the Prony series automatically. The normalized shear is defined as:

$$j_s(t) = G_0\, J_s(t) \text{ with } J_s(t) = \frac{\gamma(t)}{\tau_0}$$

where $J_S(t)$ is the shear compliance, γ (t) is the total shear strain, and τ_0 is the constant shear stress in a shear creep test. At time t = 0, $J_S(0) = 0$.

The creep data are converted to relaxation data through the convolution integrals:

$$\int_0^t g_R(s)\, j_s(t-s)ds = t$$

Abaqus then uses the normalized shear modulus $g_R(t)$ in a nonlinear least-squares fit to determine the Prony series parameters

4.2.4. Temperature Effects: Williams-Landel-Ferry Law (WLF)

The effect of temperature, T, on the material behaviour is introduced through the dependence of the instantaneous stress, τ_0, on temperature and through a reduced time concept. The expression for the linear-elastic shear stress is rewritten as:

$$\tau(t) = G_0(T)\left(\gamma - \int_0^t \dot{g}_R(\xi(s))\,\gamma(t-s)ds\right) \; with \; \dot{g}_R(\xi) = \frac{dg_R}{d\xi}$$

with G_0 the instantaneous shear modulus dependent of the temperature and □(t) is the reduced time.

$$\xi(t) = \int_0^t \frac{ds}{A(T(s))}$$

with $A(T(s))$: a shift function at time t.

This reduced time concept for temperature dependence is usually referred to as thermo-rheologically simple (TRS) temperature dependence. Often the shift function is approximated by the Williams-Landell-Ferry (WLF) form.

The integral formulation for linear isotropic viscoelasticity is:

$$\sigma(t) = \int_0^t 2G(\tau - \tau')\,\dot{e}dt' + I\int_0^t K(\tau - \tau')\,\dot{\phi}dt'$$

with e the mechanical deviatoric

ϕ the volumetric strains

$K(\tau)$ the bulk modulus functions of the reduced time "τ"

G(□) the shear modulus functions of the reduced time "τ"

The reduced time τ is related to the actual time through the integral differential equation:

$$\tau = \int_0^t \frac{dt'}{A_t(T(t))} \frac{d\tau}{dt} = \frac{1}{A_t(T(t))}$$

where T is the temperature and A_T is the shift function, therefore, if $A_T = 1$, $\tau = t$. A commonly used shift function is the Williams-Landel-Ferry (WLF) equation, which has the following form:

$$-\log A_T = h(T) = \frac{C_1^g(T - T_g)}{C_2^g(T - T_g)}$$

where C_1^g and C_2^g are universal constants and T_g is the glass transition temperature as mentioned above. If $T \leq (T_g - C^2{}_g)$, deformation changes will be elastic. C^1g and $C^2{}_g$ were once thought to be "universal" constants whose values were obtained at T_g, but these constants have been shown to vary slightly from polymer to polymer.

The finite element code allows the WLF equation to be used with any convenient temperature, other than the glass transition temperature, as the reference temperature. The form of the equation remains the same, but the constants are different. Namely:

$$\log A_{\theta T} = h(T) = \frac{C_1\ (T - T_g)}{C_2\ (T - T_g)}$$

where T_0 is the reference temperature at which the relaxation data are given, and C_1 and C_2 are the calibration constants at the reference temperature. The "universal" constants $C^1{}_g$ and $C^2{}_g$ are related to C_1 and C_2 as follows:

$$C_1 = \frac{C_1^g}{1 + \frac{(T_0 - T_g)}{C_2^g}} \quad and\ C_2 = C_2^g + T_0 - T_g$$

4.3. Finite Element Models: Results and Analysis

In this part, different finite element model have been used for calculate the internal stresses develop in the pipeline coating during the process. The approach is to write a finite element model based on simple constitutive laws (i.e. with few parameters in the law) able to quantify internal stress. For this purpose, three models have been compared:

Model 1: Linear elastic model with all parameters constant;
Model 2: Linear elastic model with E=f(T);
Model 3: Viscoelastic model.

4.3.1. Description of Finite Element Models

A steel pipeline of 20 meters length and 15 cm diameter, coated with the three layers of epoxy, adhesive and polyethylene have been considered. An axisymmetric model has been chosen for pipeline modelling. For symmetry reasons along longitudinal axis, only a half pipe has been modelled, and calculations have been done on one meter, i.e. on a 2 meters length pipeline.

It has been previously shown that withdrawals calculated with these models are proportional to the pipe length considered [30]. In "complex" model, the adhesive layer has been integrated in the PE layer since its properties are very similar to the polyethylene topcoat layer. Indeed, the adhesive is a maleic anhydride grafted PE with same physical properties (measured by DSC).

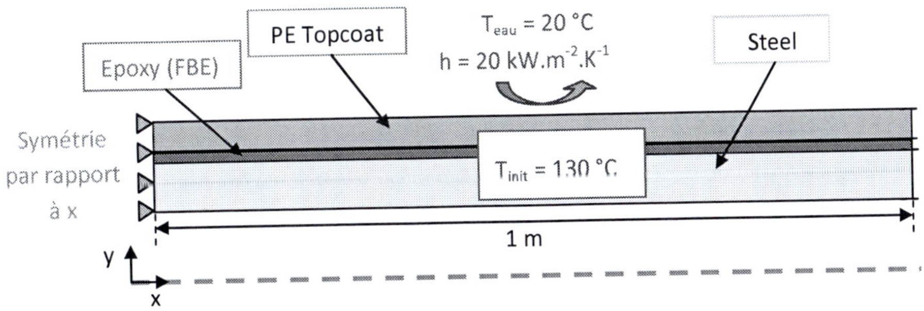

Figure 8. Numerical model description: load, boundary and initial conditions.

The water quenching has been simulated by application of a convection h (h = 20 kW.m^{-2}.K^{-1}, which corresponds of the classical value for shower quenching [37]) on the external surface of pipe, which describes thermal exchange with water at 20°C. No condition has been applied on the internal surface of pipeline. The initial temperature of the pipeline has been fixed at 130°C, which corresponds to polyethylene crystallization temperature. Over this temperature, it has been supposed than no stress can be stored. Each interface has been defined as a permanent contact « Tied ». The quenching time was determined from industrial specifications for application which give a steel temperature of about 55°C at the end of cooling stage. The meshing was realized by quadrilateral element with a quadratic interpolation using the structured meshing technique. The meshing has been optimized in such a way that the elements were concentrated on both sides of the interfaces where stress levels will be measured.

4.3.2. Linear Thermoelastic Behaviours: All Physical Properties Independent of Temperature

Chang [21, 22] has realized a finite elements modelling study on three-layers coatings, using very simple hypothesis. Physical properties were all considered as independent of temperature and the model was based on linear thermoelastic laws for all materials. He obtained stress values around 40 MPa.

A similar finite element model have been realised by Legghe [30]. She has used a linear thermoelastic model, where all parameters are taken constant with temperature. The value considered has been taken at room temperature from literature or from experimental data when available. The numerical model has been validated with two industrial results: temperature distributions and shrinkage values.

4.3.2.1. Temperature Distribution

Temperature field in each layer after coating application and cooling by water quenching during 190 s have been described on Figure 9. The temperature gradient is the same along the whole length of the pipeline. As polyethylene is a good thermal insulating material, cooling of the internal layers is quite difficult. We can note a thermal gradient, mainly in the polyethylene layer from 20°C on the surface to 47°C in internal part. Epoxy and steel reach a temperature of about 55°C (Figure 10).

The linear thermoelastic model with all parameters constant has done a calculated temperature of about 55°C in the steel ring inner radius, which corresponds quite well to the temperature observed on application line after 3 minutes of cooling. So the cooling modelling is coherent with industrial observations in terms of thermal exchanges.

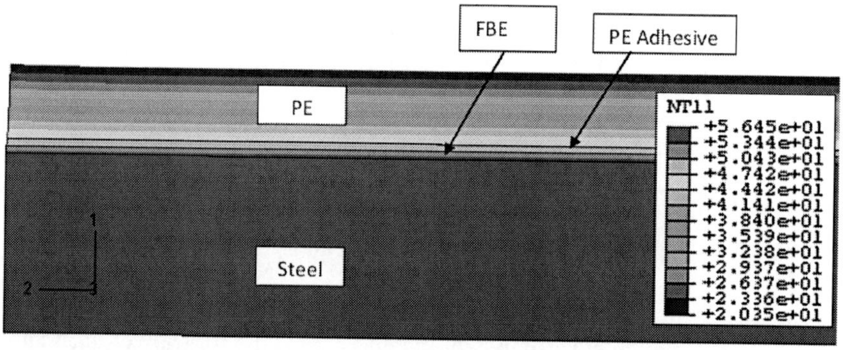

Figure 9. Temperature field isovalues cartography in °C (after 190 s water quenching).

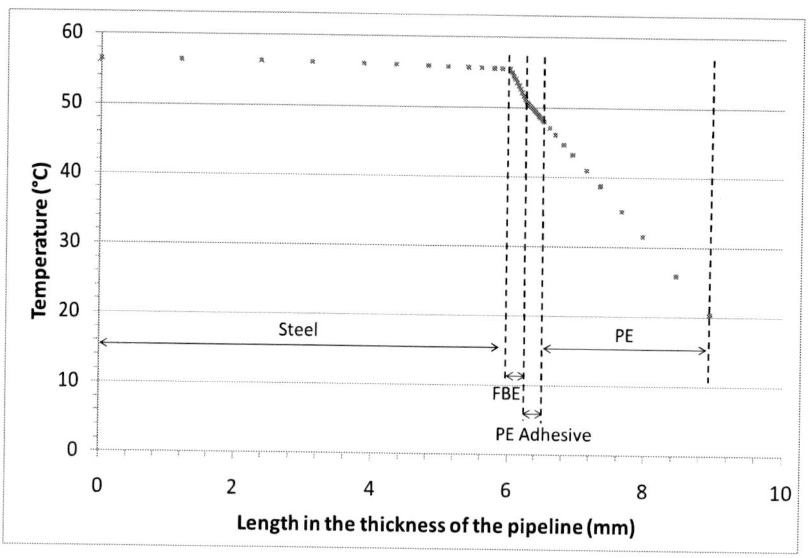

Figure 10. Temperature profile along thickness of each material (d=0 : external surface of steel pipeline and T= 20°C : external surface of PE coating).

4.3.2.2. Shrinkage Values

The displacement field in the longitudinal direction at the edge of the pipeline for each element of the system has been described on Figure 11. A withdrawal of about 1.2 mm is observed at each end of the pipe of 2 m length.

The shrinkage value of PE coating (1.2 mm at each end of the pipe in case of 2 m pipeline length) which corresponds to 12 mm by linear extrapolation for a 20 m length pipeline is coherent with the order of magnitude around 1 cm observed at the end of application on polyethylene. Moreover a radial contraction around the metallic cylinder with a withdrawal of about 0.3 mm for the external diameter with a clear edge effect has been calculated (Figure 12). A 30 % difference between contraction at the edge of the pipe and in the middle has been calculated.

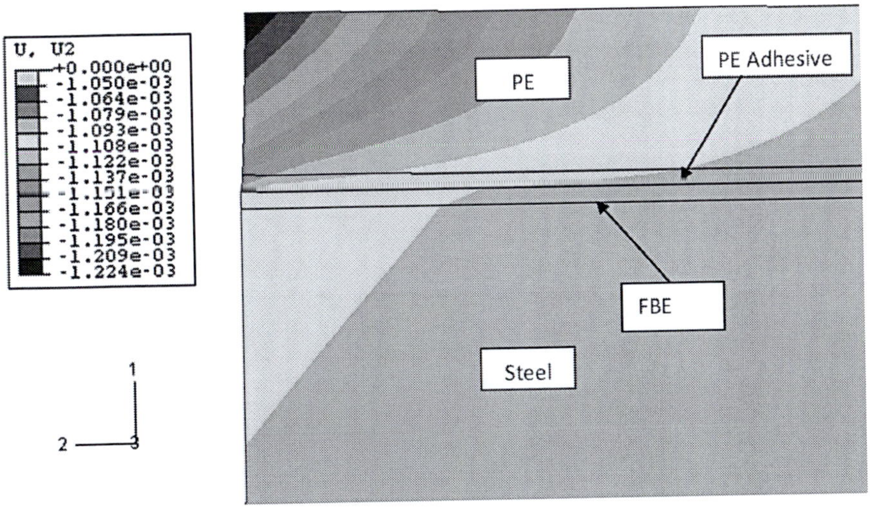

Figure 11. Displacement field isovalues cartography (in m) in longitudinal direction (direction 2).

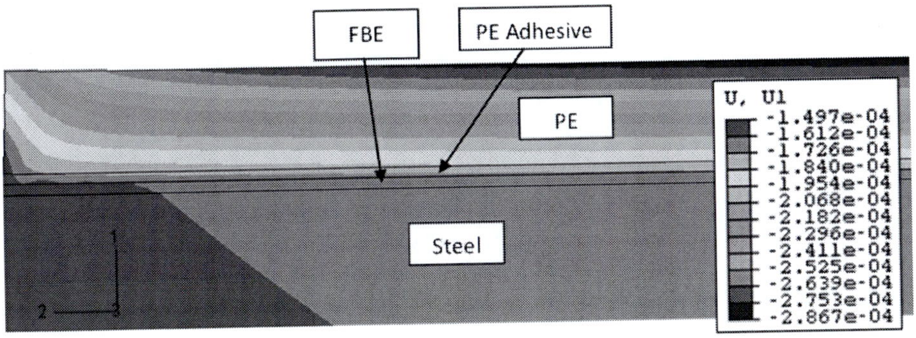

Figure 12. Displacement field (in m) in radial direction (direction 1).

4.3.2.3. Internal Stresses

For highlight interfacial stresses generated by cooling, shear stresses must be considered. The use of Tresca criteria which maximises shear stresses seems to be the most adapted and has consequently been used. More precisely the shear value in PE adhesive layer has been followed. It corresponds to the weakest layer in the assembly during storage. Numerical calculi have shown that stresses were located at the interfaces, mainly epoxy/adhesive interface, at the edge of the pipe. The value of Tresca criteria was near to 70 MPa.

The stress field calculated with this finite element model had given very high values, far from realistic values. The overestimation results from materials parameters, have been considered as independent on temperature and taken at 20°C, where the system was the most rigid. Moreover, no dissipation has been considered to take into account viscoelastic behaviour of polyethylene and epoxy. This model leaded to a temperature field coherent with cooling times and to withdrawals similar to those observed experimentally. However, it clearly overestimates stresses, which has shown its limits.

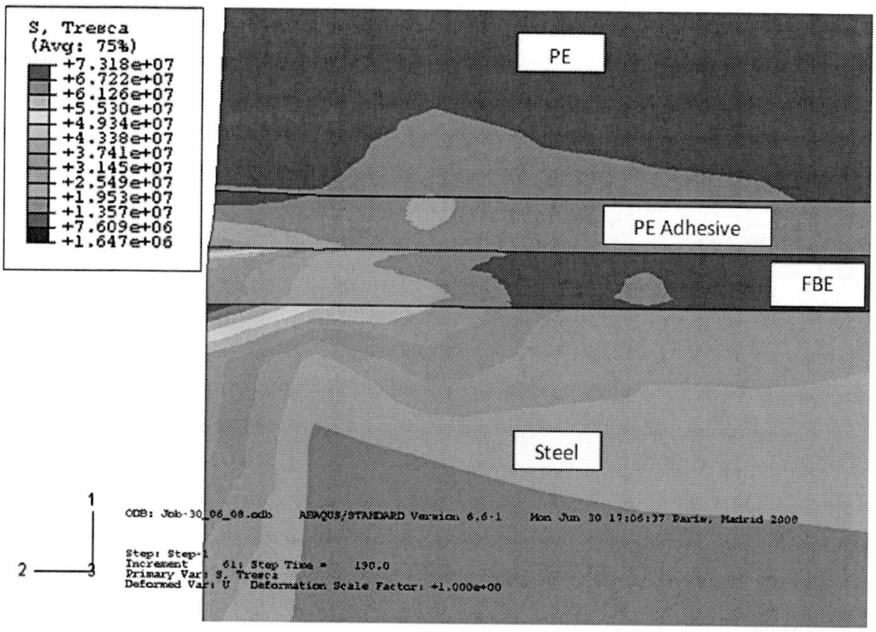

Figure 13. Stress field isovalues cartography (in Pa) Tresca criteria (maximum shear).

The next part of this chapter tries to improve the model by introducing Young modulus dependent on temperature. This parameter drives the system rigidity, and thus plays an important role on internal stresses.

4.3.3. Linear Thermoelastic Behaviours: Young Modulus Dependent of Temperature

The second model describes in this chapter is still a linear elastic model identical to the first model, except that the young modulus of polyethylene and epoxy were temperature dependant. Calculi result have shown than the displacement and the temperature fields were the same as those obtained with the first model. Tresca stress field obtained with this model have been describes on Figure 14.

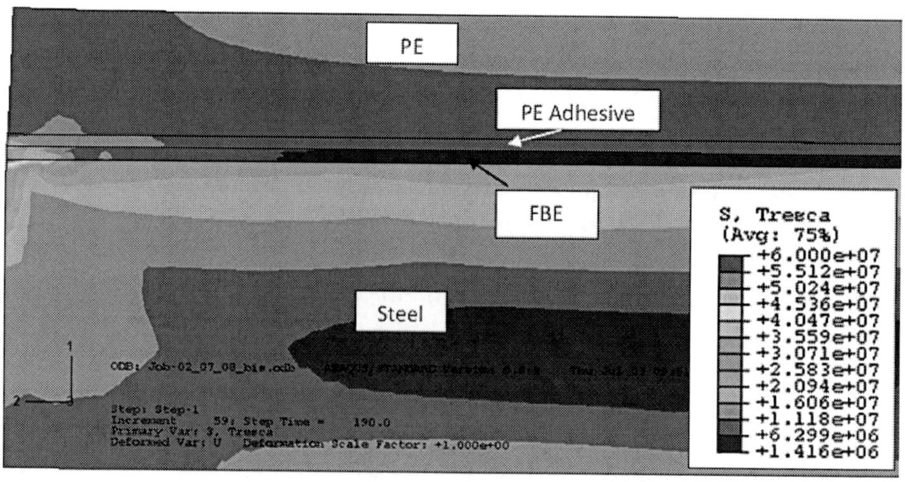

Figure 14. Tresca Stress field isovalues cartography (in Pa).

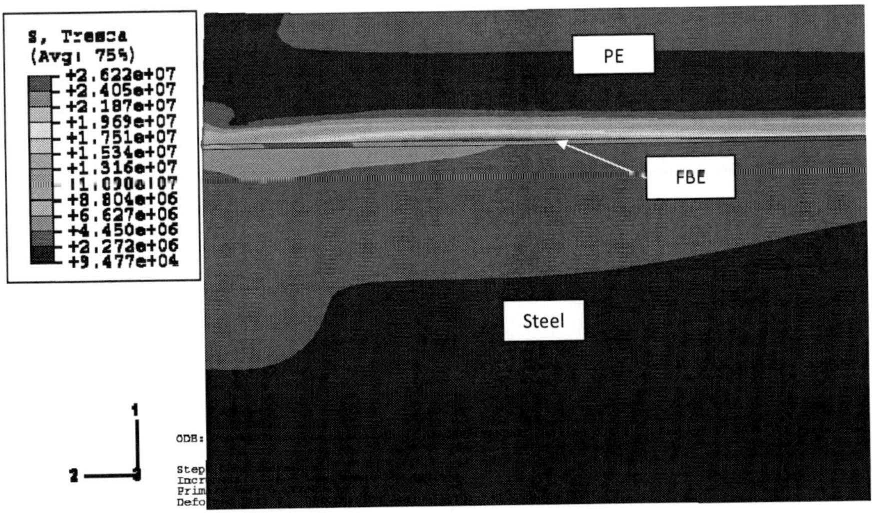

Figure 15. Tresca Stress field isovalues cartography of three-layer model (in Pa) – Thermoviscoelastic model.

Introducing the variation of Young modulus with temperature had reduced estimated stresses, with a maximum value of 60 MPa instead of 70 MPa (i.e. 15% reduction). But this second model still overestimated stresses, probably because of the absence of viscoelastic dissipation. Maximal stresses were still located at epoxy/adhesive interface and were still much higher than the polymers tensile strengths.

To complete this model, the variation with temperature of thermal expansion coefficient and of calorific capacities has been introduced. A divergence of the model at the ends of pipe, due to the high thermal expansion at high temperature has been obtained which was not compensated by high viscous dissipation in this range of temperature.

In the next part viscous dissipation behaviours have been introduced for polyethylene and epoxy layers.

4.3.4. Thermoviscoelastic Behaviours

The viscoelastic dissipation for epoxy and polyethylene layers has been introduced in finite element model. The entry parameters were similar to those of the second model, but viscoelasticity parameters have been introduced through shear creep test data and WLF law coefficients.

Numerical results have shown than the temperature and displacement fields calculated with were the same as those obtained with the previous models, which were coherent with experiment. A stress concentration has been observed at the interfaces, especially for the epoxy/PE interface (Figure 15). In terms of stress distribution, the maximal stress values at polyethylene/epoxy interface with 26 MPa, which was still quite high. The stress at the epoxy/steel interface was about 10 MPa.

The evolution of Tresca stress values along the steel/epoxy interface as a function of the distance from the end of pipeline has been described on Figure 16. From Tresca criteria, a maximum value of 26 MPa has been found at 4 mm from the edge of pipe, along epoxy/polyethylene interface. Above 3 cm from the edge, the value of stress was less than 5

MPa, which shown a clear edge effect. Moreover, the radial compression stress was smaller at the end of the pipe.

A value of about 6 MPa in the epoxy layer has been calculated, near the interface with steel, which represents a realistic value of stress. It was indeed of the same order of magnitude as residual adherence measured for epoxy on steel after ageing [18,38]. This model must be completed and developed, but it has shown realistic behaviour with coherent values and has shown the validity of this type of modelling for the estimation of internal stresses in pipeline coating.

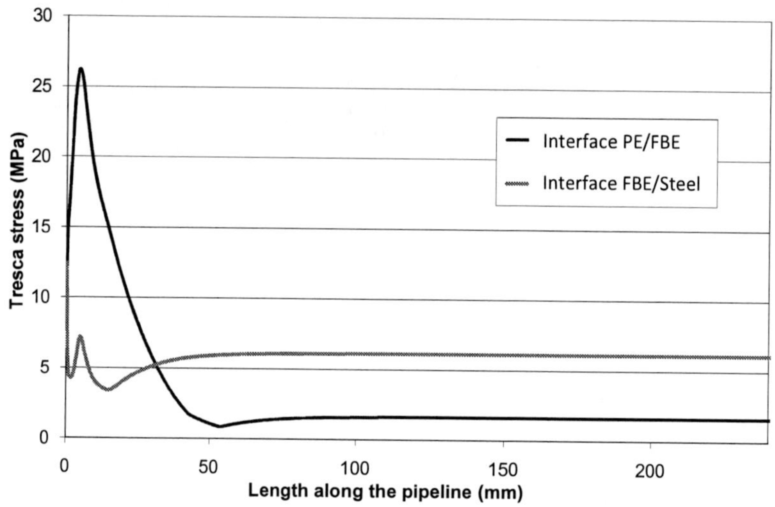

Figure 16. Tresca criteria evolution along the pipe at epoxy/PE and at steel/epoxy interfaces.

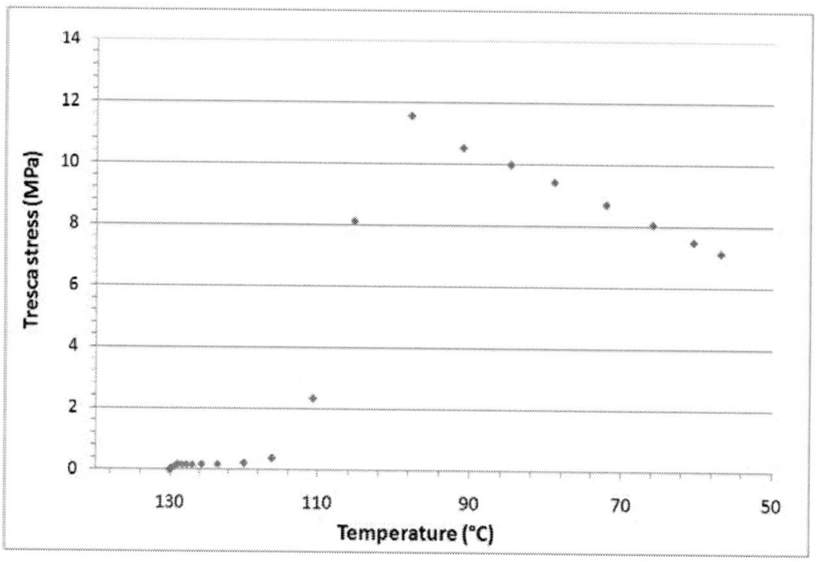

Figure 17. Stress evolution from Tresca criteria as a function of temperature at steel/epoxy interface.

Above some centimetres from the edge, stress value has decreased down to 5 MPa, which can explain the disbondings observed in the adhesive at the edges of the pipes. All models show a clear edge effect due to coating properties.

Stress evolution at epoxy/steel interface, during cooling, as a function of steel temperature has been represented on Figure 17. It shows an increase of internal stress from the temperature of 120°C, which corresponds to polyethylene crystallisation. This stress increases up to a tens of MPa, and then begins to decrease below 100°C.

The epoxy, which was near its glass transition temperature, can then relax a high part of the stored stresses. It has been observed the same behaviour for the two polyethylene thicknesses, with a higher stress for the thicker one. Moreover, it had a higher temperature after cooling because of the insulating behaviour of the polyethylene.

Some points must be improved to reach more accurate stress values. Viscoelastic parameters of polyethylene and epoxy should be investigated with experimental measurements to validate literature values.

The WLF coefficients, which were estimated here from literature, could be calculated from isothermal frequency sweep tests on polyethylene and epoxy. Plasticity of polyethylene could also been considered. Conductivity variations with temperature for the different materials should be investigated, because it plays an important role in thermal transfers occurring during application process.

5. Conclusion

By a finite element approach, to evaluate stresses generated during coating application. The use of a quite complete model, which takes into account viscous dissipation, allows obtaining realistic behaviours and stress values, giving interesting perspectives for this type of studies. Stresses are concentrated near the interfaces, in the adhesive layer, at the edges of the pipe.

This is coherent with the experimental observations which indicate current failure in this layer at the end of pipes. Coatings are generally eliminated on several centimetres at the end of pipes to limit this stress concentration. Epoxy/steel interface also presents significant stresses, of several MPa, on the whole length of the pipe. This value is similar to the residual adherence value of epoxy primer after humid ageing which could explain the disbonding observed on three-layer coatings.

This coupled modelling/experiment approach showed here its interest and should be continued to obtain more accurate results and to improve the understanding of stress generation during application and of the behaviour during ageing.

Part B: Durability in Humid Environments

The use of 3-layers polyolefines coatings for pipeline protection is motivated by their mechanical properties and resistance to abrasion, as well as their inertness to many environmental factors. Their chemical and physical stability also add to their value as protective coatings.

Nevertheless, the observation of coating disbonding on three layers systems with no apparent initial defect in wet environments suggest a water diffusion process through the different polymer layers. Most disbondments have been notices when the operating temperature of the pipeline is about 50°C-60°C but loss of adhesion exists for temperatures as low as 35°C. The loss of adhesion in the reported cases were related to exposure to soil conditions, and more especially to water diffusion [6].

1. Definitions

Water diffusion kinetics and coatings permeability must then be precisely known for each polymer layer to determine the time necessary for water molecules to reach the polymer/metal interface. The case of three layers coating systems, which should offer the best barrier properties to water vapour, is considered here.

The permeability defines the grade of transmissibility of the coating. It corresponds to an amount of penetrant in a specific time, under a given pressure and temperature, through a given coating thickness and area. Then, the permeability unit for water vapour for example, is g.µm/(m^2.24h.bar). It is obtained from permeation measurements [39]. It depends on several parameters such as the macromolecular mobility [40], the polymer ability to form hydrogen bonds, the crystalline proportion in polyolefin topcoat for example, the plasticizing effect of the solvent…

The diffusion coefficient represents the ability for a water molecule to move between two equilibrium states [41]. In permanent flow [42]:

$$\boldsymbol{J = D\frac{c}{h}}$$

with D: the diffusivity/diffusion coefficient (m^2/s); J: the flow of water by unit of time through a material of thickness h; c: the water concentration (mol/m^3).

The permeability P is used when the pressure gradient p' is considered rather than the concentration:

$$J = P\frac{p'}{h}$$

The solubility is then defined by:

$$s = \frac{c}{p'}$$

with c the equilibrium solvent concentration for a given pressure p'.

Consequently, the permeability can be simply determined from diffusion and solubility with:

$$P = D.S$$

2. Diffusion Kinetics

2.1. Fickian Diffusion

Two approaches are generally proposed to describe transport phenomenon of low-molecular weight substances in polymers [43,44]. One approach is based on the principles of irreversible thermodynamics that employs phenomenological coefficients to correlate the diffusion gradient across the membrane [45]. The other approach considers that the moisture absorption behavior of polymeric materials can been treated as a steady-state or diffusion-controlled process as described by Fick's first law [46]:

$$J = -D\frac{\partial c}{\partial x}$$

where J is the flow of water in the *x* direction which is along the thickness of a specimen and the only direction in which mass transfer is taking place.

The conservation of species equation in one dimension is given by:

$$\frac{\partial c}{\partial t} = -\frac{\partial}{\partial x}J$$

Fick's second law is used in non-steady or continually changing state diffusion, i.e., when the concentration within the diffusion volume changes with respect to time. The second Fick's law in one dimension is described by:

$$\frac{\partial c}{\partial t} = -D\frac{\partial^2 c}{\partial x^2}$$

assuming the temperature and the diffusion constant are constant inside the material.

A solution of Fick's second law describes the water uptake of a thin film in immersion when side effects can be neglected (film thickness h much lower than lateral dimensions):

$$\frac{\Delta M_t}{M_0} = \frac{\Delta M_\infty}{M_0}\left\{1 - \frac{8}{\pi^2}\sum_{m=0}^{\infty}\frac{1}{(2m+1)^2}exp\left[-\frac{D(2m+1)^2\pi^2 t}{h^2}\right]\right\}$$

with ΔM_t, the absorbed water at t, M_0 the initial weight and ΔM_∞ the water uptake at saturation.

The diffusion coefficient can then be calculated from the approximation for short times when the saturation stage is reached:

$$\frac{\Delta M_t}{M_0} = \frac{\Delta M_\infty}{M_0}\sqrt{\frac{16D}{\pi}}\frac{\sqrt{t}}{h}$$

It is generally agreed that the transport of water vapour through polyethylene obeys Fick's laws [47].

2.2. Non Fickian Diffusion

In many cases, deviations from fickian behavior are observed, especially in epoxy matrix [48,49].

The observed deviations are partly attributed to the time-dependent viscoelastic response of polymers [50]. Indeed, many glassy polymers are not at equilibrium and relax slowly, especially during absorption [51]. In these cases the diffusion coefficient becomes dependent not only on temperature, but also on exposure time and concentration [52]. Furthermore, moisture uptake does not reach saturation, but increases slowly with time.

In many epoxy adhesive systems, the water diffusion kinetics is well described by the two-phase model [48]. The theory assumes the existence of mobile and bound diffusing molecules. The existence of these phases were confirmed from deuterium NMR analysis [49] and measurements on absorbed water crystallization and melting by DSC [18] during water absorption of epoxy. The authors showed that the mobile phase corresponds to diffusing molecules, whereas the bound phase corresponds to clusters.

If n(t) represents the mobile water molecules and N(t) the bound molecules per unit volume at t, with γ (respectively β), the probability per unit time for free (respectively bound) molecules to become bounded (respectively free) then at equilibrium:

$$\gamma . n_\infty = \beta . N_\infty$$

Then, the one-dimensional diffusion case becomes:

$$\frac{\partial n}{\partial t} + \frac{\partial N}{\partial t} = D\frac{\partial^2 n}{\partial x^2}$$

and (a)

$$\frac{\partial N}{\partial t} = \gamma n - \beta N$$

For a sample thickness h, the boundary and initial conditions are:

$n(x,0) = 0$ and $N(x,0) = 0$ for $x < {}^{h}/_{2}$

$n\left({}^{h}/_{2}, t\right) = n_\infty$ and $N\left(-{}^{h}/_{2}, t\right) = n_\infty$

Defining $\boldsymbol{k = \pi^2 \frac{D}{h^2}}$ (b)

an approximate solution of equation (a) is obtained considering that β or $\gamma << k$

$$m_t = m_\infty \left\{ \frac{\beta}{\gamma+\beta} e^{-\gamma t} \left[1 - \frac{8}{\pi^2} \sum_{l=1}^{\infty} \frac{e^{-k(2l+1)^2 t}}{(2l+1)^2} \right] + \frac{\beta}{\gamma+\beta} \left(e^{-\beta t} - e^{-\gamma t} \right) + \left(1 - e^{-\beta t} \right) \right\} \quad \text{(c)}$$

which corresponds to the Fick's model when γ=0.

For short immersion times, (c) can be approximated by:

$$m_t \approx \frac{4}{\pi^{3/2}}\left(\frac{\beta}{\gamma+\beta}m_\infty\right)\sqrt{kt}$$

For medium times, a pseudo-equilibrium moisture content can be defined by:

$$m_1 \approx \frac{\beta}{\gamma+\beta}m_\infty$$

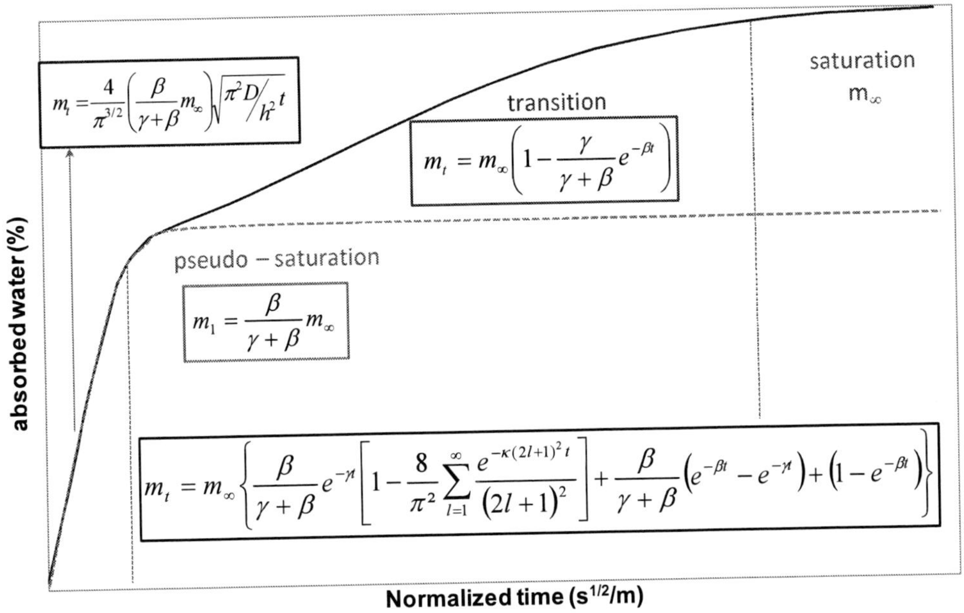

Figure 18. Identification of the parameters of a Langmuir-type diffusion model [48].

For long immersion times, (c) becomes:

$$m_t \approx m_\infty\left(1-\frac{\gamma}{\gamma+\beta}e^{-\beta t}\right) \quad \text{(d)}$$

From (b) to (d) and experimental data, generally reporting absorption as a function of $t^{1/2}$, it is then possible to determine κ and then D, γ and β as soon as saturation is reached. As a conclusion, the two-phase model describes correctly water diffusion in many epoxy primers or adhesives, especially in FBE used in three layers coating systems (Figure 19). It can then be used to calculate the diffusion coefficient for such systems.

The earliest stage of diffusion is governed by the interactions between the polymer and water molecules which form weak bonds with polar sites of the network. The pseudo equilibrium corresponds to the saturation of these polar sites.

The on-going water uptake up to the saturation is probably due to the accumulation of water in microvoids inside the matrix or fillers or at filler/matrix interfaces [53,54].

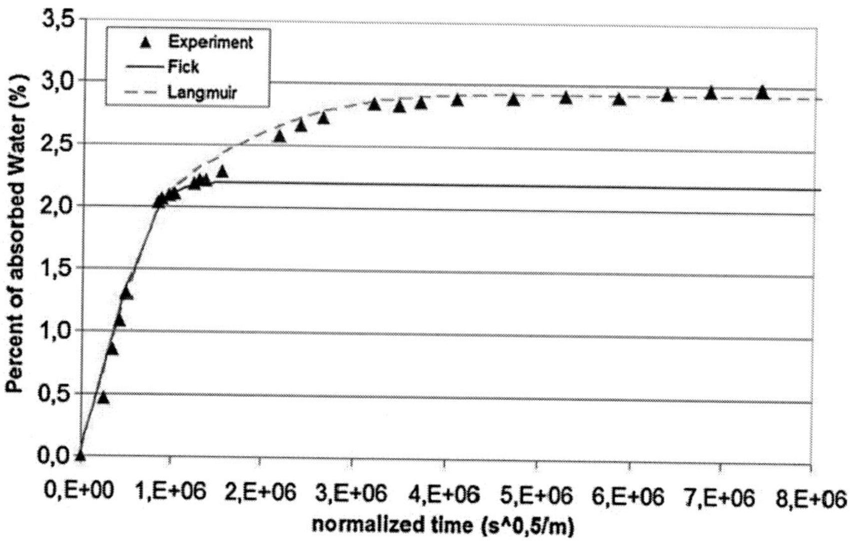

Figure 19. Water uptake vs. $t^{1/2}/h$ for an epoxy film in deionised water at 60°C [18].

3. Parameters Influencing Water Diffusion

3.1. Influence of Microstructure

3.1.1. Influence of Crosslinking

An under-crosslinked network generally increases the water absorption and diffusion coefficient. Such epoxy networks generally show a non Fickian absorption due to the on-going crosslinking during immersion [55].

On one other hand, the activation energy for the diffusion coefficient increases with crosslink density, due to the low accessibility to microvoids.

On the other hand, the mass at saturation of the under-crosslinked network increases with immersion temperature while it is nearly constant for the full crosslinked network. In that later case, the free volume is low and it is more difficult for water molecules to reach hydrophilic sites [56].

At low immersion temperature, water diffusion process is controlled by thermodynamics, with hydrogen-bonds formation. This is compatible with a higher chemical affinity of the high crosslinked network (etherification of epoxy, free reactive secondary amines).

At higher immersion temperature, the diffusion is controlled by kinetics and water diffuses via free volume, which is more important in the under crosslinked network.

3.1.2. Influence of Resin Nature

The absorption kinetics and equilibrium moisture content of epoxy systems are strongly dependent on the chemical structure of the resin and hardener.

It is generally admitted that the microstructure of the final epoxy resin systems in terms of free volume and size of voids influences the water sorption [57]. But it was also shown in several studies on various epoxy-amine systems that the absorption kinetics is independent of the free volume present in the resin and that it is mainly governed by polar interactions

[58,59]. The structure and functionality of the curing agent and prepolymer influence the hydrophilic properties of epoxy-amine systems.

Tertiary amines increase water bonding with the network, especially with hydroxyl groups in the β position [60]. Many Fusion Bonded Epoxy contain dicyandiamide (DDA) as hardener. It was also shown that under hygrothermal ageing conditions, DGEBA cured with DDA systems are sensitive to plasticization and hydrolysis [61,62,63].

Moreover, the ratio amine hardener / DGEBA prepolymer also influences the water diffusion kinetics [64]. Indeed, an excess of amine plasticizes the resin and then increases the molecular mobility as well as the number of potential H-bonds sites. Both phenomena lead to an increase in water uptake.

The lack of amine also lead to an under crosslinked network and then, to an intermediate Tg, but this system contains few reactive sites with water molecules. This system shows consequently the lowest water uptake. The system with stoechiometric proportion shows the highest Tg but an intermediate water uptake due to residual reactive amine sites. The nature and amount of reactive sites able to form H-bonds with water molecules highly influence water diffusion.

3.2. Influence of Filler Nature

The addition of filler may increase the water diffusion paths and can then limit water diffusion. Interactions between fillers and matrix strongly influence water absorption. Depending on filler surface nature, the matrix-filler interface can be more or less resistant to water uptake [53]. The water absorption of a glass-beads filled polymer is much lower when the glass surface is treated with an appropriate silane treatment. Some oxides such as MgO, ZnO, CaO can react with the penetrant, which decreases its diffusion rate. Conversely, neutral fillers such as TiO_2 or carbon black are reported to increase the diffusion of acid at matrix/filler interface [65].

Table 4. Water vapour permeability and diffusion coefficient of varied polyolefins (topcoat) and FBE (primer)

	Temperature (°C)	P (WV) (g.25μm/m²/24h/bar)	D (WV) (m²/s)
LDPE	25	45-222	$0.5\text{-}7\times10^{-12}$
	40	170-260	3×10^{-11}
	50	590	-
HDPE	25	45-135	12×10^{-12}
	50	230	-
HDPE CB	25	130	-
	50	270	-
FBE	25	590	2×10^{-13}
	50	870	12×10^{-13}

3.3. Influence of Polyolefine Density

Polyolefine density increases with their crystalline fraction. The permeability of low-density polyethylene is then higher than the one of high-density polyethylene due to the higher amorphous proportion [41,47,66,67,68]. High-density polyethylenes have permeabilities from 1/3 to 1/6 that of low-density polyethylene. The permeability of polypropylene is intermediate to high- and low-density polyethylenes. A proportional relationship is sometimes proposed between permeability and amorphous phase fraction.

3.4. Influence of Temperature

The permeability evolution with temperature is well described by an Arrhenius –type equation [69, 47]. Activation energy of around 33kJ/mol are given for water vapour permeability of polyolefins.

4. Experimental Data from Literature

The water solubility in polyolefines is low and the fickian diffusion coefficient is independent on water concentration. Several values are given in literature [41,18,66,67,70 71].

From permeametric measurements of water vapour (WV):

From gravimetric measurements at different immersion temperatures [18,67,74,72,73, 54]:

The permeability of polyolefins and epoxy primer must then be taken into account, especially at high operating temperatures. Water diffusion kinetics parameters show that water molecules reach the metal/polymer interface very rapidly taking account of the different polymer layers thicknesses.

Table 5. Solubility and diffusion coefficient of water in varied polyolefins (topcoat) and FBE (primer)

	Immersion temperature (°C)	D (m^2/s)	Solubility (%)
LDPE	25	$0.5\text{-}10\times10^{-12}$	0.01-0.09
	40	30×10^{-12}	0.02-0.19
HDPE	20	3.6×10^{-13}	0.03-0.1
	40	1.3×10^{-12}	
	60	$1.5\text{-}5\times10^{-12}$	
Epoxy novolac	20	$1\text{-}2\times10^{-13}$	2
	60	$0.5\text{-}2.5\times10^{-12}$	2.2-5
FBE	20	$1.8\text{-}2.5\times10^{-13}$	1-2.5
	60	$0.5\text{-}1.5\times10^{-12}$	2-14

The water solubility of LDPE generally varies between 0.03 and 0.13 mg/cm^3 [18,67, 74]. These values depend on the oxygenation state of the polymer. The activation energy values are about 20 to 50 kJ/mol for epoxy adhesives in general [56,75] and between 21 to 45 kJ/mol for FBE in particular [18]. The service temperature of the pipeline is an accelerating factor which must then be considered. For example, in the case of a three layer coating pipeline at 60°C in contact with water, with a standard FBE primer of 200 µm and a 3 mm PE topcoat, the first water molecules will reach the epoxy-metal interface in less than 4 days.

Table 6. Arrhenius parameters for various polyolefins

	Activation energy (kJ/mol)	P_0 (g.25µm/m².day)
LDPE	33	$1.3x10^8$
HDPE	20	$3.5x10^5$
CB filled HDPE	21	$6.9x10^5$

5. Wet Disbonding

5.1. Description of the Phenomenon

The high surface energy of metals associated to the diffusion of water molecules through the different polymer layers favor the access of water at metal/epoxy adhesive interface. It is known that the adherence of polymers to metal in weakened by a humid environment, especially at high temperatures [13,76,77,78,79,80,81,]. Indeed, a cohesive fracture inside the polymer often becomes adhesive at metal/polymer interface after a humid ageing.

This weakening may be explained by the sensitivity of interfacial bonds to water. When no surface treatment is applied on steel, very few covalent bonds can be formed between epoxy coatings and steel substrates [79,81]. Indeed, in most cases, metal/polymer bonds are secondary bonds (acid-base interactions), which can be easily replaced by hydrogen bonds with water molecules, forming a weak layer between the polymer and the metal surface [71,78, 82]. The metal oxides hydratation makes them more basic which limits the formation of acid/base interactions with the polymer.

This weakening may also be provoked by the dissolution of metal oxides in the presence of water [81,83]. In any case, it is necessary to have a good metal surface preparation to limit the development of such water sensitive oxides. Moreover, if corrosion occurs close to coating defect, hydroxyl ions resulting from metal dissolution can hydrolyze the polymer leading to a cathodic delamination process. If the polymer can form chemical bonds with the metal substrate, the system generally presents a good resistance to wet disbonding, except if the covalent bonds are sensitive to hydrolysis [78].

A partial adhesion reversibility has been observed after drying [71,79,84]. This is especially observed when adhesion results from secondary bonds. This reversible phenomenon can only take place when the distance between polymer molecules and metal atoms are small enough. For long immersion times, these conditions cannot be fulfilled and the loss of adhesion is irreversible. This adhesion is thus at least partly reversible by drying [85,86].

Various studies about the influence of surface topography on adhesion have improved the understanding of this mechanism of adhesion [87, 88, 89, 90] This type of adhesion is not really sensitive to water, unless plasticization of the coating modifies its mechanical properties [85].

5.2. Improvement of Adherence

As mentioned above, an optimization of surface roughness [87,88,89,91, 92, 93] can increase the mechanical adhesion which consists in mechanical interlocking of the coating into the substrate asperities [94,95]. The addition of zinc or chrome based pigments into the primer limits increases corrosion resistance and then the adherence in wet environments [77]. Chemical treatments such as acidic, anodizing or phosphating treatments can create an oxide layer on the substrate to increase the adhesion with primers [96]. Silane-based primers may increase the wettability and create covalent bonds between the metal and the polymers [81,86,97].

5.3. Adhesion Measurements of Coatings

As shown in Part A, stresses are accumulated in the coating during the process. It is assumed that the stored elastic energy associated with this stress will reduce adherence [10,98,99]. Moreover, during service time in humid environments, tensile hygroscopic stresses can develop inside the coating due to water absorption [99]. Globally, the internal, hygroscopic or thermal effects generate an elastic energy β in the coating, which can be described as follows:

$$\beta \cong h.\sigma.\varepsilon \tag{e}$$

where, h is the coating thickness, σ and ε are the total stress and strain.

Then, the energy balance necessary to remove a coating from a substrate is given by:

$$W_T = \gamma + W_P - \beta \tag{f}$$

where W_T is the work associated to the stress applied to remove the coating, γ the interfacial work of adhesion and W_P the work expended in plastic deformation.

Practical adhesion tests give access to these physical data. The externally applied stress σ_z required to remove the coating from its substrate will decrease with increasing coating stress σ (or the stored elastic energy β).

Among these numerous techniques [100] , the Pull-Off test [99, 101, 102, 103] is particularly relevant to follow the evolution of wet adhesion during the course of ageing on simplified coatings (FBE/steel), to focus on adhesive/metal interface [38].

The tests must be performed rapidly after sample removal from immersion bath to avoid the coating desorption. The pull-off stress and the failure type (adhesive at steel/polymer interface, cohesive in polymer) are reported. In simplified systems, with FBE thicknesses of several hundred micrometers, the standard deviation is about 1 MPa. The stress values given by this technique can be linked to the adherence energy [10,99] as followed:

$$\boldsymbol{\sigma_z \cong \left[\frac{2K}{h}(\gamma + W_p - \beta)\right]^{1/2}}$$

with K, the bulk modulus of the coating.

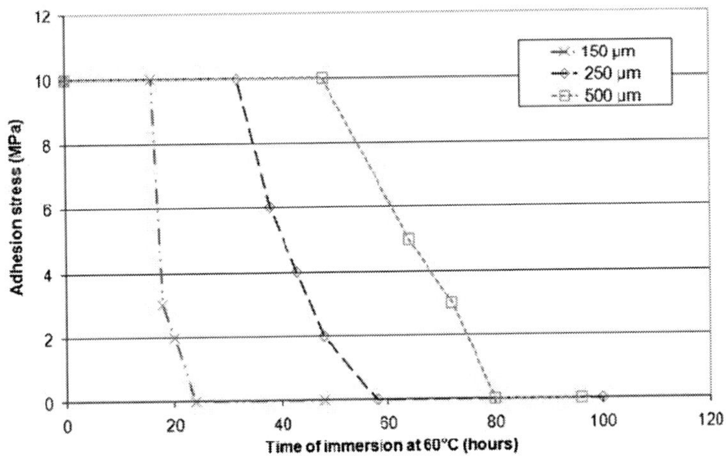

Figure 20. Adhesion stress vs. time of immersion in deionised water at 60°C polished steel coated with different thicknesses of epoxy [38].

The test is nevertheless limited by the adhesion of the dolly to the coating. Indeed, if the coating adhesion to steel is higher than this value, the failure takes place at dolly/coating interface and no adhesion value of the coating to steel is available. It is the case in the following example (Figure 20) were the interfacial strength was then considered superior to 10 MPa (mean dolly/primer interfacial strength).

5.4. Correlation with Diffusion Data

The pull off test shows a rapid loss of adhesion with immersion time on simplified FBE/polished steel systems. The time needed to observe this loss of adhesion depends on the coating thickness, which can be correlated with diffusion properties. Indeed, the time necessary for water to saturate the supported coating can be estimated from gravimetric data on free films, which gives the following relationship for a coating thickness h:

$$t_{sat} = \frac{\pi h^2}{4D}$$

This time is estimated for different coating thicknesses and then, compared to the time necessary to have a complete loss of adhesion from pull off test measurements. The comparison is reported in Figure 21.

A good correlation is observed below 250µm, suggesting that the loss of adhesion is controlled by the presence of water at metal/polymer interface, as soon as the polymer is saturated. In that case, the steel sample were mirror-polished, to de-correlate physical or chemical adherence from mechanical adherence (anchorage).

As mentioned above, in FBE/steel interactions are sensitive to water because due to their polar nature. The loss of adhesion is then directly linked to the arrival of water at the interface. When the coating is dried, this adhesion is partially reversible as long as polar interactions can be reformed [71,79,84,38].

Above a critical coating thickness (250 µm in the example), disbondment occurs before the saturation of the coating. This deviation increases with coating thickness, suggesting that the contribution of the stored elastic energy β in the coating is no longer negligible. Indeed, this value takes account the internal stresses generated from the process (e) and hydroscopic effects which both increase with polymer thickness. From (f) it is clear that the stored elastic energy β reduces the global adherence [27'99]. Moreover, the diffusion coefficient is generally higher when a material is under tensile stress [104,105]

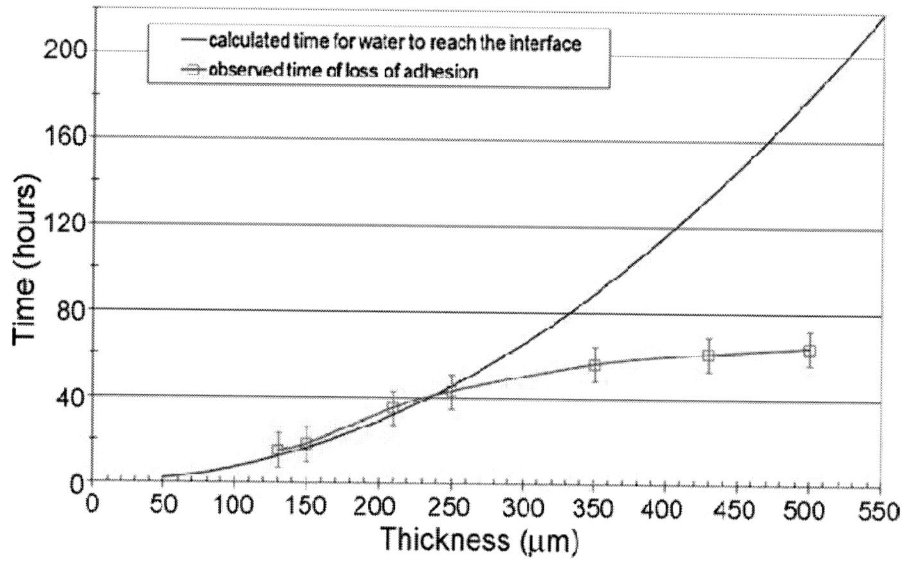

Figure 21. Saturation time and time of loss of adhesion vs. thickness [38].

Wet disbonding of thick coatings on metal substrates must then be analysed in terms of coating and interfacial properties evolutions. Above a critical thickness, the negative effects of the elastic energy generated in the coating increase the weakening effects of water molecules on interfacial bonds.

6. Conclusion

The disbonding of pipeline coatings must be avoided during service time to protect the steel structure from corrosion. Numerical studies supported by practical observations show that internal stresses are generated in the coating during the application process, especially in three layers systems.

During storage or in service situations, the coated pipelines are then exposed to humid environments associated in some cases to high temperatures. In such conditions, water diffusion through the organic coatings can be very fast, even in three layers coating systems. Without specific metal surface treatments, the adhesion between primers (FBE) and steel substrates results from polar interactions, which are water sensitive.

In one hand, increasing the coating thickness delays the time of arrival of water molecules at metal/polymer interface, but on the other hand, it decreases the global adherence by increasing the internal and hygroscopic stresses.

7. Future Trends

Several ways can be explored to improve the durability of three layers systems. One way is to correctly achieve surface preparation before coating application during industrial process. The metal/polymer interface could be strengthened using adequate surface treatments to create water resistant covalent bonds.

Moreover, it is necessary to calculate precisely the lifetime for a given system. As shown in part A, the internal stresses must first be calculated. Then, water diffusion kinetics can be determined for each layer. From these initial data, a coupled mechanical-diffusion model approach can be proposed. For example, Shanati et al. [106] have proposed a finite element model for coupled diffusion and stress in polymer materials. Nonlinear diffusion and time-dependent deformation (creep) have been considered for the 2D and 2D ½ models. Comparisons with experimental data for diffusion in stressed polymers have given encouraging results. Abdel Wahab et al. [107] have proposed a coupled stress-diffusion finite element analyses for study the durability of adhesively bonded joints aged in hot/wet environment. This work have been realised under commercial finite element code (Ansys). Numerical results have allowed to explain the experimental results of the effect of moisture on joint strength. Leger et al. [108] have proposed a thermo-diffuso-mechanical modelling based on a thermodynamic framework to solve the diffusion problem in industrial adhesive. The numerical model has been implanted in a commercial finite element code (Abaqus) to simulate diffusion behaviour of the adhesive. Numerical results have been compared with experimental values and confirm these hypotheses.

A global approach of the problem must then be considered, taking into account the internal stress state and diffusion phenomena with potential synergy effects. The coupled mechanical-diffusion model approach aims in particular at predicting the lifetime of three-layer coated pipelines and, as a consequence, could enable to choose rationally between the monolayer and the three layer protection.

REFERENCES

[1] Harris, G.H.; Lorenz, A. New coatings for the corrosion protection of steel pipelines and pilings in severely aggressive environments. *Corrosion Science*, 1993 35, 1417-1423.

[2] Guidetti, G.P.; Rigosi, G.L.; Marzola, R. The use of polypropylene in pipeline coatings. *Progress in Organic Coatings*, 1996 27, 79-85.

[3] Kamimura, T.; Kishikawa, H. Mechanism of cathodic disbonding of three polyethylene coated steel pipes, *Corrosion*, 1998 54, 979-987.

[4] Leng, A.; Streckel, H.; Stratmann, M. The delamination of polymeric coatings from steel. Part 2: first stage of delamination, effect of type and concentration of cations on delamination, chemical analysis of the interface. *Corrosion Science*, 1998 41, 579-597.

[5] Leng, A.; Streckel, H.; Hofmann, K.; Stratmann, M. The delamination of polymeric coatings from steel. Part 3: effect of the oxygen partial pressure on the delamination reaction and current distribution at the metal/polymer interface. *Corrosion Science*, 1998 41, 599-620.

[6] Melot, D.; Paugam G.; Roche, M. Disbondments of pipelines coatings and their consequences on corrosion risks. *17th International conference on Pipeline Protection, Edinburgh: BHR Group*, 2007.

[7] Guermazia, N.; Elleuchb, K.; Ayedia, H.F.; Kapsac, P. Aging effect on thermal, mechanical and tribological behaviour of polymeric coatings used for pipeline application, *Journal of materials processing technology*, 2008 203, 404-410.

[8] Kehr, J.A.; Enos, D.G. FBE, a foundation for pipeline corrosion coatings. *Corrosion*, 2000. 00757(NACE international Conference).

[9] Coeuille, F. Revêtements polymères de canalisation de fluide: caractérisation et évolution de l'adhésion en milieu agressif. *Ph.D. Dissertation*, Ecole des Mines de Paris, France, 2002.

[10] Perera, D.Y. On adhesion and stress in organic coatings. *Progress in Organic Coatings*, 1996 28, 21-23.

[11] Abdelkader, A.; White, J., Comparison of internal stresses in coatings cured on rigid substrates and on unrestrained thin substrates. *Progress in organic coatings*, 2002 44, 121-129.

[12] Yan, G.; White, J.R. Residual Stress Development in a Bi-Layer Coating. *Polymer Engineering and Science*, 1999 39, 1856-1865.

[13] Schmidt, R.G.; Bell, J.P. Epoxy adhesion to metals. *Advances in Polymer Science*, 1986 75, 33-71.

[14] Kamarchik, P. Crosslink density and the development of internal stress in organic coatings. Macromolecular. Symposia, 2001 175, 411-419.

[15] Ramani, K.; Zhao, W. The evolution of residual stresses in thermoplastic bonding to metals. *International Journal of Adhesion and Adhesives*, 1997 17, 353-357.

[16] Wang, H.; Siow, K.S. Measurement of Tg in Epoxy Resins by DSC-Effects of Residual Stress. *Polymer Engineering and Science*, 1999 39, 422-429.

[17] Perera, D.Y.; Oosterbroek, M. Hygrothermal stress evolution during weathering in organic coatings. *Journal of coatings technology*, 1994 66, 83-88.

[18] Legghe, E. Etude des paramètres et des mécanismes conditionnant l'adhérence entre un primaire époxy et un substrat acier: Cas des revêtements tricouche de pipelines. *Ph.D. Dissertation*, Université de Toulon et du Var, 2009.

[19] Timoshenko, S.P.; Goodier, J.N. *Theory of elasticity*, 1st edition, Mc Graw-Hill Book Company, p 399.

[20] Chu, J.L.; Sanboh, L. Thermal stresses in a long coaxial cylinder of two materials: Zero axial force. *Mechanics of Materials*, 1995 20, 105-123.

[21] Chang, B.; Sue, H.-j.; Wong, D.; Kehr, A.; Pham, H.; Siegmund, A.; Snider, W.; Jiang, H.; Browning, B.; Mallozzi, M.; Guo, S. Integrity of 3LPE Pipeline Coatings - Residual Stresses and Adhesion Degradation. *7th international Pipeline Conference*, Calgary, 2008.

[22] Chang, B.T.A.; Jiang, H.; Sue, H.J.; Guo, S.; StJean, G.; Pham, H.; Wong, D.; Kehr, A.; Lo K.H. Disbondment mechanism of 3LPE pipeline coatings. *17th International conference on Pipeline Protection*, Edinburgh, 2007.

[23] Corcoran, E. M. Determining Stresses in organic coatings using plate beam deflection. *Journal of Paint technology*, 1969 41, 635-640.

[24] Perera, D.Y.; Eynde, D.V. Considerations on a cantilever (Beam) Method for measuring the internal stress in organic coatings. *Journal of Coatings Technology*, 1981 53, 39-41.

[25] Negele, O.; Funke, W. Internal stress and wet adhesion of organic coatings. *Progress in organic coatings*, 1996 28, 285-289.

[26] Tsui, Y.C.; Howard, S.J.; Clyne, T.W. The effect of residual stresses on the debonding of coatings-II. An experimental study of a thermally sprayed system. *Acta Metallurgica et Materialia*, 1994 42, 2837-2844.

[27] Howard, S.J.; Tsui, Y.C.; Clyne, T.W. The effect of residual stresses on the debonding of coatings-I. A model for delamination at a bimaterial interface. *Acta Metallurgica et Materialia*, 1994 42, 2823-2836.

[28] Saarnak, A.; Nilsson, E.; Kornum, L.O. Usefulness of the measurement of internal stresses in paint films. *J. Oil. Col. Chem. Assoc.*, 1976 59, 427-432.

[29] Legghe, E.; Joliff, Y.; Belec, L.; Aragon, E. Computational analysis of a three-layer pipeline coating: Internal stresses generated during the manufacturing process. *Computational Materials Science*, 2011 50, 1533-1542.

[30] Legghe, E.; Joliff, Y.; Belec, L.; Aragon, E.; Margaillan, A. Computational analysis of internal stresses generated during the manufacturing process of a monolayer or three-layer pipeline coating. *Computational Materials science*, 2010 48, 360-365.

[31] Cardarelli, F. *Materials Handbook - A Concise Desktop Reference*, 2nd Edition, Springer-Verlag London Limited, 2008.

[32] Astruc, A. Microstructure et perméabilité de revêtements anticorrosion : influence des contraintes hygrothermiques et rôle des interphases. *Ph.D. Dissertation*, Université du Sud-Toulon-Var, France, 2007.

[33] Nitta, K.H.; Ishiburo, T. Ultimate Tensile Behavior of Linear Polyethylene Solids. *Journal of Polymer Science: Part B - Polymer Physics*, 2002 40, 2018-2026.

[34] ABAQUS. *Analysis User's Manual, Contact formulation for ABAQUS/Standard contact pairs*, chapter 18, section 18.7, 2007.

[35] ABAQUS. *Analysis Manual, Viscoelasticity*, chapter 4, section 4.8, 2007.

[36] Arzoumanidis, G. A.; Liechti, K.M. Linear Viscoelastic Property Measurement and Its Significance for Some Nonlinear Viscoelasticity Models. *Mechanics of Time-Dependent Materials*, 2003 7, 209-250.

[37] Absi, J.; Glandus, J.C.; Improved method for severe thermal shocks testing of ceramics by water quenching. *Journal of the European Ceramic Society*, 2004 24, 2835-283.

[38] Legghe, E.; Aragon, E.; Bélec, L.; Margaillan, A.; Melot, D. Correlation between water diffusion and adhesion loss: Study of an epoxy primer on steel. *Progress in organic coatings*, 2009 66, 276-280.

[39] ISO 1663 – 1999. Rigid cellular plastics - Determination of water vapour transmission properties.

[40] Kumins, C.A. Transport through polymer films. *Journal of Polymer Science Part C: Polymer Symposia*, 1965 10, 1-9.

[41] Klute, C.H.; Franklin, P.J. The permeation of water vapor through polyethylene. *Journal of Polymer Science*, 1958 32, 161-176.

[42] Park, G.S. Transport principles: solution, diffusion and permeation in polymer membranes. *Congres - North atlantic treaty organization. Advanced study institute* 1986 181, 57-107.

[43] Peppas, N. A.; Meadows, D. L. Macromolecular structure and solute diffusion on membranes: An overview of recent theories. *Journal of Membrane Science*, 1983 16, 361-377.

[44] Frisch, H.L. Sorption and transport in glassy polymers – A review. Polymer Engineering & Science, 1980 20, 2-13.

[45] Vieth, W.R. *Diffusion in and through polymers*. Munich: Hanser Publishers, 1991.

[46] Crank, J. *The Mathematics of Diffusion*. 2nd Ed. New York: Oxford University Press, 1975.

[47] Hamilton, R.L. Water vapor permeability of polyethylene and other plastic materials, *The Bell System Technical Journal*, 1967 46, 391-415.

[48] Carter, H.G.; Kibler, K.G. Langmuir-type model for anomalous moisture diffusion in composite resins. *Journal of Composite Materials*, 1978 12, 118-131.

[49] Popineau, S.; Rondeau-Mouro, C.; Sulpice-Gaillet, C.; Shanahan, M.E.R. Free/bound water absorption in an epoxy adhesive. *Polymer*, 2005 46, 10733-10740.

[50] Weitsman, Y. A continuum diffusion model for viscoelastic materials. *J. Physical Chemistry*, 1990 94, 961-968.

[51] De Kee, D.; Liu, Q.; Hinestroza, J. Viscoelastic (Non-Fickian) Diffusion. *The Canadian Journal of Chemical Engineering*, 2005 83, 913-929.

[52] Apicella, A.; Nicolais, L.; Astarita, G.; Drioli, E. Effect of thermal history on water sorption, elastic properties and the glass transition of epoxy resins. *Polymer*, 1979 20, 1143-1148.

[53] Nicaise, E.; Gandini, A.; Cheradame, H. Behaviour of Composites Used in the Confinement of Nuclear Wastes 2 - Water Absorption in Particle-filled Epoxide Networks. British Polymer Journal, 1986 18, 251-255.

[54] Sauvant-Moynot V., Kittel J., Melot D., and Roche M. Three layer polyolefin coatings: how the FBE primer properties govern the long term adhesion. in 17th *International conference on Pipeline Protection*. 2007. Edinburgh: BHR Group.

[55] Wong, T.C.; Broutman, L.J.. Moisture Diffusion in Epoxy Resins - Part I -Non-Fickian sorption processes. *Polymer Engineering and Science*, 1985 25(9): 521-528.

[56] Nogueira, P.; Ramirez, C.; Torres, A.; Abad, M.J.; Cano, J.; Lopez, J.; Lopez-Bueno, I.; Barral, L. Effect of Water Sorption on the Structure and Mechanical Properties of an Epoxy Resin System. *Journal of Applied Polymer Science*, 2001 80, 71-80.

[57] Jeffrey, K.; Pethrick, R.A. Influence of chemical structure on free volume in epoxy resins: A positron anninilation study. *European Polymer Journal*, 1994 30, 153-158.

[58] Soles, C.L.; Chang, F.T.; Gidley, D.W.; Yee, A.F. Contributions of the nanovoid structure to the kinetics of moisture transport in epoxy resins. *Journal of Polymer Science Part B: Polymer Physics*, 2000 38, 776-791.

[59] VanLandingham, M.R.; Eduljee, R.F.; Gillespie, J.W.Jr. Moisture diffusion in epoxy systems. *Journal of Applied Polymer Science*, 1999 71, 787-798.

[60] Morel, E.; Bellenger, V.; Verdu, J. Structure-water absorption relationships of amine-cured epoxy resins. *Polymer*, 1985 26, 1719-1724.

[61] De Neve B.; Delamar M.; Nguyen T.T.; Shanahan M.E.R. Failure mode and ageing of steel epoxy joints. *Applied Surface Science*, 1998 134, 202-212.

[62] De Neve, B.; Shanahan, M. Physical and chemical effects in an epoxy resin exposed to water vapour. *Journal of Adhesion*, 1995 49, 165-176.

[63] Xiao, G.Z.; Shanahan, M.E.R. Swelling of DGEBA/DDA epoxy resin during hygrothermal ageing. *Polymer*, 1998 39, 3253-3260.

[64] Carfagna, C.; Apicella, A.; Nicolais, L. The effect of the prepolymer composition of amino-hardened epoxy resins on the water sorption behavior and plasticization. *Journal of Applied Polymer Science*, 1981 27, 105-112.

[65] Kalenda, P.; Kalendova, A. Improved Chemical resistance of Epoxy resin-based coating composition. *Dyes and Pigments*, 1995 27, 305-312.

[66] Lasoski, Jr.S.W.; Cobbs Jr.W.H. Moisture permeability of polymers. I. Role of crystallinity and orientation. *Journal of Polymer Science*, 1959 36, 21-33.

[67] Klute, C.H. Diffusion of small molecules in semicrystalline polymers: Water in polyethylene. *Journal of Applied Polymer Science*, 1959 1, 340-350.

[68] Alter, H. A critical investigation of polyethylene gas permeability. *Journal of Polymer Science*, 1962 57, 925-935.

[69] Brandrup, J.; Immergut, E.H. *Polymer handbook*, 3rd edition, John Wiley & Sons, New York, 1989.

[70] Kamal, M.R.; Jinnah I.A. Permeability of Oxygen and Water Vapor Through Polyethylene/ Polyamide Films. *Polymer Engineering and Science*, 1984 24, 1337-1347.

[71] Rouw, A.C. Model epoxy powder coatings and their adhesion to steel. *Progress in Organic Coatings*, 1998 34, 181-192.

[72] Li, L.; ShuYong Zhang; Chen, Y.; Liu, M.; Ding, Y.; Luo, X.; Pu, Z.; Zhou, W.; Li, S. Water Transportation in Epoxy Resin. *Chemical Materials*, 2005 17, 839-845.

[73] Al-Harthi, M.; Loughlin, K.; Kahraman, R. Moisture diffusion into epoxy adhesive: testing and modeling. *Adsorption*, 2007 13, 115-120.

[74] McCall, D.W.; Doughlass, D.C.; Blyler Jr.L.L.; Johnson, G.E.; Jelinski, L.W.; Bair, H.E. Solubility and diffusion of water in low-density polyethylene. *Macromolecules*, 1984 17, 1644-1649.

[75] Zhou, J.; Lucas, J.P. Hygrothermal effects of epoxy resin. Part I: the nature of water in epoxy. *Polymer*, 1999 40, 5505-5512.

[76] Popineau, S.; Shanahan, M.E.R. Simple model to estimate adhesion of structural bonding during humid ageing. *International Journal of Adhesion & Adhesives*, 2006 26, 363-370.

[77] Bajat, J.; Dedi, O. Adhesion and corrosion resistance of epoxy primers used in the automotive industry. *Journal of Adhesion Science and Technology*, 2007 21, 819-831.

[78] Funke, W. Thin-layer technology in organic coatings. *Progress in organic coatings*, 1996 28, 3-7.

[79] Allen, K.W. A Review of Contemporary Views of Theories of Adhesion. *The Journal of Adhesion*, 1987 21, 261-277.

[80] Munger, C.G. Surface, adhesion, coatings. *Materials Performance*, 1983 22, 33-37.

[81] Schmidt, R.G.; Bell; J.P. Investigation of Steel/Epoxy Adhesion Durability Using Polymeric Coupling Agents. II. Factors Affecting Adhesion Durability. *The Journal of Adhesion*, 1988 25, 85-107.

[82] Nguyen, T.; Martin, J.W. Modes & Mechanisms of degradation of epoxy coated reinforcing steel in a marine environment. *Durability of building materials and components*, 1996 1, 491-502.

[83] Venables, J.D. Adhesion and durability of metal polymer bonds. *Adhesion*, 1983 7, 87-93.

[84] McKnight, M.; Seiler, J.F.; Nguyen, T.; Rossiter, W.J. Measuring Peel Adhesion of Coatings. *Journal of Protective Coatings and Linings*, 1995 12, 82-89.

[85] Shaw, G.S.; Rogers, C.E.; Payer, J.H. The effect of immersion on the breaking force and failure locus in an epoxy/mild steel system. The Journal of adhesion, 1992 38, 255-268.

[86] Walker, P. The effect of silanes on paint adhesion and sites of failure. *Surface and Interface Analysis*, 1991 17, 465-470.

[87] Keisler, C.; Lataillade, J. L. The effect of substrate roughness characteristics on wettability and on the mechanical properties of adhesive j oints loaded at high strain rates. *Journal of Adhesion Science and Technology*, 1995 9, 395-411.

[88] Sancaktar, E.; Gomatam, R. A study on the effects of surface roughness on the strength of single lap joints. *Journal of Adhesion Science and Technology*, 2001 15, 97-117.

[89] Zhang, S.; Panat, R.; Hsia, K.J. Influence of surface morphology on the adhesion strength of epoxy–aluminum interfaces. *Journal of Adhesion Science and Technology*, 2003 17, 1685-1711.

[90] Meine, K.; Klob, K.; Schneider, T.; Spaltmann, D. The influence of surface roughness on the adhesion force. Surface and Interface Analysis, 2004 36, 694-697.

[91] Namkanisorn, A.; Chaudhury, M.K. An Arrhenius method to study the effect of surface roughness in polymer. *Metal adhesion*,1998, 415-417.

[92] Sargent, J.P. Adherent surface morphology and its influence on the peel strength of adhesive joints bonded with modified phenolic and epoxy structural adhesives. *International journal of adhesion and adhesives*, 1994 14, 21-30.

[93] Zhai, L.; Ling, G.; Li, J.; Wang, Y. The effect of nanoparticles on the adhesion of epoxy adhesive. *Materials Letters*, 2006 60, 3031-3033.

[94] Packham, D.E. The mechanical theory of adhesion. Changing perceptions 1925-1991. *The Journal of Adhesion*, 1992 39, 137-144.

[95] Kinloch, A.J. The science of adhesion. *Journal of materials science*, 1980 15, 2141-2166.

[96] Fourche, G. An overview of the basic aspects of polymer adhesion - Part II: Application to surface treatments. *Polymer Engineering and Science*, 1995 35, 968-975.

[97] Cayless, R.A. Interfacial chemistry and adhesion: The role of surface analysis in the design of strong stable interfaces for improved adhesion and durability. *Surface and Interface Analysis*, 1991 17, 430-438.

[98] Bierwagen, G.P. Film formation and mudcracking in latex coating . *Journal of coatings technology*. 1979 51, 117-129.

[99] Piens, M.; Deurwaerder, H.D. Effect of coating stress on adherence and on corrosion prevention. *Progress in Organic Coatings*, 2001 43, 18-24.

[100] Roche, A.A.; Bierwagen, G.P.; Dolde, P.; Bouzziri, M. *Measurement of the practical adhesion of paint coatings to metallic sheets by the pull-off and three-point flexure tests*. Adhesion measurement of films and coatings, K.L. Mittal , 1995, 299-321.

[101] BS EN ISO 4624: 2003 - Paints and varnishes. Pull-off test for adhesion

[102] Roche, A.A.; Dole, P.; Bouzziri, M. Measurement of the practical adhesion of paint coatings to metallic sheets by the pull-off and three-point flexure tests. *Journal of Adhesion Science and Technology*, 1994 8, 587-609.

[103] Turunen, M.P.K.; Marjamaki, P.; Paajanen, M.; Lahtinen, J.; Kivilahti, J.K. Pull-off test in the assessment of adhesion at printed wiring board metallisation/epoxy interface. Microelectronics Reliability, 2004 44, 993-1007.

[104] Fahmy, A.A.; Hurt, J.C. Stress Dependence of Water Diffusion in Epoxy Resin. Polymer Composites, 1980 1, 77-80.

[105] Yaniv, G.; Ishai, O. Coupling Between Stresses and Moisture Diffusion in Polymeric Adhesives. *Polymer Engineering and Science*, 1987 27, 731-739.

[106] Shanati, S.; Ellis, N.S.; J. Randall, T; Marshall, J.M. Coupled diffusion and stress by the finite element method. *Applied Mathematical Modelling*, 1995 19, 87-94.

[107] Abdel Wahaba, M.M.; Crocombea, A.D.; Beeversb, A.; Ebtehaj, K. Coupled stress-diffusion analysis for durability study in adhesively bonded joints. *International Journal of Adhesion & Adhesives*, 2002 22, 61-73.

[108] Leger, R.; Roy, A.; Grandidier, J.C. Non-classical water diffusion in an industrial adhesive. *International Journal of Adhesion & Adhesives*, 2010 30, 744-753.

In: Pipelines: Design, Applications and Safety
Editors: M. G. Rivero et al. pp. 155-171
ISBN 978-1-62100-178-2

Chapter 5

NONDESTRUCTIVE EVALUATION OF STRESSED STATES OF PIPELINES BY ULTRASOUND

Nadezhda Ye. Nikitina
Blagonravov Institute of Mechanical Engineering, Nizhny Novgorod branch, Russian Academy of Sciences, Belinskogo, 85, Nizhny Novgorod, 603024, Russia
"ENCOTES" Ltd., Engineering Company, Brinskogo, 6, Nizhny Novgorod, 603950, GSP-76, Russia

ABSTRACT

This chapter contains results of theoretical and experimental investigations of the principal problems of the acoustoelasticity phenomenon application for the nondestructive testing of stressed (strained) state (SSS) of steel pipes and pipelines spent by the author and collaborators for the last 8 years. The foundations of the method of acoustoelasticity for biaxial stress evaluation with the help of bulk waves propagated normally to the plane of stress acting are established.

Specific features of stress evaluation in engineering materials by using precise measurements of velocities of elastic waves of a millimeter range of wavelengths are revealed. The results of experimental evaluation of acoustic-elastic coefficients which determine the sensitivity of ultrasonic method to actual stresses in pipe steels used in gas and oil industry are given. The opportunities of precise measurement of phase velocity of ultrasound (principal informative characteristic of the acoustoelastic effect) with use of a pulse-echo method, realized in a special portable device IN-5101A produced by "ENCOTES" Ltd, are discussed. The calculative algorithms are included into the computational module of device and enable one to evaluate in-plane principal stresses in real object. So, the experimental arrangement provides evaluation of uni- or biaxial, tensile or compressive stresses in pipes and constructive details of compressing and pumping stations, including long-term operated materials at different force and climatic actions.

The main problems of determination of stressed states of real engineering constructions in a case of the acoustical tensometry, that is monitoring of stresses during manufacturing and exploitation of pipes; and estimation of stresses «in situ», in pre-stressed industrial objects, are considered. The reliability of acoustoelastic manner for biaxial stress evaluation in main pipelines of large diameter was justified experimentally, during the loading of closed pipe by inner pressure of water. The results of the investigations show that observed data are quite close to analytical solution of the problem which was given by Gabriel Lame in 19th century.

Practical example of the acoustoelasticity phenomenon application for biaxial stress evaluation is described, namely, measurement of non-design axial stresses in technological pipelines of dust collectors of compressor station. The opportunity of application of “zero-less” ultrasonic tensometry for determination of a stressed state of objects in action has been experimentally confirmed.

INTRODUCTION

The great importance of nondestructive evaluation of stressed states of compressor stations, gas and oil pipelines is evident because the failure of those civil engineering constructions leads to great amount of human victims and deterioration of environment. In time revealing of weak high-stressed parts of engineering constructions provides the earlier prediction of their destruction and elimination of further damage. It is strongly important for such objects as main and technological pipelines, pressure-boosting and pumping equipment, oil refining.

Because of complexity of processes of manufacture of pipes and a wide range of loadings to which metals and alloys are exposed on a way to a finished detail during processes of cutting, forging, forming and other operations, exact calculation of technological stresses in this or that part of a volume of pipe wall, not always is obviously possible. The operations of mounting and exploitation of pipelines leads to additional deformation and stressing of the pipe body.

High efficiency of modern numerical methods of computation of real stressed states of main and technological pipelines, using computers, doesn't eliminate necessity of usage of experimental methods for direct measurement of stresses (deformations) in a material of a pipeline.

Calculative methods of design of SSS of industrial objects are based on use of some assumptions, as a rule, idealizing real conditions. Imperfection of methods of calculation, incomplete correspondence of design schemes to the actual conditions of details work, incorrectness of the definition of boundary conditions and operating loadings, as a rule, changing eventually, give essential errors in an estimation of stresses and strains.

Sometimes the problem of evaluation of Stress or Strain an operating engineering construction at all doesn't give in to the theoretical solution. Therefore the problem of an experimental estimation of a real tension (compression) of a pipe at various fabrication stages, assembling of pipeline and the further exploitation is rather urgent now.

Some problems of relevant prevision of dangerous stressed states of large-sized engineering constructions may be successfully solved on a base of investigations of nonlinear acoustic effects in engineering materials, for example, the acoustoelastic effect. Taking into account the nonlinear terms in the wave equation, one must find the dependence between the velocity of shear or longitudinal wave and the values of the mechanical stresses or strains in the medium. This dependence, i. e., the acoustoelastic effect, may be the physical foundation of the nondestructive manner of the evaluation of stressed states of solids [1, 2].

MAIN ADVANTAGES OF ACOUSTOELASTIC METHOD OF STRESS EVALUATION

The practical realization of the acoustoelasticity phenomenon for the investigation of SSS without destruction of an engineering material may be made by exciting waves of the amplitude essentially smaller than the effective stresses and evaluating their propagation speed changes in comparison with the unstressed material with a relative error of at least 0.01 %. Relative cheapness and safety of an acoustical method in comparison with x-ray, a wider choice of materials and their stressed states in comparison with electric strain gauging (tensometry) or magnetic methods render it attractive enough for achievement of such purposes.

The acoustic methods connected with propagation of elastic oscillations, are widely applied to studying of physical and mechanical properties of solids. The elastic wave extending in a solid medium is characterized by amplitude of oscillations, their phase and frequency, propagation and polarization vectors. So, it is possible to estimate physical properties and a technical condition of engineering materials by measuring parameters of elastic wave propagation. High getting ability of ultrasonic oscillations (in comparison with x-ray and magnetic methods) has caused wide use of ultrasound for nondestructive investigations of industrial metals and alloys.

The acoustic method of measurement of mechanical stresses is based on use of laws of propagation of elastic waves in preliminary stressed (pre-stressed) bodies. This method allows determining separately the principal stresses not only in models, but also directly in engineering designs without additional procedures of unloading. It possesses enough high resolution, provides efficiency of testing and allows spending measurements not only surface stresses, but also the stresses and residual stresses in material's volume.

Numerous constructive elements include details with one size significantly less then the other two. Frequently such details experience biaxial stress under loading, or it is possible to consider the stressed state as locally plane stress in the area of the acoustical testing by ultrasonic pulse-echo method.

For evaluation of in-plane stress of a pipe wall it is the most convenient to use the waves propagating along a normal to a plane of action of stress. In an isotropic material without stresses propagation velocity of a shear wave will be same at any orientation of a plane of its polarization. Acoustic anisotropy of pre-stressed material is shown in an interdiction for linear polarization of those shear waves which on an input on the medium aren't polarized along directions of action of the principal stresses. The difference of speeds of the "resolved" shear waves will be proportional to a difference of the principal stresses.

In order to investigate anisotropic thin structures such as pipe's envelopes by nondestructive manner shear waves of ultrasonic frequency can be used rather effectively. Two "pure" modes with mutually perpendicular polarization directions are known to exist in an anisotropic material for each direction of shear wave propagation. When shear oscillations are exited on the detail's surface at some angle with respect to the directions mentioned, the acoustical double refraction similar to that in optics can be observed. Symmetry axes directions of engineering material, which are not so evident at first sight, can be revealed by rotating the shear wave transducer. Using this method it is possible to determine technological symmetry axes directions, corresponding to maximum strength during tension or compression of materials of pipes produced by forging, forming or drawing.

Unlike many other methods of an estimation of the SSS of engineering materials where dependences of informative parameters on stresses (deformations) are phenomenological, that is are only by practical experience, here dependencies between the speeds of elastic waves and stresses can be found mathematically, in approach of the five-constant theory of elasticity. In 1953 for the first time dependences of speeds of bulk waves on uniaxial stress and all-round compression were revealed "on a feather tip" [3]. For a case of elastic waves propagation along the normal to the plane of stress acting it is possible to present the formulas of Hughes and Kelly in a form:

$$\frac{V_1^2}{V_{01}^2}-1=2\left(\frac{V_1}{V_{01}}-1\right)=\frac{1}{3K_0}\left(\lambda+2\mu+m+\frac{\lambda}{\mu}\frac{n}{4}\right)\sigma=2k_1\sigma,$$

$$\frac{V_2^2}{V_{02}^2}-1=2\left(\frac{V_2}{V_{02}}-1\right)=-\frac{1}{3K_0}\left(2\lambda-m+\frac{n}{2}+\frac{\lambda}{\mu}\frac{n}{2}\right)\sigma=2k_2\sigma, \quad (1)$$

$$\frac{V_3^2}{V_{03}^2}-1=2\left(\frac{V_3}{V_{03}}-1\right)=-\frac{1}{3K_0}\left[\frac{2\lambda}{\mu}(\lambda+2\mu+m)-2l\right]\sigma=2k_3\sigma,$$

where V_1, V_2 - speeds of the shear waves polarized along and across to a direction of action of stress, V_3 - speed of a longitudinal wave; λ, μ, $K_0=\lambda+\frac{2}{3}\mu$ – modules of linear elasticity; m, n, l – modules of nonlinear elasticity of Murnaghan [4]. Corresponding combinations of these constants are designated k_1, k_2, k_3 and named "coefficients of acoustic-elastic connection" [2].

The research is not only a subject of academic interest.

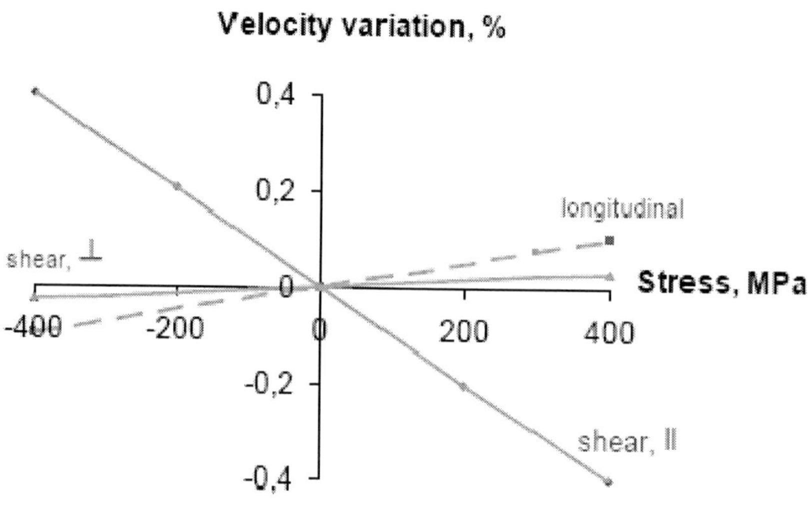

Figure 1. Velocity-stress dependencies for pipe steel.

For example, Figure 1 demonstrates connection between the values of velocity of shear waves V_1, V_2 designated || and ⊥ correspondingly, longitudinal wave V_3 and uniaxial stress, as a result of my own experiments with the samples of Sweden stainless steel cut from the old pipe of large diameter. The coefficients of acoustic-elastic connection k_1, k_2, k_3 represents the tangencies of corresponding angles and depend only on linear and nonlinear properties of the material (no any phenomenological presumptions!).

ACOUSTOELASTIC INVESTIGATION OF IN-PLANE PRINCIPAL STRESSES

Theoretic foundations of nonlinear propagation of elastic waves in stressed solid are developed in the middle of the 20th century, first experimental investigations of acoustical birefringence were carried out at the end of the 50-s [5]. The scientists of former Soviet Union contributed much to acoustoelastic phenomenon study at the end of the 60-s [6, 7]. Thus, this method youngish, but has already proved as one of reliable and perspective.

It is known [8, 9], that three waves with orthogonal vectors of polarization and with various, in general, velocities can be propagated in a hyperelastic prestressed material along the definite direction. The deformable state of such a material can be adequately described by potential function of deformation. Speed values depend on the main lengthenings in three orthogonal directions, making the orthogonal basis [8]. If there are no shear stresses and the tensions (compressions) are imposed along the symmetry axes of the material, the directions of main lengthenings and main stresses will coincide with the symmetry axes. The orthogonal basis is deduced by choosing one of these directions as wave propagation direction. Consequently neither propagation vector nor vectors of wave polarization will not turn under stress action (there won't be any nonlinear refraction of elastic waves).

However, the wide practical application of the ultrasonic method of stress evaluation is connected with some principal problems. Probable ways and perspectives of acoustoelastic phenomenon using to determine biaxial stressed states of engineering materials with the help of bulk waves traveling normally to the plane of stress action are considered here. The schematic illustration of shear waves propagation in pre-stressed material is shown on Figure 2. Longitudinal wave propagates in the same direction.

It is clear that it is not so easy to transit from a mathematical model of isotropic hyperelastic solid, where acoustoelastic effect exists "in the pure form", to media of more complex structure such as metals and alloys, in particular, pipe steels, if precise measurements of elastic waves velocities are carried out in a millimeter range of wavelengths. The effect of intrinsic material's anisotropy is not the only one influence the acoustical birefringence of ultrasonic shear waves and their interaction between each other. Due to the acoustoelasticity phenomenon the relative value of the shear waves velocities difference is also proportional to the uniaxial stress value or to the difference of in-plane principal stresses in initially isotropic material. However, the effect is principally nonlinear. So, the value of the velocity difference due to existence of stress is essentially smaller than the difference due to own material's anisotropy, which may be explained in the frames of linear elasticity of anisotropic solid.

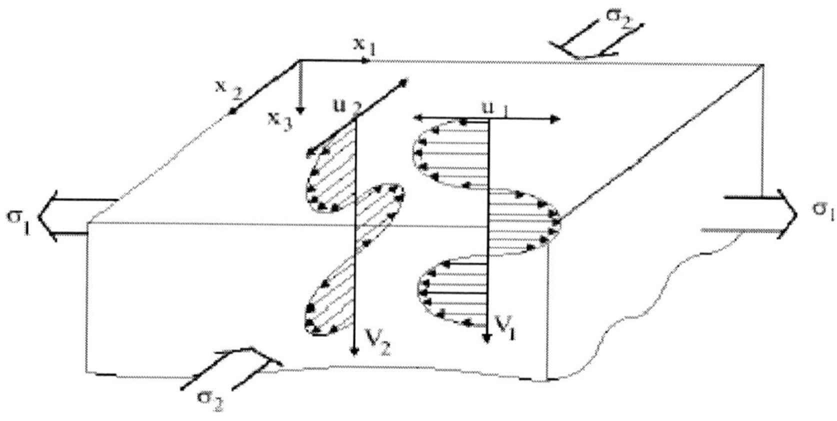

Figure 2. Scheme of shear waves propagation in pre-stressed material.

There are two statements of the problem of nondestructive evaluation of mechanical stresses in an engineering material.

A. Values of measured parameters appropriate to the initial unstressed material are known (or it is possible to measure them).
B. The values are unknown (the material is already stressed or an essential change of its internal condition occurred in process of stresses onset).

In both cases the initial (unstressed) material can differ greatly from the hyperelastic material.

Results of theoretical and experimental investigations of the acoustoelastic phenomenon practical application in order to determine applied and residual stresses in multi-sectioned special-purpose engineering constructions have been carried out by author for many years may be generalized in the following manner. In my view, it is possible to suggest a step-by-step procedure of design and application of some simple and exact algorithms of biaxial stress calculations on basis of precise measurement of informative characteristics of acoustoelasticity.

- Consider that a rolled material is ortotropic [10] and tensile or compressive stresses acting along its symmetry axes. This point of view is explained by the fact that it is really true for large quantity of practical cases. For example, axial and circumferential stresses are commonly observed in pipes produced from rolled sheets.
- Take as a basis some simple and logical relationships similar to photoelasticity equations between acoustic and mechanic parameters, as it is made in the work [11], at the same time keeping anisotropic properties of the material in the values of speeds, and considering coefficients of proportionality between stress and speed independent from the direction of loading [12].
- As far as it is possible, use in equations directly measured parameters and dimensionless (relative) quantities.

- Propose practical techniques of experimental evaluation of coefficients of proportionality between mechanical stresses and ultrasonic velocities (i.e. some algebraic combinations of linear and nonlinear elasticity constants of a material), using the material samples whose loading conditions are known exactly.
- Suggest quit adequate methods of informative ultrasonic parameters measuring, in accordance with strong requirements of accuracy and reliability of measurements, typical to practical use of such "weak" effects, as acoustoelasticity phenomenon in a real engineering material.
- Whenever it is possible, consider in calculations internal and external factors influencing essentially the accuracy and reliability of stress evaluation final outputs.
- Remember about the restrictions imposed on an opportunity of acoustoelastic identification of the stressed state in each specific situation at a stage of algorithmic maintenance, let alone accuracy of ultrasonic measurements in real engineering materials.

In practice, the velocity isn't a directly measured ultrasonic wave parameter. Such a parameter could be propagation time, and the thickness of testing material also varies with stress changing (see Figure 3). So there must be three (not two) acoustoelastic equations determining two principal stresses. To solve this problem we use longitudinal wave propagated in the same direction as shear waves (across the plane of stress acting). Longitudinal waves are used together with shear waves as a specific "thickness-meter" to provide simultaneous monitoring of the "acoustical path" for ultrasonic pulses.

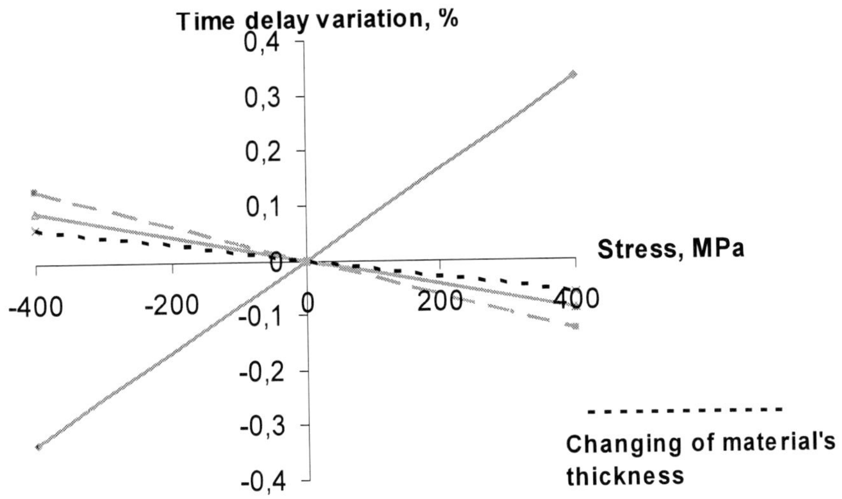

Figure 3. Variation of the "acoustical path" of ultrasonic pulses.

Therefore, it is necessary to excite in a material all three bulk waves which can propagate on a normal to a plane of action of stresses – two mutually perpendicularly polarized shear waves and a longitudinal wave.

Thus, the combined equations for biaxial stress evaluation on the base of precise time-of-flight measurements of shear and longitudinal ultrasonic waves may be written in a form [2, 13]:

$$\frac{L}{t_1}\frac{t_{01}}{L_0}-1=k_1\sigma_1+k_2\sigma_2,$$
$$\frac{L}{t_2}\frac{t_{02}}{L_0}-1=k_2\sigma_1+k_1\sigma_2,$$
$$\frac{L}{t_3}\frac{t_{03}}{L_0}-1=k_3\sigma_1+k_3\sigma_2. \qquad (2)$$

Here L_0, L - a distance of wave propagation before and after the occurrence of stresses, proportional to a thickness of a material; t_{0i}, t_i - times for which this distances were passed (i=1,2,3); k_1, k_2, k_3 - coefficients of acoustic-elastic connection. (it is supposed that elastic-acoustic properties of a material are identical in all directions).

In case of uniaxial stress (for example, $\sigma_2 = 0$) the equations (1) and (2) coincide.

Having excluded variation of a material thickness $\frac{h}{h_0}=\frac{L}{L_0}$ from the combined equations (2) and neglecting terms of the second order in relation to values $\Delta_i = \left(\frac{t_3}{t_i}\frac{t_{0i}}{t_{03}}-1\right)$ (i=1,2), we will receive simple enough algorithm of determination of a biaxial stresses by results of precise time-of-flight measurements of shear and longitudinal ultrasonic waves [2, 13]:

$$\sigma_1 = K_1\Delta_1 - K_2\Delta_2,$$
$$\sigma_2 = K_1\Delta_2 - K_2\Delta_1, \qquad (3)$$

Here $K_1 = \frac{k_1-k_3}{(k_1-k_3)^2-(k_2-k_3)^2}$, $K_2 = \frac{k_2-k_3}{(k_1-k_3)^2-(k_2-k_3)^2}$ - coefficients of elastic-acoustic connection [2], which depend only on linear and nonlinear properties of the material, have dimension of stresses (of modules of elasticity) and can be calculated theoretically or by the acoustic-mechanic testing of material samples. Dimensionless quantities $\Delta_1 = \left(\frac{t_3}{t_1}\frac{t_{01}}{t_{03}}-1\right), \Delta_2 = \left(\frac{t_3}{t_2}\frac{t_{02}}{t_{03}}-1\right)$ contain only relative values of acoustic parameters. t_{0i}, t_i (i = 1,2,3) are time delays of elastic waves composing the unique orthogonal basis providing two shear and one longitudinal waves propagation normally to the plane of stress action without turning of the plane of polarization. The parameters Δ_1, Δ_2 do not depend on any change of the acoustical path during material deformation.

It may be known in advance that the stressed state of a detail is uniaxial. Then, bearing our assumptions in mind, knowledge of propagation time of only two types of waves is sufficient for stress evaluation. Considering the greatest sensitivity to stress of the wave polarized along a direction of its action, the formula for stress evaluation may be writing down in a form:

$$\sigma_1 = \frac{1}{k_1 - k_i}\left(\frac{t_1}{t_i}\frac{t_{0i}}{t_{01}} - 1\right), \ i\text{=2 or 3.} \tag{4}$$

By neglecting the terms of the second order with respect to relative values of time-of-flight variations one can receive a well known formula for determination of uniaxial stress acting in the direction 1 of shear wave polarization on basis of acoustical birefringence phenomenon:

$$\sigma_1 = D(a - a_0)\,. \tag{5}$$

So, the value of uniaxial stress appearing in the plane perpendicular to shear waves propagation direction, may be evaluated directly using the elastic-acoustic coefficient $D = \frac{1}{k_1 - k_2} = K_1 + K_2$ and the values of acoustical anisotropy parameters of the material after and before the stress occurrence:

$$a = \frac{t_2 - t_1}{t_2}, \ a_0 = \frac{t_{20} - t_{10}}{t_{20}}\,. \tag{6}$$

Here a_0 is the value of material intrinsic anisotropy. In isotropic hyperelastic medium $a_0 = 0$, the coefficient $D = 8\mu^2/4\mu + n$, where μ - shear modulus, n – Murnaghan modulus of nonlinear elasticity [4].

Experimental Equipment

A portable device IN-5101A (Figure 4) was developed by “ENCOTES” Ltd. for practical realization of opportunities of the acoustoelasticty method. The device provides measuring of uni- or biaxial, tensile or compressive stresses in engineering objects, including long-term operated pipelines at different force and climatic actions.

Three piezoelectric plates radiating two mutually orthogonally polarized shear waves and a longitudinal wave (a three-component piezoelectric transducer on Figure 4) are installed in a rigid case. An acoustic contact between piezoelectric plates and metal surface is executed using an epoxy resin without a hardener. The calculative algorithms (3) are included into the computational module of device and enable one to evaluate in-plane principal stresses in real object.

Ultrasonic Stress meter IN-5101A represents a new generation of devices based on use of acoustoelasticity phenomenon, provided reliable measurements of uni- and biaxial stresses in potentially dangerous sections of main or technological pipelines. The operation of IN-5101A is user-friendly in comparison to devices based on other physical principles (e.g. X-rays).

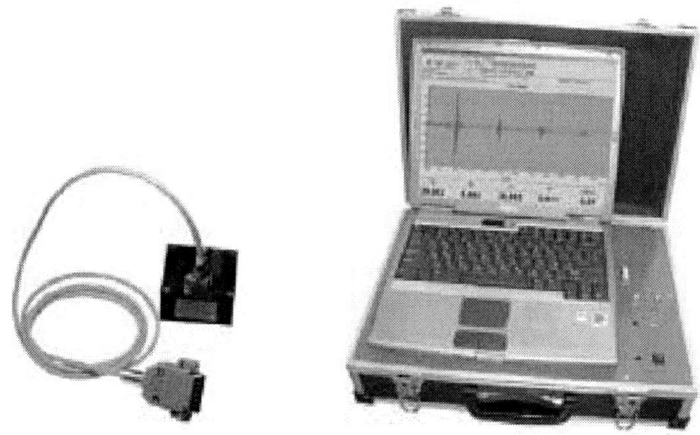

Figure 4. Three-component transducer and device IN-5101A.

Main Practical Purposes of IN-5101A

1. Measurement of applied and residual stresses appearing as a result of manufacture, assemblage and exploitation of large-size engineering constructions.
2. Monitoring of stressed states in the process of stress relaxation and material softening in the vicinity of weld seams.
3. Evaluation of the dangerous residual stresses value in constructions operated for a long time under heavy loads and unfavorable internal and external conditions.
4. Appropriate estimation of stress-strain states of loaded parts and details of constructions for their timely repair or substitution.
5. Timely revealing of damaged parts and prediction of destruction of pipelines, gas and oil refinery.

Nondestructive Evaluation of Biaxial Stress in a Closed Pipe

Pipes and pipelines represent the details with one size significantly less then the other two. The ultrasonic pulse-echo method can be used rather effectively for nondestructive inspection of such details [15]. Frequently pipes experience essentially biaxial stress under loading, or it is possible to consider the stressed state as locally biaxial stress in the area of the acoustical testing by ultrasonic pulse-echo method. Combined equations (3) provide the evaluation of in-plane principal stresses in accordance with the acoustoelastic properties of testing material and the results of precise ultrasonic measurements.

The reliability of acoustoelastic manner for biaxial stress evaluation in linear pipelines of large (820–1420 mm) diameter was justified experimentally. A terminated steel pipe cut out from main pipeline was closed by two bottoms and exposed to inner pressure of water of 70 atm (previously). Pressure of 25 and 50 atm were applied to closed pipe with simultaneous nondestructive stress evaluation by acoustoelastic method.

Gabriel Lame founded the analytical solution of the problem in 19th century. For a case of the thin pipe walls, the elasticity theory predicts that the stress normal to the pipe surface is essentially smaller than axial (σ_z) and circumferential (σ_θ) stresses. So, the stressed state of a small part of a thin envelope of the pipe is performed as in-plane stress:

$$\sigma_z \approx p\frac{d}{4h},\ \sigma_\theta = p\frac{d}{2h}. \tag{7}$$

The diameter d of steel pipe used in our experiments was 1020 mm and thickness h was 9 mm. The pipe was made from Sweden stainless steel (its domestic analogue is steel 09G1FB); the yield stress was equal to 500 MPa. Precise measurements of time-of flight of shear and longitudinal waves propagated across the pipe surface were carried out at 4 testing points (1, 2, 3, 5), before loading and after the first (p = 2,45 MPa) and second (p = 4,9 MPa) steps of the loading. Special compact device IN-5101A designed by "ENCOTES" Ltd has been used in the experiments. IN-5101A realizes the acoustoelastic effect, e.g. linear dependences between elastic wave velocity and stress. The advanced technology for nondestructive testing of biaxial stresses in engineering materials was used in our investigations. The values of stresses acting along and across the pipe axes (axial and circumferential stresses, correspondingly) were automatically evaluated in real time by the special computational block, based on the theoretical study [2, 13] and placed inside the experimental equipment. Figures 5, 6 demonstrate the results of comparison of theoretical and experimental values of stresses in the closed pipe under inner pressure.

The results of our investigations show that the observed data are quite close to Lame predictions. The average difference between the stresses evaluated by ultrasound and by theoretical calculations is equal to 11 MPa for σ_z , 25 MPa for σ_θ and does not exceed 5% of the steel yield point equal to 500 MPa.

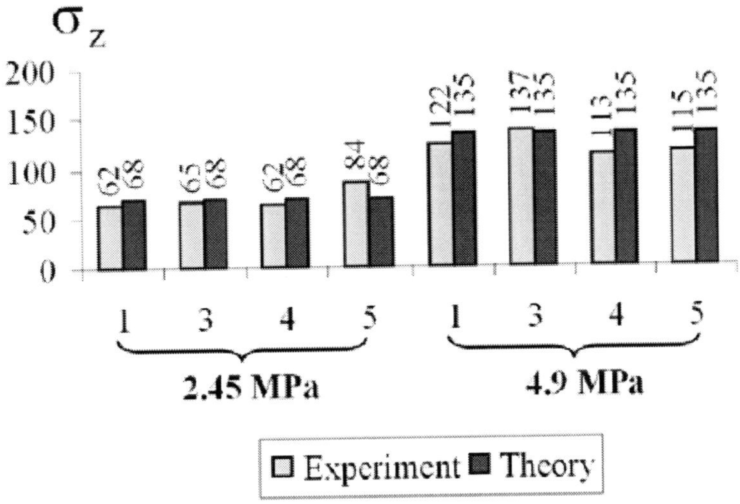

Figure 5. Values of axial stresses in pipe under pressure.

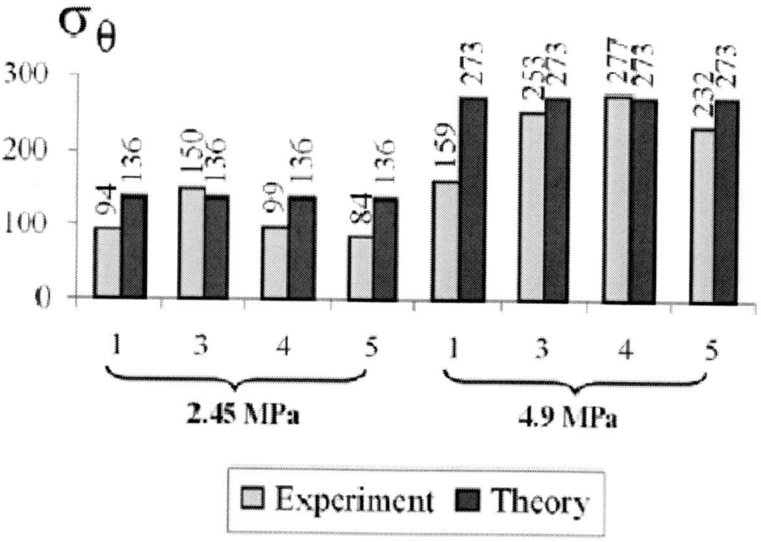

Figure 6. The same, for circumferential stresses.

It is very good accuracy for nondestructive stress evaluation. So, the acoustoelastic method, realized with the help of the device IN-5101A, has sustained check by an arbitrary analytic method with good results. This fact allows hope for successful introduction of this rather new perspective method in practice of the nondestructive evaluation of biaxial stresses in gas or oil pipelines.

EVALUATION OF NON-DESIGN AXIAL STRESSES IN TECHNOLOGICAL PIPELINES

Opportunities of the nondestructive ultrasonic method for evaluation of stresses «in situ» are shown by an example of measuring of axial stresses due to bending of technological pipelines. The results of application of acoustoelasticity effect are given for carrying out direct measurements of mechanical stresses in pipelines of sole timbers of the main equipment of compressive station. Features of application of the acoustoelasticity method for measuring stresses in these objects and prospects of its application for solving problem of the periodic testing of a stress state of pipeline systems are considered.

Relative cheapness and safety of an ultrasonic method make the acoustoelastic effect attractive enough to the purposes of nondestructive revealing of applied or residual stresses in main and technological pipelines. However, the determining of stresses "in situ", when the own acoustical anisotropy a_0 is unknown, may be connected with large mistakes. In earlier 1963 R. T. Smith remarks that the measurements of birefringence (double refraction) of shear waves "give no information about the origin of anisotropy" [14].

For example, the value of intrinsic acoustical anisotropy due to rolling and working of pipe steels may reach 5-8%. In the same time, the relative value of shear waves velocity difference in steels under yield stress is not exceed 0,3-0,5%. Even the average value of a_0 is preliminary known for the engineering material, it may be dispersed among the material volume, and the value of the dispersion may exceed the parameter $a(\sigma_{yield})$.

Here we try to realize the acoustoelastic manner of measuring stresses in seamless pipes which are used in technological pipelines and are manufactured without using rolled sheets which are characterized by high degree of own acoustical anisotropy.

Laboratory Experiments

The gas inlet and output pipes of dust collectors (DC) of compressor station were the object of our investigation. These are seamless pipes with diameter d = 426 mm and wall thickness h = 16 mm produced from imported steel (its domestic analogue is steel 09G2S). As a result of the technology of their manufacturing, seamless pipes are usually characterized by small acoustic anisotropy, which is lower than or comparable to that determined by the presence of stresses. On the contrary, in domestic longitudinal electric-welded pipes, the value of intrinsic acoustic anisotropy reaches several percent and is distributed very non-uniformly over the pipe circumference.

The use of the emergency reserve of pipes from the same batch as the pipes of the studied object is the simplest method for acquiring information on the unstressed state of the pipe steel. Studies of the acoustical properties of seamless and straight-seam pipes used in pipeline systems have shown that these properties slightly differ along the pipe axis and may substantially vary along its circumference. Therefore, to find the values of the initial acoustic parameters for evaluation stresses "in situ" in accordance with combined equations (3) it may be sufficient to cut a coil from a spare or used pipe.

Our laboratory investigations were realized by using 180-mm-wide coil (Figure 7) cut from the pipe of the emergency supply of pipes, the same as the object under study. Time-of-flight measuring for shear ultrasonic pulses propagated normally to the pipe surface was provided. The propagation times of shear waves of polarization parallel (t_1) and perpendicular (t_2) to the pipe axis were measured in 8 points of ring circumference.

Measurements of time delays of pulses of shear waves of mutually perpendicular polarizations allow determine the value of intrinsic anisotropy a_0 corresponding to unstressed state of pipe material. The observed data are shown in Figure 8.

Figure 7. The view of tested pipe coil.

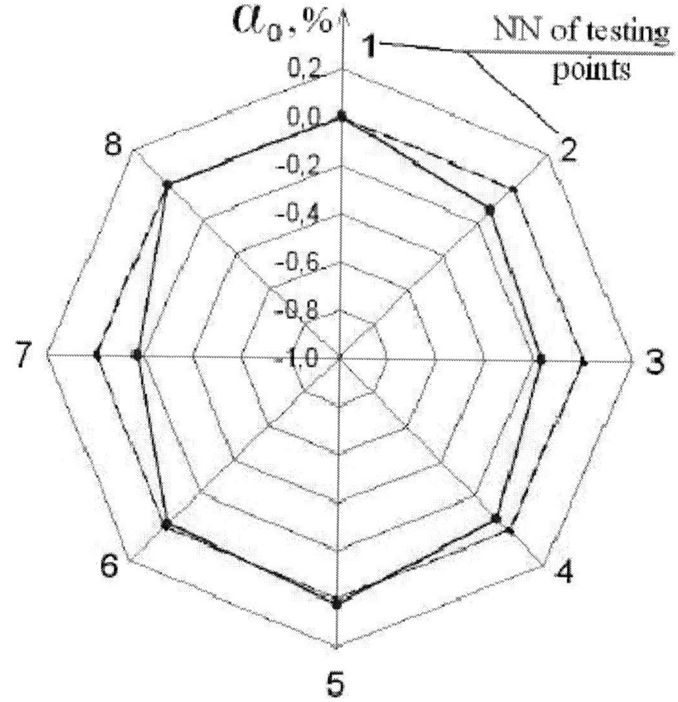

Figure 8. Intrinsic acoustic anisotropy of pipe steel.

So, the own acoustic anisotropy of pipe material is a_0 = -0.07% with a spread of ±0.05% relative to the average value. Taking into account that the coefficient D for low-carbon low-alloyed steels is -121000 MPa, this spread converted into stresses may provide an error of ±60 MPa, i.e., about 17% of the yield stress of pipe steel is assumed to be 360 MPa.

Evaluation of Stressed State of Pipelines

The possibility of determining of non-designed axial stresses of technological pipelines of dust collectors of an operated compressor station was studied experimentally by ultrasonic pulse-echo method, using special portable device IN-5101A. Figure 9 shows the reference points at which the propagation times of shear waves of polarization parallel (t_1) and perpendicular (t_2) to the pipe axis were measured. We performed precise ultrasonic measurements in gas outlet pipes of 12 dust collectors.

It was mentioned above, that measurements of time delays of pulses of shear waves of mutually perpendicular polarization allows to determine in an isotropic material a difference of in-plane principal stresses acting in directions of polarization of waves. In our case, one can evaluate that difference by use formulas (5, 6) and parameter a_0 = -0.07% - the average value known from the laboratory testing.

The table presents the observed and calculated data and gives the results of nondestructive evaluation of axial stresses σ_1 in pipeline system. The circumferential stress

σ_2 was calculated from the well known formula (7) for thin-wall vessel under pressure (p = 2.8 MPa is the operated pressure in the pipe).

Non-design axial stresses evaluated by the ultrasonic method are varying from compressive (dust collector no 2) to tensile (dust collector no 12). It may be explained by probable horizontal displacement of underground collectors during the repair of pipeline structure (Figure 10).

The data of direct measurements of stresses were used to refine the computation of the stressed-strained state of the pipeline structure by the finite-element method, based on the observed data of geodesic measurements of vertical displacements of reference points of the object. Taking into account the horizontal displacements of underground collectors which were revealed no directly, by ultrasonic testing, we provided the satisfactory coinciding of the results of finite-element calculations and data of nondestructive stress evaluation.

Table of results of axial stress evaluation

No DC	1	2	3	4	5	6	7	8	9	10	11	12
$a \cdot 10^4$	-0.6	17.8	2.5	2.0	9.2	-5.5	-8.4	-7.3	-5.2	-8.5	0.4	-22.1
$(a - a_0) \cdot 10^4$	6.4	24.8	9.5	9.0	16.2	1.5	-1.4	-0.3	1.8	-1.5	0.4	-15.1
$\sigma_1 - \sigma_2$, MPa	-77	-300	-115	-109	-196	-18	17	4	-22	18	-90	183
σ_1, MPa	**-40**	**-263**	**-78**	**-72**	**-159**	**19**	**54**	**41**	**15**	**55**	**-40**	**220**
σ_2, MPa	37	37	37	37	37	37	37	37	37	37	50	37

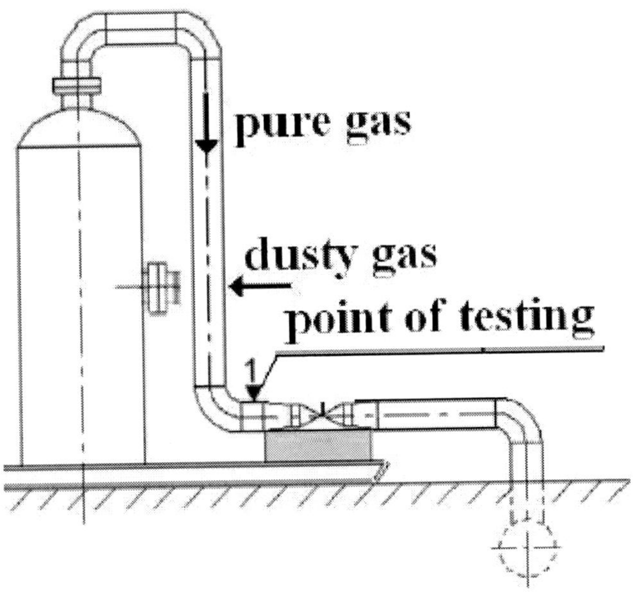

Figure 9. Arrangement of points of ultrasonic testing.

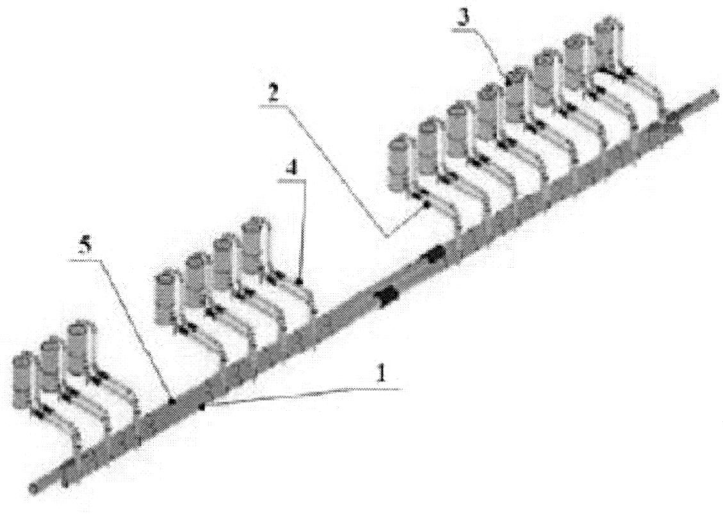

Figure 10. General view of piping system: 1 – collector of incoming gas; 2 – inlet pipe; 3 – housing of a dust collector; 4 - outlet pipe; 5 - collector of outgoing gas.

Conclusions and Recommendations

1. The opportunity of application of "zero-less" ultrasonic tensometry for evaluation of a stressed state of objects in action has been experimentally confirmed by an example of measuring of non-designed axial stresses in technological pipelines of dust collectors of gas-transferring stations.
2. The method has good prospects for determining stresses in weld-less pipes, which are manufactured without using rolled sheets that frequently possesses a high degree of an intrinsic acoustic anisotropy.
3. The proposed manner can be also used for other objects, which material has the intrinsic acoustic anisotropy not exceeding 0.1%, and there is an opportunity of testing an analogous material in the unstressed state.

Acknowledgments

The work was supported by Russian Foundation for Basic Research (Grant No 10-08-01108).

Great acknowledgements to Engineering and Technical Centre of "SEVERGAZPROM" Ltd. for the help in organization and realization of experiments.

References

[1] N. Ye. Nikitina and L. A. Ostrovsky, *Ultrasonics* 35, 605-610 (1998).

[2] N. Ye. Nikitina, *Acoustoelasticity. Experience of practical application,* N. Novgorod: Talam, 2005, 208 p., in Russian.
[3] D. S. Hughes and J. L. Kelly, *Phys. Rev.* 92, 1145-1159 (1953).
[4] F. D. Murnaghan, *Amer. Journ. Math.* 59, 236-260 (1937).
[5] R. W Benson and V. J. Raelson, *Product. Eng.* 30, 56-59 (1959).
[6] G. A. Budenkov and Zh. G. Nikiforenko, *Defektoskopija* No 3, 59-63 (1967), in Russian.
[7] O. I. Guscha and V. K. Lebedev, *Applied mechanics* No 2, 89-92 (1968), in Russian.
[8] C. Truesdell, *Arch. Ratl. Mech. Anal.* 8, 263-296 (1961).
[9] R. Thurston and K. Bragger, *Phys. Rev.* 133(6A), A1604-A1610 (1964).
[10] T. D. Shermergor, *Theory of elasticity of micro-inhomogeneous media.* Moscow: Nauka, 1977, in Russian.
[11] H.-D. Tietz and D. Weigt, D, *Feingeratetechnik* No 11, 501-503 (1979), in German.
[12] A. N. Guz', F. G. Makhort and O. I. Guscha. *Introduction to acoustoelasticity,* Kiyev: Naukova doomka, 162 p., 1977, in Russian.
[13] N. Ye. Nikitina, *Russian Journ. of Nondestruct. Testing.* 35(No 1), 41-46 (*Defektoskopija* 35(No 1), 1999).
[14] R. T. Smith, *Ultrasonics* No 1, 135-147 (1963).
[15] N. Ye. Nikitina, *"Shell Structures. Theory and Applications", Proc. of 8th SSTA conference.* London: Taylor& Frencis/Balkema plc., 399-402 (2005).

In: Pipelines: Design, Applications and Safety
Editors: M. G. Rivero et al. pp. 173-188
ISBN 978-1-62100-178-2

Chapter 6

AERIAL ALTITUDE GAS PIPELINE

Alexander A. Bolonkin*
1310 Avenue R, Suite 6-F,
Brooklyn, New York 11229, USA

ABSTRACT

Design of new cheap aerial pipelines, a large flexible tube deployed at high altitude, for delivery of natural (fuel) gas over a long distance is delineated. The main component of the natural gas is methane, which has a specific weight less than air. The lift force of one cubic meter of methane equals approximately 0.5 kg. The lightweight film flexible pipeline can be located in air at high altitude and, as such, does not damage the environment. This aerial pipeline dramatically decreases the cost and the time of construction relative to conventional pipelines of steel, which saves energy and greatly lowers the capital cost of construction.

The article contains a computed project for delivery of 24 billion cubic meters of gas per year.

Keywords: Gas pipeline, aerial pipeline, cheap pipeline, altitude pipeline, inflatable pipeline.

1. INTRODUCTION

1.1. Natural Gas

Natural gas is a gaseous fossil fuel consisting primarily of methane (CH4) but including significant quantities of ethane, propane, butane, and pentane—heavier hydrocarbons removed prior to use as a consumer fuel —as well as carbon dioxide, nitrogen, helium and hydrogen sulfide. Before natural gas can be used as a fuel, it must undergo extensive processing to remove almost all materials other than methane. The by-products of that

* aBolonkin@juno.com, http://Bolonkin.narod.ru

processing include ethane, propane, butanes, pentanes and higher molecular weight hydrocarbons, elemental sulfur, and sometimes helium and nitrogen.

Natural gas is not only cheaper, but burns cleaner than other fossil fuels, such as oil and coal, and produces less carbon dioxide per unit energy released. For an equivalent amount of heat, burning natural gas produces about 30% less carbon dioxide than burning petroleum and about 45% less than burning coal. The major difficulty in the use of natural gas is transportation and storage because of its low density. Natural gas conventional pipelines are economical, but they are impractical across oceans. Many existing pipelines in North America are close to reaching their capacity, prompting some politicians representing colder areas to speak publicly of potential shortages.

With 15 nations accounting for 84% of the world-wide production, access to natural gas has become a significant factor in international economics and politics. The world's largest gas field by far is Qatar's offshore North Field, estimated to have 25 trillion cubic meters (9.0×10^{14} cu ft) of gas in place—enough to last more than 200 years at optimum production levels. The second largest natural gas field is the South Pars Gas Field in Iranian waters in the Persian Gulf. Connected to Qatar's North Field, it has estimated reserves of 8 to 14 trillion cubic meters (2.8×10^{14} to 5.0×10^{14} cu ft) of gas.

It is difficult to evaluate the cost of heating a home with natural gas compared to that of heating oil, because of differences of energy conversion efficiency, and the widely fluctuating price of crude oil. However, for illustration, one can calculate a representative cost per BTU. Assuming the following current values (2008):

For natural gas.

- One cubic foot of natural gas produces about 1,030 BTU (38.4 MJ/m^3).
- The price of natural gas is \$9.00 per thousand cubic feet (\$0.32/m^3).

For heating oil.

- One US gallon of heating oil produces about 138,500 BTU (38.6 MJ/l).
- The price of heating oil is \$2.50 per US gallon (\$0.66/l) .

This gives a cost of \$8.70 per million BTU (\$8.30/GJ) for natural gas, as compared to \$18 per million BTU (\$17/GJ) for fuel oil. Of course, such comparisons fluctuate with time and vary from place to place dependent on the cost of the raw materials and local taxation.

1.2. Current Pipelines

Natural Gas Pipelines

The long-distance transportation of natural gas became practical in the late 1920s with improvements in pipeline technology. From 1927 to 1931 more than ten major gas pipeline systems were built in the United States. Gas pipelines in Canada connect gas fields in western provinces to major eastern cities. One of the longest gas pipelines in the world is the Northern Lights pipeline, which is 3,400 miles (5,470 kilometers) long and links the West Siberian gas fields on the Arctic Circle with locations in Eastern Europe.

Oil and Petroleum Products Pipelines

Pipelines are used extensively in petroleum handling. In the field, pipes called gathering lines carry the crude oil from the wells to large storage depots located near the oil field. From these depots the oil enters long-distance trunk lines, which may carry it to an intermediate storage point or directly to refineries.

By the last quarter of the 20th century, there were about 250,000 miles (400,000 kilometers) of oil pipeline in operation in the United States. About one third of the total mileage consisted of crude-oil trunk lines. Pipelines carry large volumes of crude oil from fields in Texas, Louisiana, and Oklahoma to refineries in the Midwest and on the East coast of the United States. The 800-mile (1,300-kilometer) north-south trans-Alaskan oil pipeline, which began operation in 1977, connects the Prudhoe Bay fields on the northern coast of Alaska to Valdez on Prince William Sound. Europe also has several crude-oil pipelines supplying inland refineries. In the Middle East large pipelines carry crude oil from oil fields in Iraq, Saudi Arabia, and other oil-exporting countries to deepwater terminals on the Mediterranean Sea.

From the refinery, large volumes of petroleum products travel to the market area by way of product lines. In areas of high consumption, several similar products gasoline, furnace oil, and diesel oil, for example may be shipped in the same line in successive batches of several tens or hundreds of thousands of barrels each. The first European product line, called the TRAPIL line, was completed in 1953 in France. In 1964 the world's largest products line began operation in the United States; the pipeline can transport 1 million barrels of products per day from Houston, Tex., to New Jersey.[1]

Table 1. A list of countries by total length of pipelines [18]

Rank	Country	Total length of pipelines (km)	Structure	Date of Information
1	United States	793,285	petroleum products 244,620 km; natural gas 548,665 km	2006
2	Russia	244,826	condensate 122 km; gas 158,699 km; oil 72,347 km; refined products 13,658 km	2007
3	Canada	98,544	crude and refined oil 23,564 km; liquid petroleum gas 74,980 km	2006
4	China	49,690	gas 26,344 km; oil 17,240 km; refined products 6,106 km	2007

The Trans-Alaskan Pipeline

The trans-Alaskan pipeline was the most expensive pipeline in the world costing 9 billion dollars to build. It is 48 inches (122 centimeters) in diameter and 800 miles (1,300 kilometers)

[1] Reference: Compton's Interactive Encyclopedia.

long. Oil moves southward through it, at a rate of 1.5 million barrels each day, from the giant Prudhoe Bay oil field on the northern coast of Alaska to the ice-free port of Valdez, where the oil is shipped by tankers to refineries on the West coast of the United States. The pipeline traverses three mountain ranges and 250 rivers and streams. For some 400 miles (640 kilometers) it is suspended on pylons above permanently frozen ground called permafrost. If the pipeline had been buried in the ground, the heat from the oil in the pipeline would have melted the permafrost, causing considerable environmental damage.

1.3. Some currently Planned Pipeline Projects

The *Nabucco pipeline* is a planned natural gas pipeline that will transport natural gas from Turkey to Austria, via Bulgaria, Romania, and Hungary. It will run from Erzurum in Turkey to Baumgarten a der March, a major natural gas hub in Austria. This pipeline is a diversion from the current methods of importing natural gas solely from Russia which exposes EC to dependence and insecurity of the Kremlin practices. The project is backed by the European Union and the United States.

The pipeline will run from Erzurum in Turkey to Baumgarten a der March in Austria with total length of 3,300 kilometers (2,050 mi). It will be connected near Erzurum with the Tabriz-Erzurum pipeline, and with the South Caucasus Pipeline, connecting Nabucco Pipeline with the planned Trans-Caspian Gas Pipeline. Polish gas company PGNiG is studying the possibility of building a link to Poland with the Nabucco gas pipeline.

In the first years after completion the deliveries are expected to be between 4.5 and 13 billion cubic meters (bcm) per annum, of which 2 to 8 bcm goes to Baumgarten. Later, approximately half of the capacity is expected to be delivered to Baumgarten and half of the natural gas is to serve the markets en-route. The transmission volume of around 2020 is expected to reach 31 bcm per annum, of which up to 16 bcm goes to Baumgarten. The diameter of the pipeline would be 56 inches (1,420 mm).

The project is developed by the Nabucco Gas Pipeline International GmbH. The managing director of the company is Reinhardt Mitschek. The shareholders of the company are: OMV (Austria), MOL (Hungary), Transgaz (Romania), Bulgargaz, (Bulgaria), BOTAŞ (Turkey), RWE (Germany).

In 2006, Gazprom proposed an alternative project competing Nabucco Pipeline by constructing a second section of the Blue Stream pipeline beneath the Black Sea to Turkey, and extending this up through Bulgaria, Serbia and Croatia to western Hungary. In 2007, the South Stream project through Bulgaria, Serbia and Hungary to Austria was proposed. It is seen as a rival to the Nabucco pipeline. Ukraine proposed White Stream, connecting Georgia to Ukrainian gas transport network.[2]

These mega-pipeline projects and others currently planned will require investment of at least $200 Billion in the next few years. We propose a much cheaper alternative, an aerial pipeline that is lifted because natural gas is lighter than air and as such a lifting gas. An inflatable pipeline pressurized with natural gas will levitate and float up as if full of helium. Each end of the pipeline will be tethered to the ground but the middle will soar above land and sea at altitudes between .1 and 6 kilometers. It can span seas and can be up to a hundred

[2] http://en.wikipedia.org/wiki/Category:Future_pipelines.

times cheaper than conventional undersea pipelines. It can be tethered only 100meters above the ground and easily monitored and repaired.

The main differences of the suggested Gas Transportation Method and Installation from current pipelines are:

1. The tubes are made from a lightweight flexible thin film (no steel or solid rigid or hard material).
2. The gas pressure into the film tube is more an atmospheric pressure (>1 atm) and usually less than 1.5×10^5 N/m^2 (<1.5 atm.). (The current gas pipelines have pressure up 220 atmospheres.).
3. The great majority of the pipeline [except compressor (pumping) and driver stations] is located in atmosphere at a high altitude (0.1-6 km) and does not have a rigid compression support (pillar, pylon, tower, bearer, etc.) though it may be tethered. The current pipelines are located on the ground, under ground or under water.
4. The compression stations of the aerial pipeline are located at intervals of 200 – 1000 km. The compression stations of the ground pipeline are located at intervals of 50 – 100 km.
5. The transported gas contained within the pipeline supports the gas pipeline at high altitude.
6. Wing devices may provide additional lift.
7. The gas pipeline can be used as an air transport system for oil and solid payloads with a high speed for each container shipped up of perhaps 250 m/sec.
8. The gas pipeline can be used as for the large scale transfer of mechanical energy.

The suggested Method and Installation have huge advantages in comparison with modern steel gas pipelines [1 -8].

2. Description of Innovation

A gas and payload delivering gas/load pipeline is shown on figurs 1 - 6.

Figure 1 shows the floating pipeline.

The installation works as follows: The compressor station pumps in gas from the system input (natural gas processing plant, storage system, etc) into the floating pipeline. The tube is made from light strong flexible gas-impermeable fire-resistant material (film), for example, from composite material (fibers, whiskers, nanotubes, etc.) [7 – 8]. The internal gas pressure is a bit higher than the outside atmospheric pressure (up 0.1- 0.2 atm) both to keep the tube inflated and to send the contents flowing toward the far end of the pipeline. Natural gas has methane as a main component with a specific density about 0.72 kg/m^3. Air, by contrast, has a specific density about 1.225 kg/cubic meters. It means that every cubic meter of gas (methane) or a gas mixture with similar density has a lift force approximating 0.5 kg. Each linear (one meter) weight of the tube is less than the linear lift force of the gas in the tube and the pipeline thus has a net lift force. The pipeline rises up and is held at a given altitude (0.1 - 6 km) by the tensile elements 3. The altitude can be changed by the winches 7. The compressor station is located on the ground and moves the flowing gas to the next compressor station, which is usually located at the distance 100 - 500 km from previous. Inside of the pipeline there are valves (Figure 4) dedicated to lock the tube in case of extreme damage. The pipeline has also the warning light 5 for aircraft.

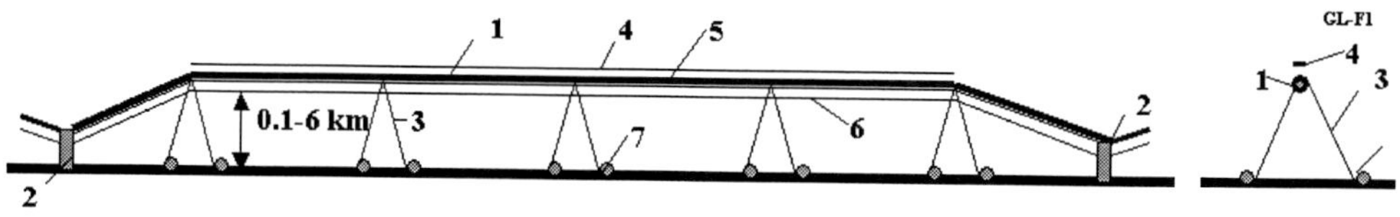

Figure 1. General view of air pipeline between two compression (pump) stations. (*a, left*) Side view, (*b-right*) front view. Notations: 1 - pipeline in atmosphere, 2 - compression station, 3 - tensile stress element, 4 - support wing device, 5 – aircraft warning light, 6 - load (container) monorail, 7 - winch.

Figure 2 shows the cross-section of the gas pipeline and support ring. The light rigid tube support ring helps intermediate the motions and stresses of, variously, the lift force from the gas tube, wing support devices, and the tension elements (tiedown tethers) 3.

Figure 3 shows the compressor (pumping) station. The station is located on the ground or water and works in the following way. The engine 32 rotates compressor 31. This may be in the form of propellers located inside a rigid ring body connected to the flexible tubes of installation 1. The pipeline tubes are located floating in atmosphere near the compressor. The propeller moves the gas in the given direction.

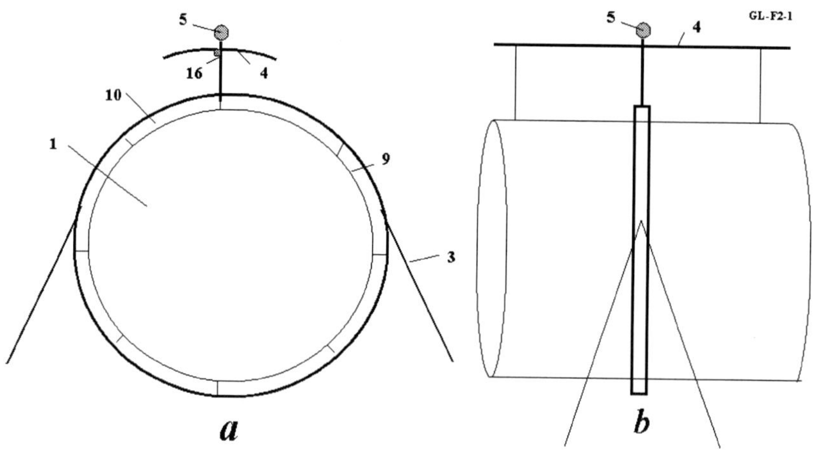

Figure 2. Cross-section and support ring of aerial pipeline. (*a-left*) front view, (*b-right*) side view. Notations: 9 - double casing, 10 -rigid ring, Other notations are same as in Figure 1.

Figure 4 shows two variants of a gas valve. The first valve version is an inflatable ball. The ball would expand out and close the gas path. The second version is a conventional light flat valve-closing disc in a circular form.

The valve works in the following way. When the tube section is damaged, the pressure decreases in a given section. The control devices measure it, the valves close the pipeline. The valve control devices have a connection with a central dispatcher, who can close or open any section of the pipeline. Figure 4a shows the spherical valve (a ball) in packed (closed) form. Figure 4b shows the spherical valve (a ball) in an open form.

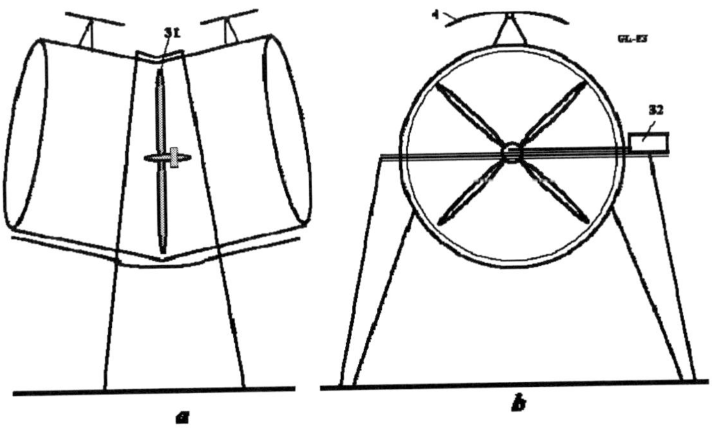

Figure 3. Compression (pumping) station. *Notations*: 31 - compressor (propeller), 32 - engine. Other notations are same as Figures 1-2.

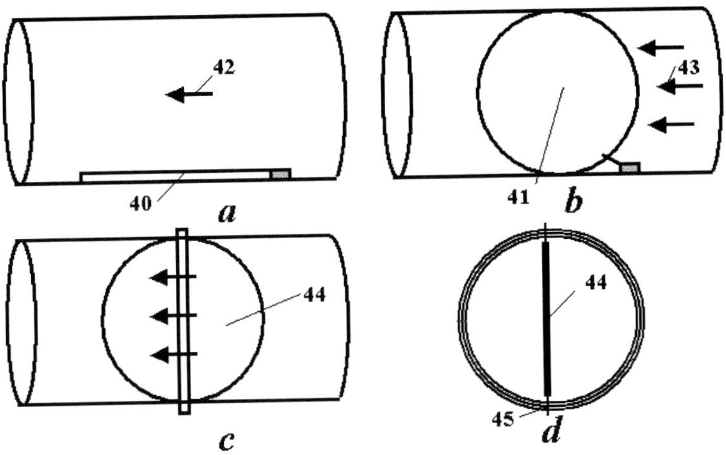

Figure 4. Gas valve. (a)-(b) Inflatable valve, (c)-(d) Flat valve. Notations: 40 - inflatable valve in compact position, 41 - inflatable valve in fill out position, 42 - gas flow, 43 - gas pressure, 44 - flat choker disc.

The tubes of pipeline have a double wall (films) and gas sealing liquid may be between them. If the walls damage the streak gas flows out and the second film closes the hole in the first film and conserves the pipeline's gas.

The wing device 4 is a special automatic wing. When there is windy weather and the side wind produce a strong deviating force, the wing devices produce a strong lift force and supports the pipeline in a vertical position.

The wind device works in the following way. When there is a side wind the tube has the wind drug and the wind device creates the additional lift force. All forces (lifts, drags, weights) are maintained toward equilibrium. The distance between the tensile elements 3 is such that the tube can withstand the maximum storm wind speed. The system can have a compensation ring. The compensation ring includes ring, elastic element, and cover. The ring compensates the temperature's change of the tube and decreases stress from wind.

Figure 5 shows the building of the offered gas pipeline. The ship uncoils the roll of the inflatable film tube and an end ground station fills it a gas. The gas lifts the tube at need altitude and ship team fixes (anchors) a tube position. Other method: Deliver the filled gas tube by helicopter (Figure 5a).

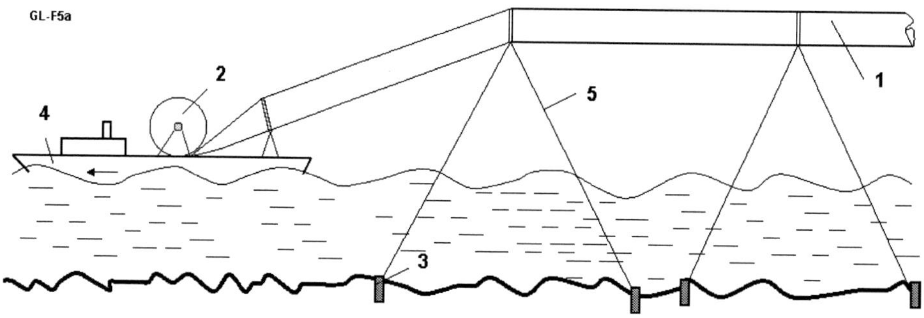

Figure 5. Building the offered gas pipeline by sea ship. *Notation*; 1 - Gas pipeline; 2 - Roll from the inflatable film tube; 3 - Load for fixing tube to sea bottom; 4 - Support ship; 5 - Tensile links.

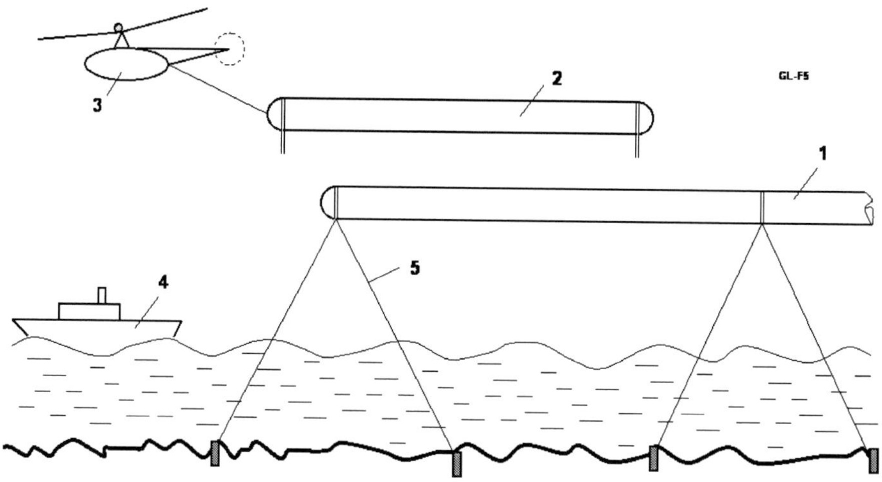

Figure 5a. Building the offered gas pipeline by helicopter. Notation: 1 – gas pipeline; 2 – delivered section of gas pipeline; 3 – deliver helicopter; 4 – support ship; 5 – tensile links.

The suggested gas pipeline has advantages over the conventional steel gas pipeline:

1. The suggested gas pipeline is made from a thin film which is hundreds of times less expensive than a current ground or sea bottom gas pipeline made from steel tubes.
2. The construction time may be decreased from years to months.
3. There is no need to supercompress a gas (over 200 atmospheres) and spend huge energy for that.
4. No large area of expensive user rights needed on ground surface and little surface environmental damage during the building and use of the pipeline.

5. No surface environmental damage in case of pipeline damage during use of the pipeline.
6. Easy to repair.
7. Less energy used in gas transit and delivery.
8. Additional possibility of payload delivery in both directions.
9. If the pipeline is located at high altitude, it is more difficult for terrorist diversion and for petty or large scale gas theft by third parties when the suggested pipeline cross the third country (for example, unknown parties in the Ukraine typically are said to steal 10-15% amount of the gas from Russian pipelines transiting the Ukraine.).
10. The suggested transportation system may be also used for a transfer of mechanical energy from one driver station to another, more cheaply than electric power could be sent.

More detailed description of the innovation may be found in references [1]-[8]. Below, projects suitable for many regions are proposed. This paper is by no means exhaustive and apparent problems in this project, such as standing up to strong winds, have been solved and the author is prepared to discuss the problems and solutions with serious organizations, which are interested in researching and developing these ideas and related projects.

3. Methods of the Estimation of the Altitude Gas Pipeline

1. Gas *Delivery Capability* is

$$G = \pi D^2 V/4 \ [\mathrm{m^3/sec}],\ M = \rho G \ [\mathrm{kg/s}], \qquad (1)$$

where: D is diameter of tube [m]; V is average gas speed [m/sec], M is mass flow [kg/s], ρ = 0.74 kg/m^3 is gas methane density. Result of computation is presented in Figure 6.

2. Increment *pressure* [N/m^2] is

$$P = \lambda L \rho V^2/2D\ , \qquad (2)$$

where: λ is dimensionless factor depending on the wall roughness, the fluid properties, on the velocity, and pipe diameter (λ= 0.01 - 0.06); L is the distance between pump station [m]; ρ is fluid density [kg/m^3].

The dimensionless factor can be taken from graph [5, p.624]. It is

$$\lambda = f(R,\varepsilon)\ , \qquad (3)$$

where: $R = VD\rho/\mu$ is Reynolds number, μ is fluid viscosity; ε is measure of the absolute roughness of tube wall. In our case λ = 0.015.

Result of computation of the gas pressure in relation to the distance between pump station and gas speed are presented in Figure 7.

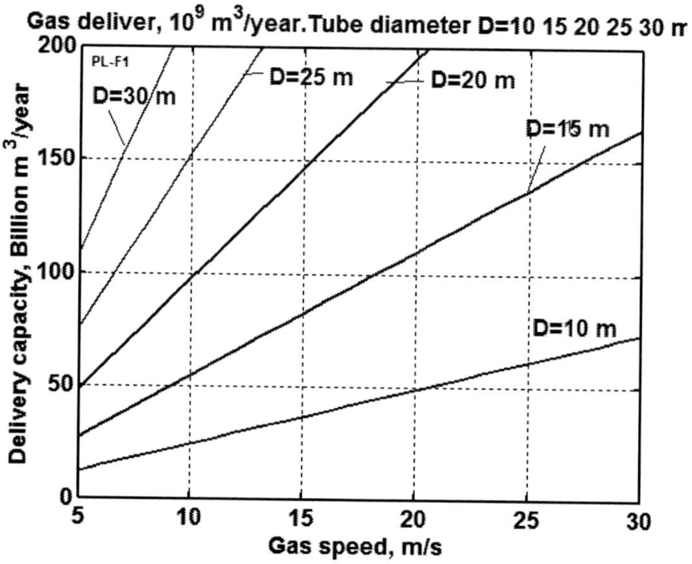

Figure 6. Delivery capacity of gas by the offered floating gas pipeline by gas speed and tube diameter.

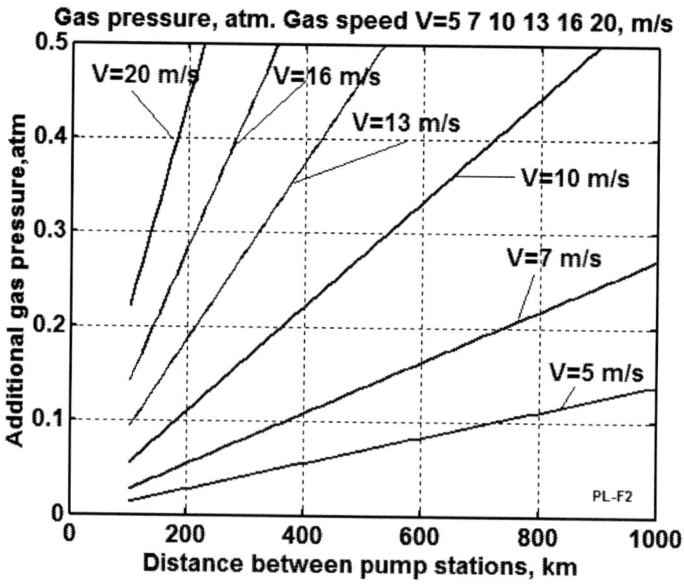

Figure 7. Gas pressure versus the distance between pump stations for different gas speeds.

3. *Lift force F* of each one meter length of pipeline is

$$F = (\upsilon_a - \upsilon_m)v\,, \qquad (4)$$

where: υ_a = 1.225 kg/m^3 is air density for standard atmosphere; υ_m = 0.72 kg/m^3 is methane density; $v = \pi D^2/4$ is volume of one meter length of pipeline, F [kg/m]. Results of computation are presented in Figure 8.

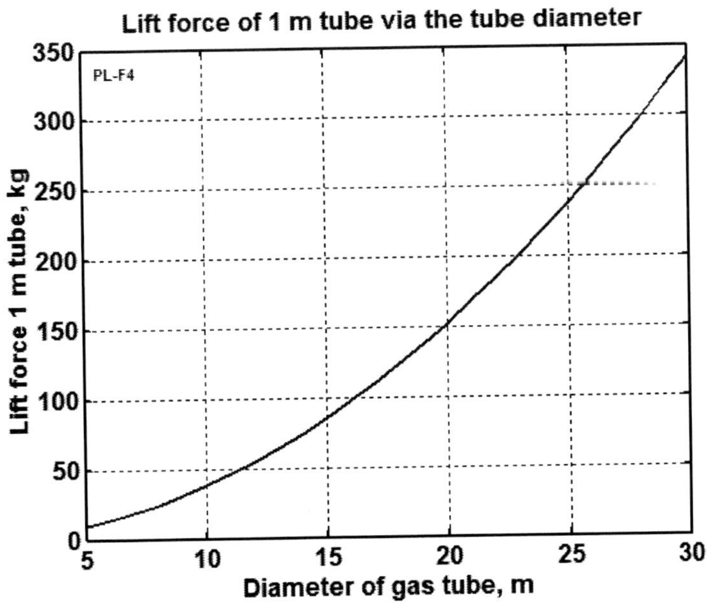

Figure 8. Lift force of each 1 meter of methane gas bearing tube in relation to tube diameter.

4. Needed *thickness* δ of tube wall is

$$\delta = PD/2\sigma, \tag{5}$$

where: P is pressure [see Eq.(2)]; $\boldsymbol{\sigma}$ is safety stress. That is equal to 50 - 200 kg/mm^2 for materials made from current artificial fibers. (Stronger materials, of course, would allow better results) Results of computation are presented in Figure 9.

5. *Weight* of one meter of pipeline is

$$W = \pi D\delta\gamma, \tag{6}$$

where: γ is specific weight of matter tube material (film, cover). That equals about

γ = 900 - 2200 kg/m^3 for matter from artificial fiber.

6. *Air drag D* of pipeline from side wind is

$$D = C_d\rho V^2S/2, \tag{7}$$

where: C_d = 0.01 ~0.2 is drag coefficient; S is logistical pipeline area between tensile elements; ρ = 1.225 kg/m^3 is air density.

7. Needed *pumping power* is

$$N = PG/\eta, \tag{8}$$

where: $\eta \approx 0.9$ is the efficiency coefficient of a compressor station.
Result of computation is presented in Figure 10.

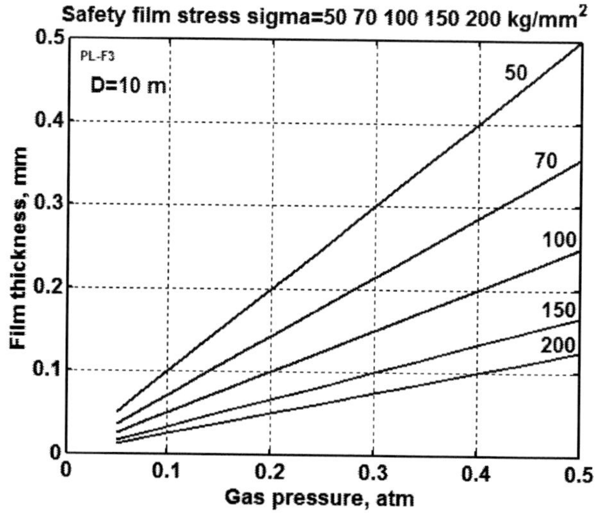

Figure 9. Film thickness via gas pressure and the safety film stress for the tube diameter D = 10 m.

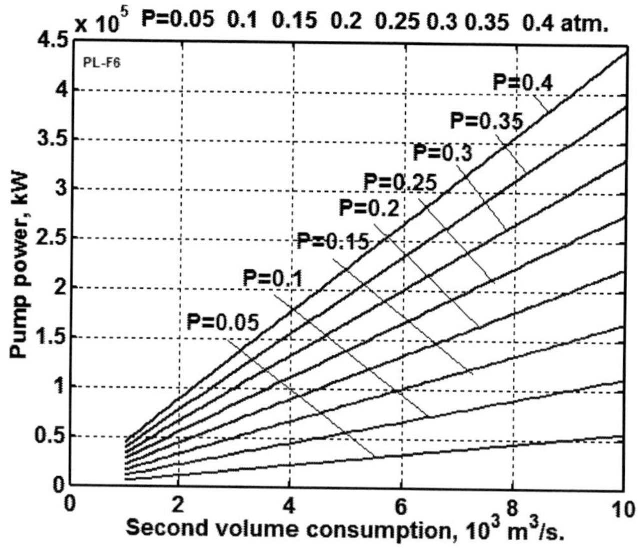

Figure 10. Pumping power versus the volume consumption and gas pressure.

We may also view that pumping power equals

$$N=2C_fMV^2L/D\eta, \tag{8a}$$

where M is mass of the gas [kg/s], L is length of pipeline between compression station [m], C_f is the friction coefficient. That means the delivery power decreases when we use the tubes of bigger diameter. In our case the chosen diameter is 10-20 m. That means the pumping power (to force the gas through the pipeline) will be about 10 times less than in a conventional high pressure pipeline having maximal diameter of 1.44 meters. The conventional steel pipeline method requests also a lot of additional pumping energy for gas compression of up to 70 – 200 atmospheres.

5. PROJECT

(Tube diameter equals D = 10 m, gas pipeline has the suspended - load transport system, the project is suitable for many regions)

5.1. Main Data

Let us take the tube diameter D = 10 m, the distance between the compressor-driver stations L = 100 - 1000 km and a gas speed V = 10 m/sec.

Gas delivery capacity is then (Eq. (1))

$$G = \pi D^2 V/4 = 800 \text{ m}^3/\text{s} = 25 \text{ billions m}^3 \text{ per year [km}^3\text{/year]}.$$

For the Reynolds number $R = 10^7$ value λ is 0.015, $P = 5\times10^3$ N/m^2 = 0.05 atm (Eq. (2)-(3) for L = 100 km). We can take V = 20 m/s and decrease delivery capacity in two (or more) times.

Lift force (Eq. 4) of one meter pipeline's length equals F = 39 kg for diameter D = 10 m. For σ = 200 kg/sq.mm the tube wall thickness equals δ = 0.0125 mm (Eq. 5 for P = 0.05). We take the thickness of wall as δ = 0.15 mm. That is enough *for a distance of 1200 km between pumping stations!*

The cover weight of each one meter of the pipeline's length is 7 kg [Eq. (6)]. The needed power of the compressor station (located at distance of 1000 km) equals N = 44,400 kW for η = 0.9 [Eq.(8)].

5.2. Cost of Altitude Gas Pipeline

Let us take the film thickness δ = 0.15 mm, specific weight of film γ = 1440 kg/m3, film cost c = \$4/kg, diameter of tube D = 10 m. Then volume of 1 m of tube is $v = \pi D\delta = 4.7\times10^{-3}$ m^3, weight is $W = \gamma v$ = 6.77 kg/m, cost of 1 m tube film is $C_1 = cW$ = \$28/m.

The tube needs one wing attached to the tube for each length 10 m. Let us take the cost of one such wing \$20. It is C_2 = \$2/m. Support cables and anchors for each tube length of 100 m cost about \$200 or C_3 = \$2/m. Let us take the cost of pumping station good for 100 km is \$100,000 and good for 1000 km the \$1 mln. That means the cost of the pumping station is C_4 = \$1/m.

The total cost of 1 m of pipeline (material) is $C = C_1 + C_2 + C_3 + C_4$ = \$33/m or C = 33K/km (K is thousand). The Labor and additional works usually costs 100-200% from material. That means each 1 km of the offered altitude pipeline may cost as little as $100K/km after the tooling and other overhead costs are paid for.

Are these figures definitive? Of course not. This is new work, not done before, not part of prior art. There are substantial safety and engineering risks, possibly including development of new tools and materials, certainly developing the specifics of handling gas in large inflatable tubes (for example, triboelectric issues involving friction of wind flow past inflatable dielectrics (the film surface) These cost estimates show the potential low side. The first versions would cost more, not least for safety engineering and prototyping.

But the cost savings potential is most substantial, which argues for its consideration.

In conventional pipelines the steel tubes alone cost 0.3 – 1 million dollars per km. The builder must also pay a lot of money for the right of way on the Earth's surface. The deliver must also annually pay fees for gas transition through the territory of other countries - transition rights. The offered pipelines are very cheap and in many cases may be located over the high seas, cutting down greatly on the right of way fees.

Compare the cost of the current and proposed ground and sea pipelines in Table 1. As you see the offered pipe may be up to ~ 10 – 100 times cheaper than conventional ground pipelines of same length. They may be built if need be more rapidly; in a year or so, with extra money to cover the expense of such rapid construction, in as little as 5-6 months, to be contrasted with 4-5 years as conventional pipelines. This time value of capital saved is a major advantage.

Such a pipeline does not require payment for transition across other countries via ground rights (air rights are usually far cheaper because the uses of the land may continue below) and they require 10 - 15 times less energy for pumping gas because there is no great compression and less gas friction.

Table 2. Design and current pipelines

#	Pipeline	Length, km	Sea part, km	Cost 1 line, $Billion	Cost 1 km, $mln/km	Min. diameter, m	Deliver capacity, km^3/year	Final year of building
1	Russia-German	1200	1200	5	4.2	1.44	27.5	2010
2	Russia-Bulgaria	900	900	10	11	?	30	2013
3	Asia-Europa	3300	-	10	3	1.42	4.5-28	2013
4	Russia-Turkey	1230	396	3.2	2.64	0.61	16	1997
5	Albania-Italy	520	-	1.9	3.65	?	10	2011
6	Turkey-Greece	296	17	?	-	0.914	7	2007
7	Grecce-Italy	800	200	1.6	2	?	8	2012
8	Proposed Levitated gas line	1000	1000	0.1	0.1	10	25	-

Source of [1- 18], Wikipedia [17].

CONCLUSION

The author suggests a new concept for a gas pipeline. The offered aerial pipeline is cheaper by mant times than a current ground (or under water) steel pipeline and may be built in short time (some months) over neutral sea surface. The offered method does not request to supercompress a gas (up 200 atmospheres) and not spend huge energy for that. The method is not needed in large ground surface and produces a little surface environmental damage during the building and using of the pipeline. The method has additional possibility of payload delivery in the both directions.

The pipeline is located at high altitude; it is more difficult for terrorist diversion and for petty or large scale gas stealing by third parties where the suggested pipeline crosses a third country. Additional research and engineering work will be required to implement such a novel idea. The first (test) pipeline must be small (some kilometers).

ACKNOWLEDGEMENT

The author wishes to acknowledge Shmuel Neuman and Joseph Friedlander for correcting the author's English and for useful technical advice.

REFERENCES

[1] A.A. Bolonkin, Patent Application "Method for Gas and Payload Transportation at Long Distance and Installations for It", Application to USPTO # 09/978,507 of 10/18/2001. http://Bolonkin.narod.ru/p65.htm

[2] A.A. Bolonkin, Patent Application "Method Transportation of Vehicles and Installations for It", Application to USPTO # 09/893,060 of 6/28/2001, patent US #6,494,143 B1.

[3] A.A. Bolonkin, Air Cable Transport System, *Journal of Aircraft*, Vol. 30, No.2, Mach-April 2003, pp. 265-259.

[4] A.A. Bolonkin and R.B. Cathcart, A Low-Cost Natural Gas/Freshwater Aerial Pipeline. http://arxiv.org search "Bolonkin".", http://arxiv.org/ftp/physics/papers/0701/0701061.pdf , presented 5 Jun. 2007.

[5] A.A. Bolonkin, Non-Rocket Space Launch and Flight, Elsevier, London, 2006, 488 pgs. http://www.scribd.com/doc/24056182.

[6] A.A. Bolonkin, New Concepts, Ideas, and Innovations in Aerospace, Technology and Human Sciences, NOVA, 2007, 502 pgs. Chapter 6. http://www.scribd.com/doc/24057071.

[7] A.A. Bolonkin, Macro-Projects: Technology and Environment. NOVA, 536 pgs., 2008. http://www.scribd.com/doc/24057930.

[8] A. Bolonkin, Suspended Structures, 2009, http://www.scribd.com/doc/25883886.

[9] K.K.Kienow (ed), Pipeline Design and Installation, 1990.

[10] Compton's Interactive Encyclopedia.

[11] M.L.Navvar (ed), The Piping Handbook, 6th ed.,1992.

[12] P.Tullis, *Hydraulics of Pipelines*, 1989.

[13] Designed oil and gas fields, main pipelines, refineries, and gas processing plants, Alberta, Canada, 1976, Map. Div.

[14] Main line natural gas sales to industrial users (microform). US Dep. of Energy Readex Microfiche E 3.11/14 7V 1979.

[15] C.W. Byrer and R.D. Malone, Preceedings of the Natural Gas Conference: Emerging technologies for the Natural Gas Industry, Houston, Texas, U.S. Department of Energy, Federal Energy Technology Center, March 1997. (Full text available on the web at: http://www.netl.doe.gov/publications/proceedings/97/97ng/ng97_toc.html.

[16] Oil and Gas R&D Programs: Securing the U.S. energy, environmental, and economic future, U.S. Department of Energy, Office of Energy, Office of Natural Gas and Petroleum Technology, March 1997.

[17] Wikipedia. Some background material in this article is gathered from Wikipedia under the Creative Commons license. http://wikipedia.org .

[18] The World Fact book accessed in June 2008.

In: Pipelines: Design, Applications and Safety
Editors: M. G. Rivero et al. pp. 189-206
ISBN 978-1-62100-178-2

Chapter 7

OUTFLOW OF GAS FROM A LIMITED VOLUME THROUGH A PIPELINE WITH FRICTION

V. I. Zvegintsev and A. Yu. Mel'nikov
Khristianovich Institute of Theoretical and Applied Mechanics,
Russian Academy of Sciences Siberian Branch, Russia

ABSTRACT

The paper describes experiments to measure characteristics of a gas flow issuing from a finite-volume reservoir to atmosphere through a pipeline of variable length. In view of the continuous variation of flow quantities in time, the process under study could be classed to unsteady processes. Nonetheless, the experimental data were found to be in excellent agreement with the results gained in one-dimensional steady-state calculations performed assuming a constant value of friction coefficient. The latter result points to applicability of quasi-steady calculations to the analysis of complex gas-dynamic processes.

INTRODUCTION

The gas flow through pipeline is generally assumed to be rather a simple, well-studied problem in industrial aerodynamics. Various engineering procedures for calculating flows through long pipelines with allowance for friction are presently available, which allows one to calculate the flow velocity, the mass rate of the gas flow, and the pressure losses [1-3]. However, all presently known calculation methods are based on the assumption that the discharged flow is steady-state whereas practical problems often require unsteady flow characteristics to be theoretically analyzed. Such problems involve, for instance, the outflow of a gas from a reservoir through an instantly opened hole. After opening of the hole, in the reservoir, in addition to wave phenomena, a continuous reduction of gas pressure and temperature occurs; as a result, the flow cannot be considered stationary at any time. If the flow leaves the reservoir through a long pipeline, then the problem becomes more complicated because of the necessity to consider the formation process of a pipe flow with time-dependent flow quantities. In numerical treatment of the problem, modern computational methods can hardly be applied since a large volume of computations is required here because of unsteadiness of the flow and a considerable extension of flow picture over one of the coordinates.

The purpose of the present study was, first, to obtain experimental data on gas outflow from a reservoir through a pipeline of varied length and, second, to check a developed quasi-steady calculation procedure for applicability in calculation of such unsteady flows.

Experimental Facility

In the experiments, we examined a flow discharged from a reservoir of volume 31 dm^3 to atmosphere through a pipeline of varied length with constant 52-mm inner diameter. Two pipeline configurations shown in Figures 1 and 2 were examined. Pipeline configuration 1 (see Figure 1) comprised six rectilinear pipe sections 1.0 to 3.0 m long and four smooth turns turning the pipeline direction through 90 degrees. The duct elements were connected together with flange connections. The digits in Figure 1 designate the location cross-sections of the flange joints. For suppressing vibrations, the pipeline was rigidly fixed on metal supports installed on a horizontal plate. The total pipeline length was 14.5 m. Pipeline configuration 2 (see Figure 2) was a rectilinear pipeline comprising six rectilinear pipe sections 2.5 to 4.0 m long. The total length of this pipeline was 13.5 m.

According to the widely accepted view, over rectilinear pipe sections the flow can be considered one-dimensional with friction losses varying in proportion to the length of particular section, and the turns, where, because of the reconstruction of the flow, sharply enhanced total-pressure losses are observed, are termed local resistances.

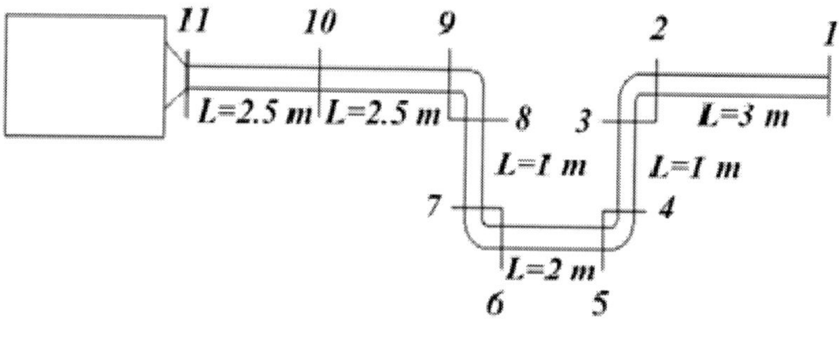

Figure 1. Pipeline configuration 1. State Standard 3262-75 water pipe. Inner diameter 52 mm. Total pipeline length 14.5 m.

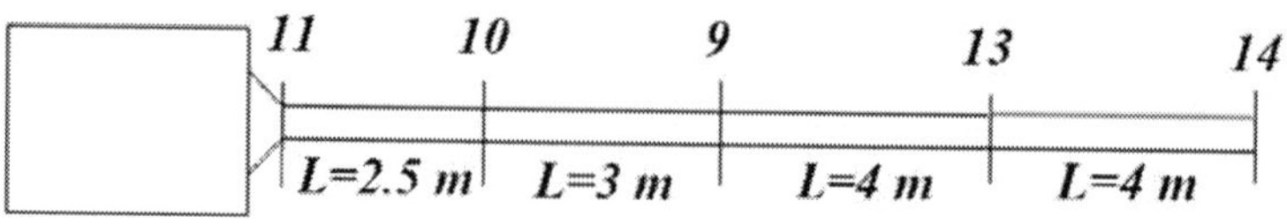

Figure 2. Pipeline configuration 2. State Standard 3262-75 water pipe. Inner diameter 52 mm. Total pipeline length 13.5 m.

Figure 3. The pneumopulse generator.

In the experiments a special apparatus, the pneumatic pulse generator, was used as the reservoir. The pneumopulse generator was a gas-dynamic apparatus first accumulating compressed air (to a pressure of 1 MPa) in a 31-dm^3 volume and, then, rapidly discharging the accumulated air into atmosphere through a 100-mm diameter quick-acting valve (opening time 3-5 ms). The design of the pneumogenerator is defended by Russian-Federation Patent No. 2023228, of August 7, 1992. The general appearance of the pneumatic generator is shown in Figure 3 and a diagram illustrating its design, in Figure 4.

The main component part of the pneumopulse generator is a compression chamber equipped with a discharge nozzle and an automatic quick-acting valve.

Inside the compression chamber a discharge nozzle is provided, shaped as a cylindrical pipe welded to an inlet flange and mounted on a support. On the outside of the flange, six pins serving the function of fixation of the generator on the object were provided. On the outlet flange, the quick-acting valve is mounted, attached to the compression chamber with the help of eight pins and a slip-on flange.

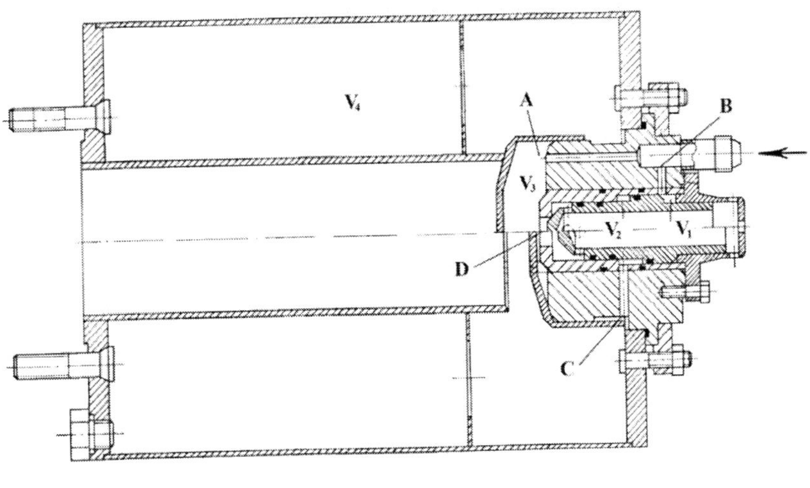

Figure 4. Diagram of the pneumopulse generator.

On initiation of gas feeding system, compressed air enters, through the inlet connection and channel B, the cavity V_1 in the control valve. Under the action of the excess pressure in this cavity, the control valve moves to the left to take position on the insert seating. Now, the air flow from the cavity V_3 of the closing valve is cut off. Simultaneously, compressed air enters the cavity V_3 through channel A. Under the action of the increasing pressure P_3, the closing valve moves to the left to close the exit from the compression chamber to the nozzle. Afterwards, the air, as it flows over in the gap between the valve and the cylindrical housing guide, starts filling the storage cavity V_4 in the compression chamber. At the moment the compression-chamber pressure P_4 reaches 90% of the pressure in the gas feeding system, the air pressure forces, acting on the control valve from the left, grow in excess of the pressure forces acting on the control valve from the right, thus letting the air from the closing-valve cavity V_3 to atmosphere. Then, the pressure P_3 in the valve cavity decreases and, under the action of the compression-volume pressure P_4 the closing valve moves to the right to open the outlet hole of the nozzle. The opening time of the valve is roughly 4 ms. Compressed air accumulated in the compression chamber flows through the discharge nozzle and the attached pipeline to atmosphere. A typical record of the pressure P_4 during the outflow of compressed air into atmosphere through a 100-mm diameter hole (without pipeline) is shown in Figure 5.

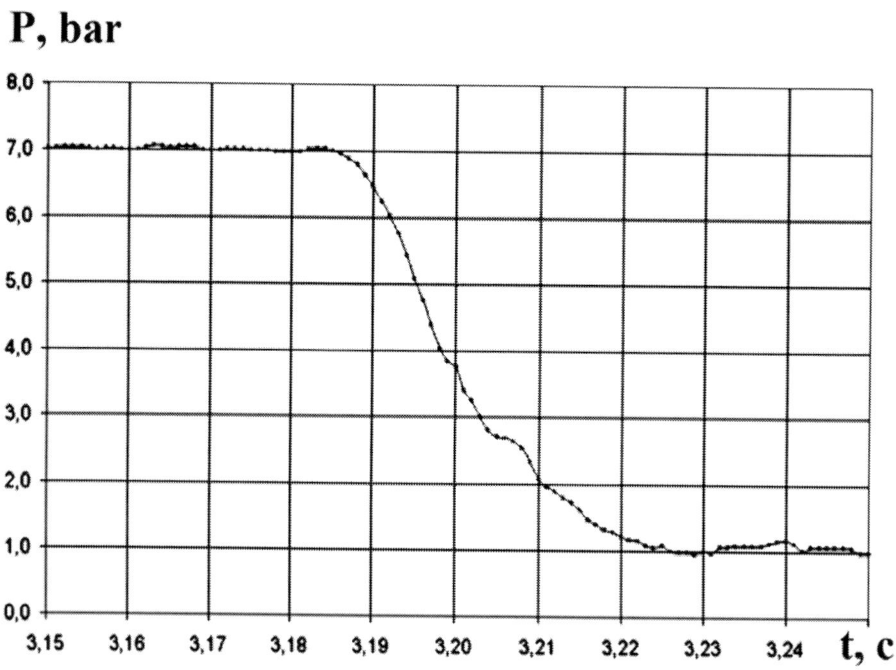

Figure 5. A typical record of air outflow from the pneumopulse generator into atmosphere at 100-mm discharge nozzle diameter.

The activation and full opening of the quick-acting valve is seen to be completed in a time less than 4 ms; after this time interval, a quasi-steady gas outflow begins, ending in full emptying of the reservoir. With a pipeline of diameter 52 mm, the gas outflow from the reservoir and the related reduction of pressure in this reservoir was a slower process, whose duration was 100 to 150 ms.

MEASUREMENT PROCEDURE AND REGISTRATION OF FLOW CHARACTERISTICS

To measure the main characteristics of unsteady air jets, measurements of the total and static pressures in the exit cross-section of the pipeline were carried out. The total pressure was measured with the help of a Pitot comb that was formed by five Pitot probes mounted on a common pylon. The static pressure was measured with one and the same cylindrical static-pressure probe. The distributions of static pressure over a measured cross-section were measured by driving the probe to different positions. The pressure measurements were performed using commercially available inductive-type pressure sensors; such sensors have small dimensions (diameter 15 mm, thickness 5 mm), and they are rather inexpensive. In measuring total-pressure values, pressure sensors with 1-MPa measurement range was used, and static-pressure values were measured using pressure sensors with measurement range 0.3 MPa. The working frequency range of the sensors was 0 to 2000 Hz.

The reference pressure was atmospheric pressure. Each Pitot was connected to the measuring cavity of pressure sensors with a pneumatic pipeline prepared from metal tubes of inner diameter 0.7 mm and length not greater than 200 mm.

The sensor signal-to-pressure conversion factor and the related measurement inaccuracies were determined using the following calibration procedure. A pressure sensor was exposed to a reference pressure whose magnitude was varied in fixed steps within a certain interval of pressures. The reference pressure was measured with a standard high-accuracy-class instrument. A proper treatment of the obtained dependence of output sensor signal on the reference pressure $U_i = f(P_i); P_i = \varphi(U_i)$ yielded a calibration curve that could be subsequently used in processing measured data. As a rule, the calibration procedure of pressure sensors was performed in the same range of external conditions in which actual measurements were carried out. The calibration curve was further refined by performing repeated calibrations during varying the pressure in direct and reverse directions. Following a preliminary analysis of calibration data, an approximating dependence had to be chosen. As a rule, a polynomial dependence of the first order was used: $P=a+bU$. The polynomial coefficients were determined by the least-squares method. Then, the formula $V_i = P_i - (a + bU_i)$ was used to calculate the residuals, presenting the deviations of measured data from the calibration curve, and also the standard dispersion

$$S_{O\Pi}^2 = \frac{1}{n-1}\sum_{i=1}^{n} V_i^2 , \tag{1}$$

where n is the total number of performed measurements.

The maximal residual gives an estimate of the absolute basic error of the pressure sensor $\Delta = \pm V_i^{\max}$ or, at 95-% probability, we have: $\Delta = \pm 2S_{O\Pi}$. The quantity

$$V = \frac{100 \cdot \Delta}{P_{\max}} \% \tag{2}$$

is called the reduced basic inaccuracy of pressure sensor (here, P_{max} is the upper-range value of the sensor). According to test data, following an individual static calibration, all the pressure sensors used had a reduced basic inaccuracy smaller than 0.5%.

EXPERIMENTAL PROCEDURE

To the pneumopulse generator, with the help of a conical reducer a pipeline was connected. The interior of the generator was filled, to a desired pressure, with compressed air. After opening of the quick-acting valve, the passage of the ejected unsteady air flow through ducts of different lengths was examined. The duration of the discharge process was 100 to 150 ms. Measurements were carried out at initial pressures in the pneumogenerator 0.3, 0.45, and 0.6 MPa.

Figure shows a diagram that illustrates the acquisition procedure of sensor signals in the experiment. Signals from sensors D1-D5, which measured the total pressure in a measured pipe exit section, and also signals from sensors D6-D7, which measured the static pressure at the pipeline outlet, were transferred to an ANCh-23 eight-channel converter. Dc voltage signals are fed to the CAMAC-based fast-operating multi-channel digital recording system to undergo subsequent computer treatment.

The data acquisition system operated as follows. The start-up of the pipe initiates the functioning of a synchronization unit and multiplexer. The sensor signals enter a normalizing amplifier, where amplification and filtration of those signals occur. Then, the amplified signals are transferred to the multiplexer to be redistributed, via channel switching, over allocated ADC memory zones. The analog-to-digital converter serves the function of converting the analog signal to a digital signal. Then, the computer reads the ADC memory and records the data in the computer memory.

The experiments were conducted as follows. First, the complete pipeline configuration was assembled, and measurements in the exit cross-section of the channel were performed. Then, the last channel element was uncoupled from the pipeline, and measurements at the exit from the remaining channel configuration were carried out. Figures 6 and 7 show the number of cross-sections in which measurements were performed. In each experiment, the total pressure at the entrance to the duct was measured with the help of a Pitot tube mounted at the joint of the duct with the conical connector (pipe cross-section 11, see Figure 6). The measured data for this pressure were used to check reproducibility of runs held under identical initial pressures in the storage capacity of the pneumogenerator.

Prior to and after a series of tests, static calibrations of all pressure sensors were carried out. The Pitots were calibrated in the pressure range from 0 to 0.6 MPa, and the static-pressure sensors, in the range from 0 to 0.2 MPa. The calibration constants of the sensors were determined using a linear approximation of the dependences of pressure versus the output signal of the data acquisition system. The mean-square inaccuracy of the sensors used was within 0.5%.

All tests were started with feeding compressed air to the pneumopulse generator. On attainment of a certain pressure in the storage capacity, the quick-operating shutoff valve was actuated. The moving element of the shutoff valve closed a special synchronization contact, and an initiating pulse was delivered to the data recording system.

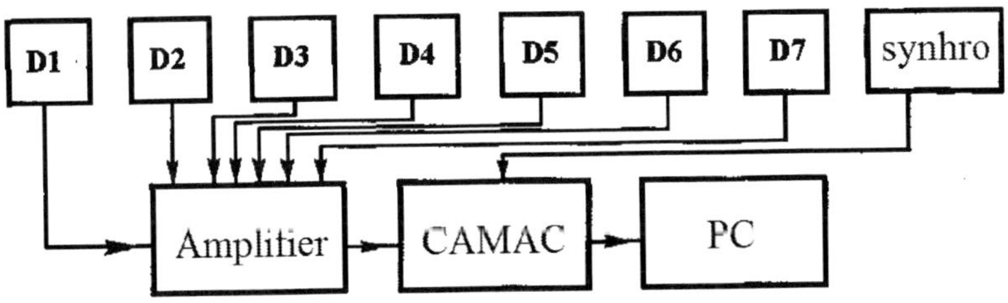

Figure 6. Diagram illustrating the acquisition procedure of measured signals.

During measurements, from each pressure sensor 512 readings were registered with a ten-bit ADC at 0.5-ms time step. For determining the main flow quantities in the pipeline outlet cross-section, the distributions of total and static pressures were measured. The total pressure was measured with the help of a Pitot comb that comprised five pylon-mounted Pitot heads. The spacing between the heads was 10 mm. For obtaining representative data, all measurements were performed at three positions of the comb relative to the cross-section of interest both in horizontal and vertical direction. Measurements in each cross-section were started with raising the comb in an upright position. The comb position relative to the outlet cross-section was changed in the following succession. In the first position, the lower Pitot head was positioned at the distance 1.4 mm from the lower edge of the cross-section. In the second position, the Pitot comb was shifted 5 mm upwards relative to the initial position. In the third position, the upper Pitot head was positioned at a distance of 1.4 mm from the upper edge of the cross-section. Measurement of pressure in the horizontal plane of the cross-section was begun from the comb position at which the leftmost Pitot was positioned at a distance 1.4 mm from the left edge of the cross-section. Then, the comb was shifted to the right by 5 mm. In the third position of the Pitot comb, the rightmost Pitot was at a distance 1.4 mm from the right edge of the cross-section. All in all, in each examined cross-section of the duct total-pressure values were measured at thirty points. This positioning sequence of the Pitot comb was strictly adhered to in each experiment; this had substantially simplified the treatment of measured data.

The static pressure was measured with one and the same cylindrical static-pressure sensor. The distribution of static pressures over the pipeline outlet cross-section was obtained by performing measurements at 12-15 different positions of the sensor in this cross-section.

For graphic representation of obtained distributions of flow quantities over jet cross-section, measured values of total and static pressures were recalculated to a special uniform grid in polar coordinate system using linear interpolation. The grid was formed by radial rays emanating from the center of the cross-section with 10° step over the azimuth angle and by equally spaced concentric circles with 0.8-mm radial separation. The grid comprised 31*37 meshes. The records of the pressures measured at various points of duct cross-section were averaged over the area, and the area-mean static and total pressures were calculated:

$$\overline{P} = \frac{\sum_i P_i \cdot F_i}{F}, \tag{3}$$

$$\overline{P}_o = \frac{\sum_i P_{0i} \cdot F_i}{F}. \tag{4}$$

From the ratio of total to static pressure, the area-mean Mach number in a given cross-section at various times could be determined. If for the ratio of measured pressures at a given point and at a given time we have $\frac{\overline{P}}{\overline{P}_0} > 0{,}528$, then the flow velocity is subsonic, and the Mach number was determined from the relation:

$$\frac{\overline{P}}{\overline{P}_0} = \left(1 + \frac{x-1}{2}\overline{M}^2\right)^{-\frac{x}{x-1}} \tag{5}$$

In the case of $\frac{\overline{P}}{\overline{P}_0} \leq 0{,}528$, the flow is either supersonic or subsonic, and the Mach number was calculated from the expression

$$\frac{\overline{P}}{\overline{P}_0} = \frac{\left(x\overline{M}^2 - \frac{x-1}{2}\right)^{\frac{1}{x-1}}}{\left(\frac{x+1}{2}\right)^{\frac{x+1}{x-1}} \overline{M}^{\frac{2x}{x-1}}}. \tag{6}$$

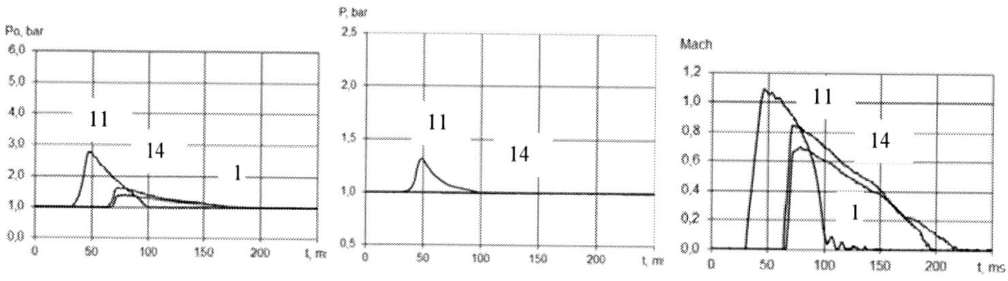

Figure 7. Typical plots of data measured at $P_0(0)$ = 3 bar.

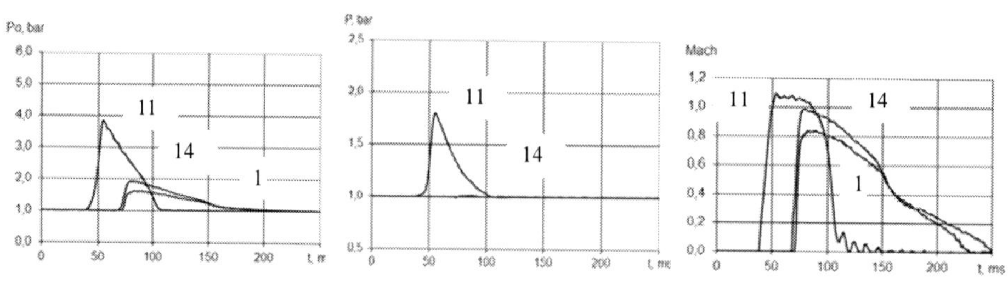

Figure 8. Typical plots of data measured at $Po(0)$ = 4 bar.

In the latter situation, the Pitot senses, instead of the total pressure, which was measured in subsonic flow, the post-shock total pressure $P_0^{`}$, denoted as P_0 for generality.

In what follows, only mean values of obtained quantities, i.e. $P = \overline{P}$; $P_0 = \overline{P}_0$; $M = \overline{M}$. will be used.

EXPERIMENTAL RESULTS

Figures 7-9 show typical graphs of total pressure, static pressure, and Mach number for the initial (11) and last (1) cross-sections of pipeline 1 and for the last cross-section (14) of pipeline 2 at various values of initial pressure.

It is seen from the graphs that the time interval from initiation of the registration system to opening of the quick-acting valve was roughly 40 ms. In this time interval, the total and static pressures in cross-section 11 both start increasing. Here, the flow Mach number reaches a value M=1.1. In 25-30 ms the leading portion of the jet reaches exit cross-section 1, and the pressure here also starts increasing. At the initial pressure Po = 6 bar the maximal total pressure at the inlet is 4.7 bar, and the pressure at the outlet is not greater than 1.9 bar. The difference between these pressures is equal to pressure losses for friction and flow turns.

The static pressure in the exit cross-section (1) remains roughly unchanged, equal to atmospheric pressure ≈1 bar. The flow velocity in this cross-section reaches the sonic velocity (M≈1), and it is subsequently retained over a time interval of 35 ms. Subsequently, a reduction of Mach number in accordance with the total-pressure drop in this cross-section is observed.

A decrease of the initial-pressure value from P_o = 0.6 MPa to P_o = 0.3 MPa had almost no effect on the duration of jet outflow (75-80 ms) and on the maximum mean Mach number in cross-section 11. Nonetheless, the shape of the Mach-number profiles proved to be time-dependent. At $Po \leq 0.3 \div 0.4$ MPa the outlet Mach number in cross-section 1 never reached the value M=1, and the flow remained subsonic throughout the whole pipe.

Figure 14 shows the experimental records of total pressure at various pipeline cross-sections obtained at one and the same start time of registration. It is seen from the graph that the pressure in cross-section 11 starts changing at t = 30 ms, simultaneously with the time of valve opening. The start time of pressure variations in subsequent cross-sections increases with increasing distance from the currently measured cross-section to the compression chamber.

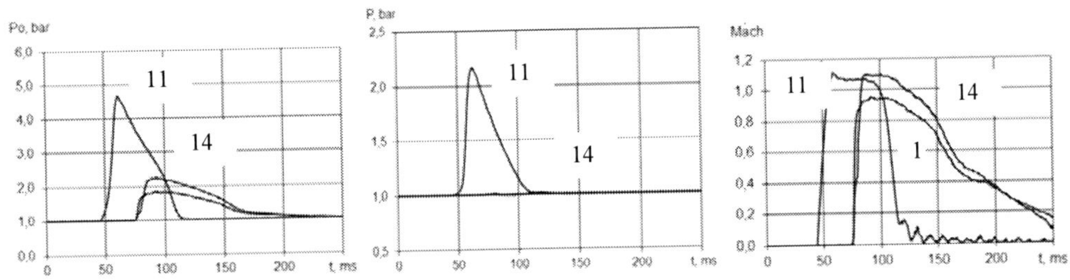

Figure 9. Typical plots of data measured at $Po(0)$ = 6 bar.

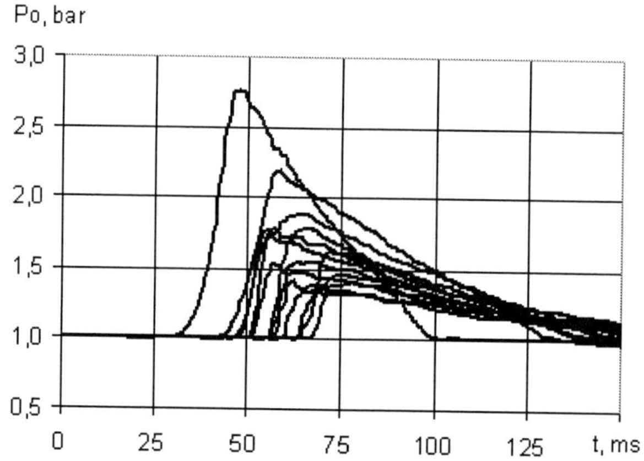

Figure 10. Pressure values measured at various pipeline cross-sections at $P_0(0)$ = 3 bar.

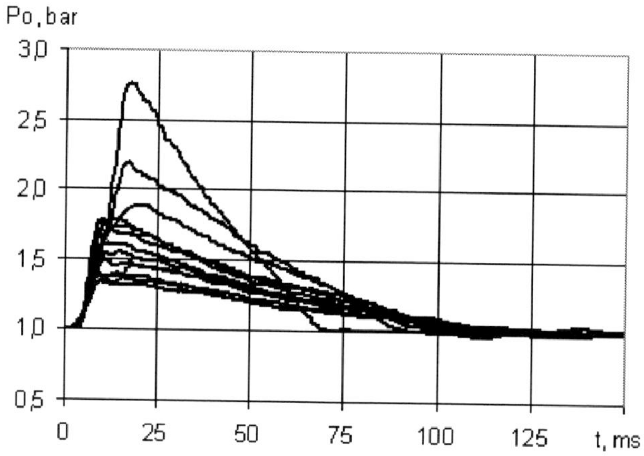

Figure 11. Matched records of time-dependent pressure in various pipeline cross-sections at $P_0(0)$ = 3 bar.

The maximum delay is about 65 ms for cross-section 1, located at the distance 14.5 m from the compression chamber. This delay comprises the time of valve opening (≈30 ms) and the time during which the gas particles flew from the compression chamber to the currently measured cross-section. Bringing the records in coincidence in terms of time origin (see Figure 15), one can identify the delay time for each cross-section and determine the time of flight of the particles.

The full circles in Figure 16 show the times of flight of particles Δt, obtained as described above, for different cross-sections (distances L) and pressures.

The presented data proved to show a good correlation with the calculated velocity of the shock wave V at a given pressure ratio $\frac{P_0}{P_{amb}}$ [4]:

$$\frac{P_0}{P_{amb}} = \frac{2 \cdot \chi \cdot M^2 - (\chi - 1)}{\chi + 1} \cdot \left\{ 1 - \frac{\chi - 1}{\chi + 1} \cdot \sqrt{(M - \frac{1}{M})} \right\}^{\frac{2 \cdot \chi}{1 - \chi}} \quad (7)$$

$$V = M \cdot a \ ; \ \Delta t = \frac{L}{V}$$

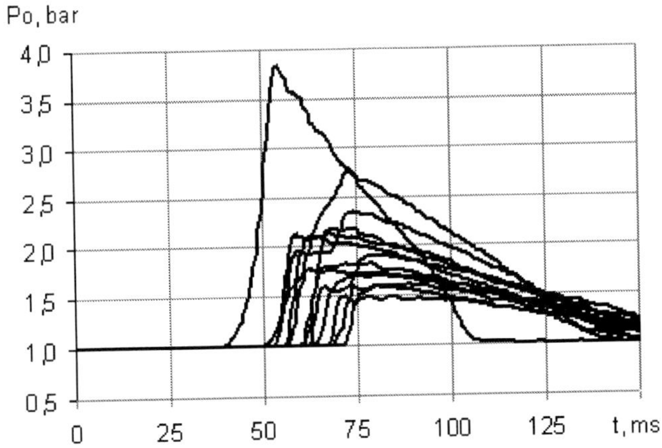

Figure 12. Pressure values measured at various pipeline cross-sections at $P_0(0) = 4$ bar.

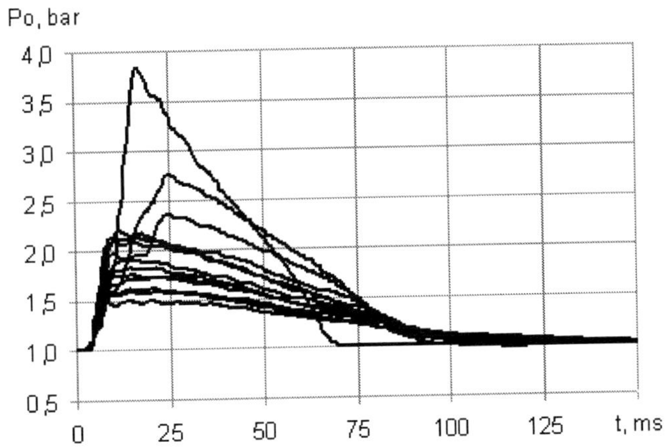

Figure 13. Matched records of time-dependent pressure in various pipeline cross-sections at $P_0(0) = 4$ bar.

The solid line shows the times required for the shock wave to pass the distance L at a given compression-chamber pressure. The calculated data are seen to be in a good agreement with the experimental data; that is why in subsequent analysis all experimental data were shifted by the time determined in the calculations. The adopted delay time was identified as the time required for the flow to become established in the various cross-sections of the pipeline. With allowance for the correction for the time of propagation of the shock wave, the flow can be considered steady-state at each moment for all pipeline cross-sections under consideration.

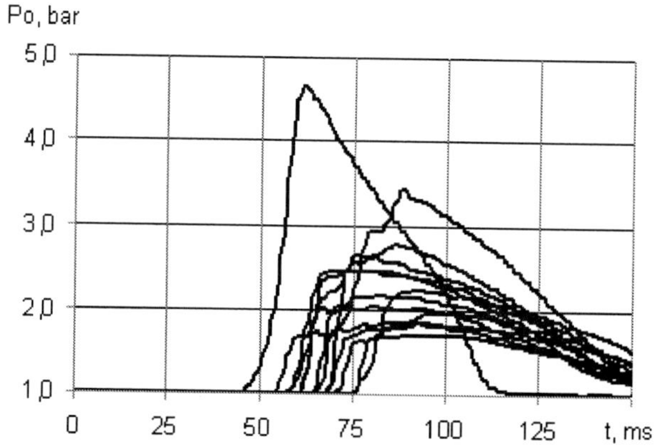

Figure 14. Pressure values measured at various pipeline cross-sections at $P_0(0)$ = 6 bar.

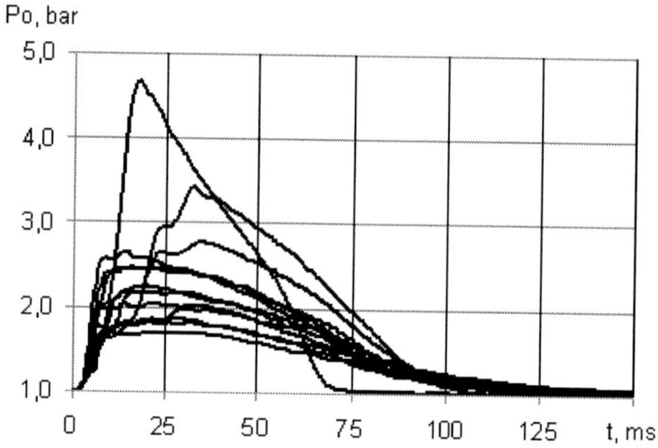

Figure 15. Matched records of time-dependent pressure in various pipeline cross-sections at $P_0(0)$ = 6 bar.

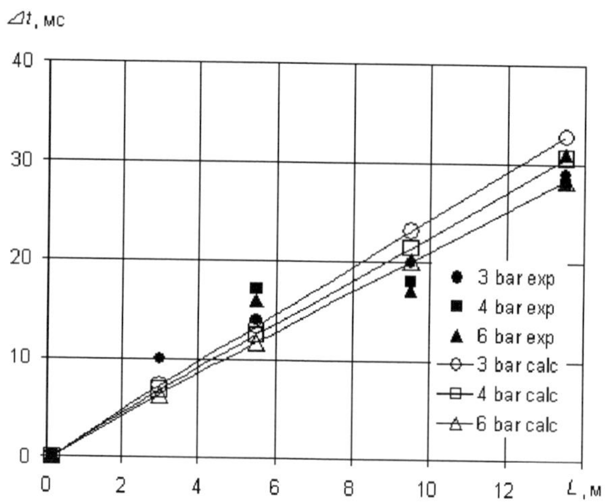

Figure 16. Time of flight of the shock wave on turning-on of the pneumopulse generator.

COMPARISON OF EXPERIMENTAL DATA WITH CALCULATIONS

In paper [5], a procedure for calculating steady-state compressible flow in a pipeline with friction was proposed. With the use of this procedure, calculations of flow quantities in various pipeline cross-sections were performed.

The calculations were carried out for individual times, with the flow assumed steady-state at those moments. In line with the proposed procedure, the increase of flow stagnation temperature and the change of flow cross-sectional area owing to the displacement thickness in boundary layer were taken into account. In the calculations, the specified parameters were the pipeline geometry, the flow quantities at the inlet to the pipeline (pressure, temperature, velocity), and the value of the friction coefficient Λ_{fr} or the roughness height h/D, used to calculate the friction coefficient by the formula

$$\Lambda_{fr} = 0{,}11 \cdot \left(\frac{h}{D} + \frac{68}{Re}\right)^{0{,}25}. \tag{8}$$

In each calculation run, the values of the total pressure, static pressure, and Mach number at the pipeline outlet were determined. At the outlet, the solution was tested for realization of the following two cases:

Case 1	M≤1	$P_2=P_{amb}$
Case 2	M=1	$P_2 \geq P_{amb}$

If none of the cases was realized, then the initial approximation was modified in terms of the velocity V_1 at the pipeline inlet, and the calculations were repeated.

At the first step, a proper choice of a single value of the friction coefficient Λ_{fr} for this cross-section was made to match the values of the total-pressure drops obtained in the calculations and in the experiments for various times. Afterwards, all flow quantities at the outlet were calculated assuming a constant value of resistance coefficient.

Figures 17-19 shows the experimental and calculated values for cross-sections 11 and 14 of pipeline 2 at different values of the compression-chamber pressure.

Figure 17 shows data obtained for the initial compression-chamber pressure $P_0(0)$ = 3 bar. Shown at the left of Figure 17 is the variation of total pressure in cross-section 11. The solid line shows the experimental values. The dotted line shows the calculated values of total pressure in the compression chamber for conditions with a quasi-steady flow issuing from a 31 dm^3 volume. The circles show the total-pressure values obtained in the calculation of a short pipeline of length 0.14 m (from the compression chamber to cross-section 11). Shown at the bottom of the same figure is the variation of static pressure at the outlet (cross-section 11). The experimentally measured pressure is shown with the solid line. The circles show the static-pressure values obtained in the calculations. Shown at the bottom is the variation of Mach number at the exit from cross-section 11. The Mach numbers calculated from experimental data are shown with the solid line.

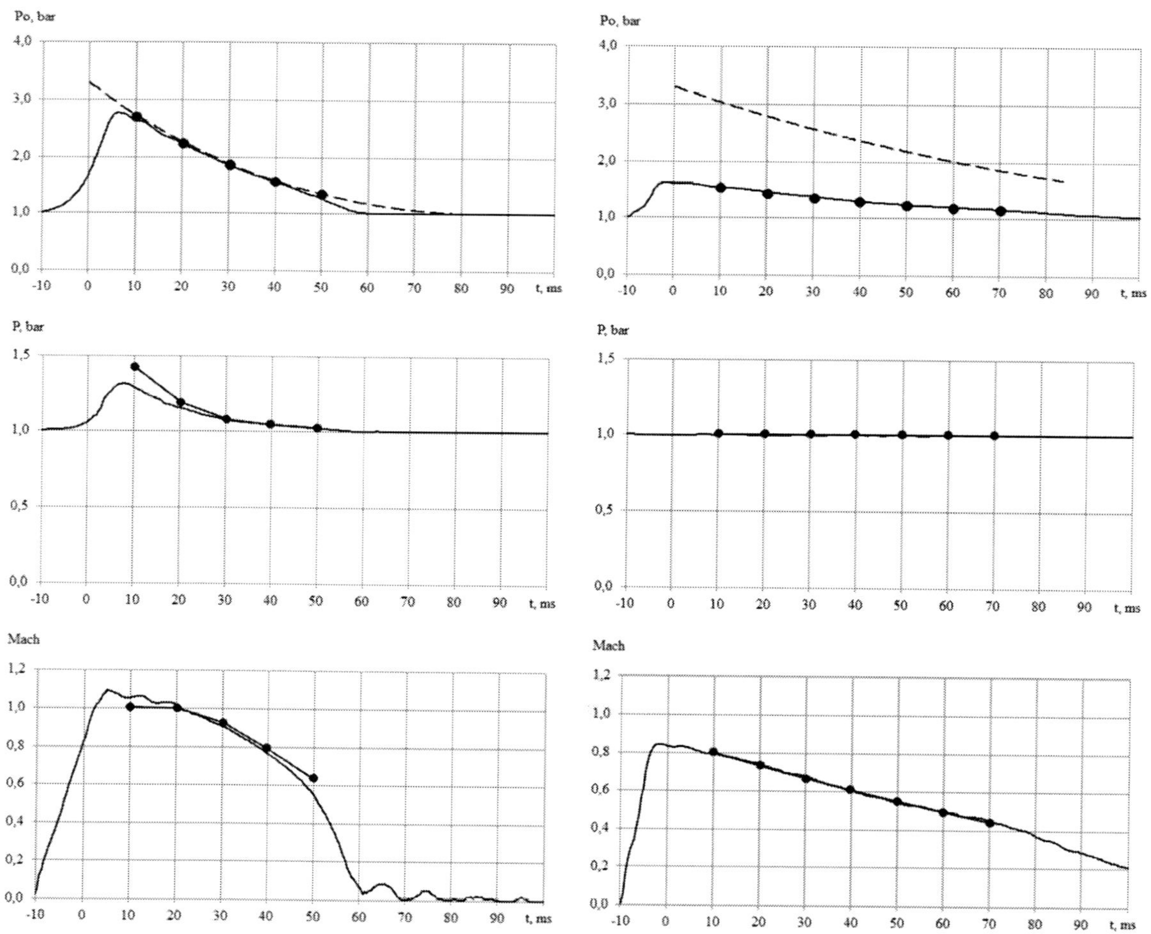

Figure 17. Measured and calculated values of flow characteristics in cross-sections 11 and 14 of pipeline 2 for $P_0(0) = 3$ bar.

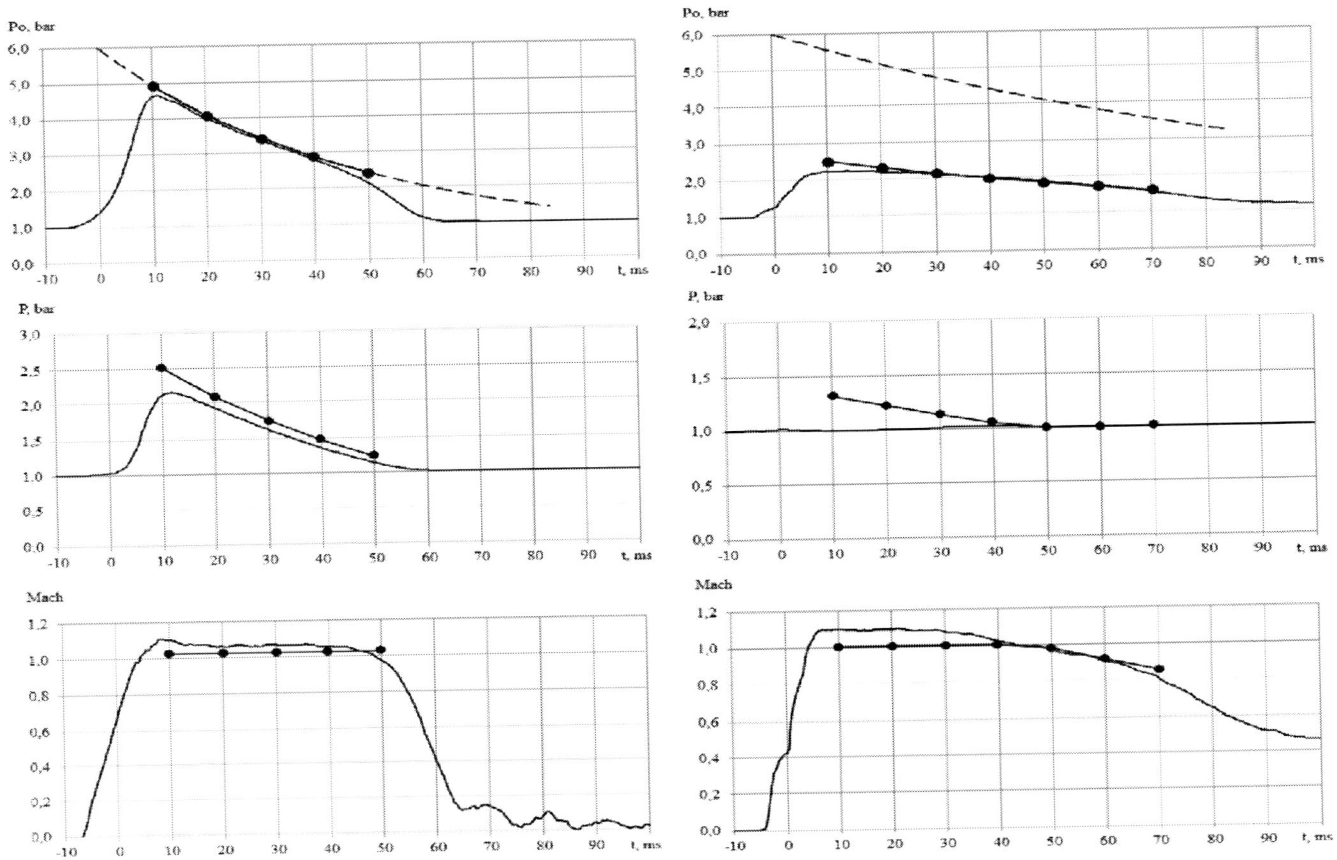

Figure 18. Measured and calculated values of flow characteristics in cross-sections 11 and 14 of pipeline 2 for $P_0(0) = 4$ bar.

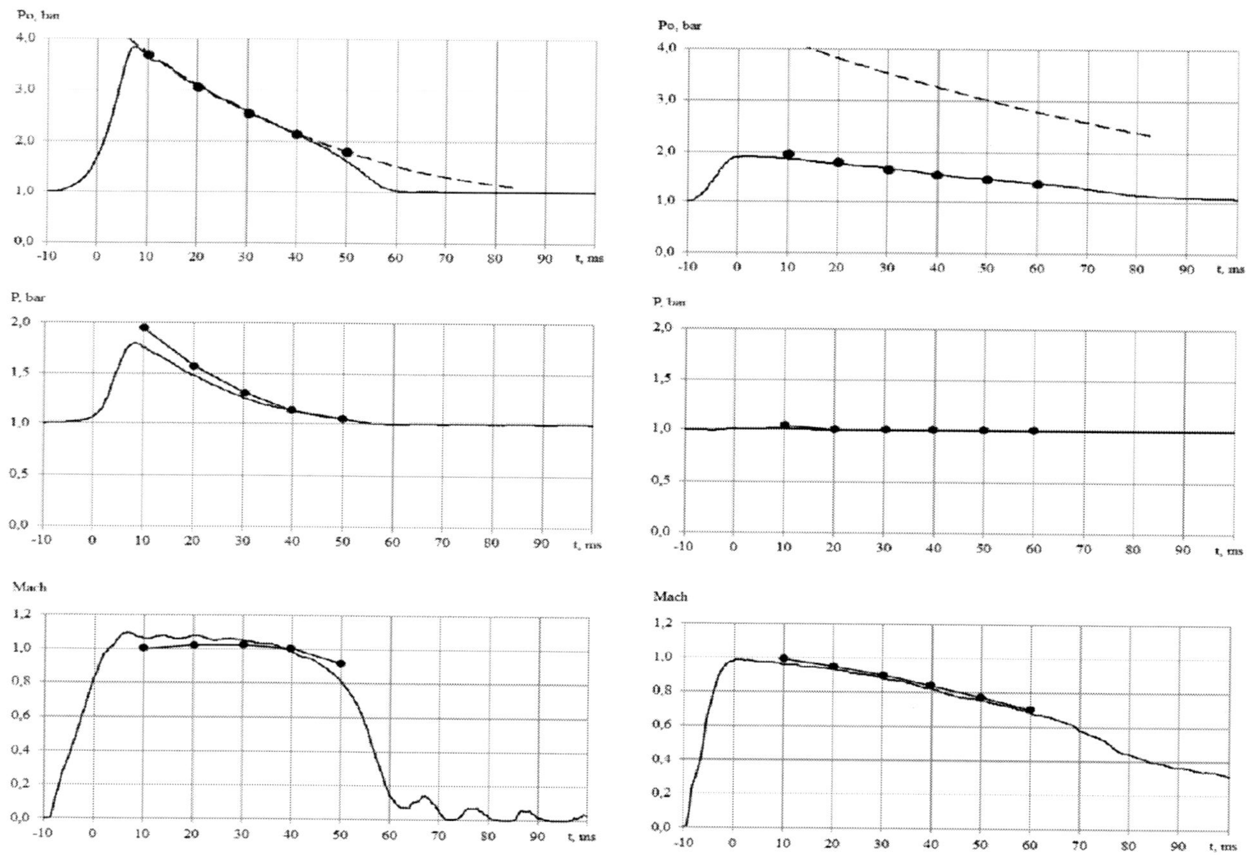

Figure 19. Measured and calculated values of flow characteristics in cross-sections 11 and 14 of pipeline 2 for $P_0(0) = 6$ bar.

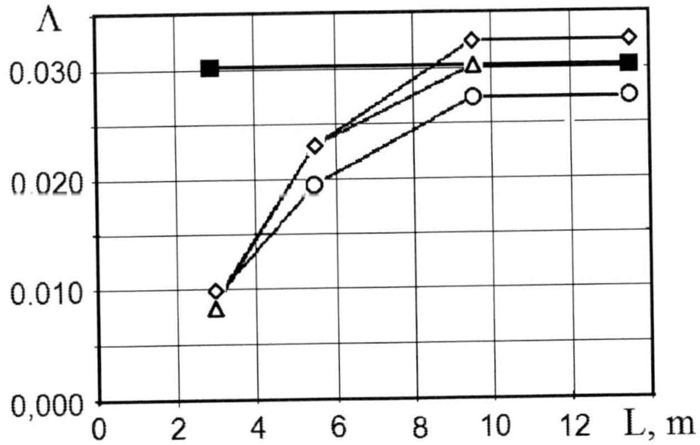

Figure 20. The friction-coefficient values adopted in the calculations.
◇ – $P_0(0)$ = 6 bar; Δ – $P_0(0)$ = 4 bar; ○ – $P_0(0)$ = 3 bar ■ – data calculated by formula (8).

The circles show the calculated Mach numbers. Here, because of the short length of the pipeline, the losses for friction are low, and the friction coefficient may have an arbitrary value. The right top part of Figure 17 shows the variation of total pressure in cross-section 14. As previously, the dotted line shows the total-pressure values calculated for the flow issuing from a compression chamber of fixed volume (31 dm^3) with specified pipeline outlet diameter (52 mm). The solid line shows the experimental values measured in cross-section 14. The circles show the total-pressure values obtained in calculation of a pipeline 13.5 m long (from the compression chamber to cross-section 14). The matching between the calculated and experimental total-pressure values was achieved by choosing a value Λ_{fr} = 0.0325 for the friction coefficient, this value corresponding to the wall roughness height h = 0.4 mm. As calculations show, in the latter case the area of the one-dimensional flow in the outlet cross-section of the pipe decreases by 11% (the flow diameter in the outlet cross-section decreases from 52 mm to 49 mm). The middle panel in Figure 17 shows the variation of static pressure at the outlet (cross-section 14). The experimentally measured pressure is shown with the solid line. The circles show the static-pressure values obtained in the calculations. It is seen that both the experimental and calculated pressure values are equal to atmospheric pressure.

The bottom part of Figure 17 shows the variation of Mach number at the outlet (cross-section 14). The Mach numbers obtained from the experimental data are shown with solid line. The circles show the Mach numbers obtained by calculations. For the most part, the flow velocity at the pipe outlet (cross-section 14) is subsonic. Analogous results are shown in Figure 18 (for initial pressure $P_0(0)$ = 4 бар) and in Figure 19 (for initial pressure $P_0(0)$ = 6 bar). It is seen from the graphs that at the chosen constant value of friction coefficient the results of steady-state calculations at each time well agree with the results of the measurements of characteristics of the complex unsteady flow with friction.

Figure 20 shows the friction-coefficient values obtained through a comparison of experimental and calculated data. The line with rectangles shows the friction-coefficient value calculated by formula (8) for 0.3-mm average pipe roughness value. It is seen from the graph that over the initial pipeline section (in the transition zone, where a developed turbulent flow has not formed yet) the measured friction coefficient proved to be much lower than the value predicted by the calculations. With increasing the pipeline length, the friction coefficient grows in value to come, at a length of 9 m (9000/52 = 160 calibers), to a constant level yielded by formula (8) at 0.3-mm roughness height. Also, it is seen from the figure that, on increasing the initial pressure in the compression chamber from 3 to 6 bar, the friction coefficient changes approximately by 10%.

REFERENCES

[83] E.I. Idel'chik, Handbook of Hydraulic Resistance, M.O. Shteinberg (editor), third revised and supplemented edition, Moscow, *Mashinostroenie,* 1992. – 672 p.

[84] G.N. Abramovich, Applied Gas Dynamics, in 2 parts, Pt. 1: Manual for Students of Technical Colleges, third edition, revised and supplemented, Moscow, Nauka, *Glavn. Red. Fiz.-Mat. Literat.,* 1991. – 600 p.

[85] L.A. Vulis, On the sonic transition in gas flow, in: *Thermodynamics of Gas Flows,* Moscow, Energoizdat, 1950.

[86] A.G. Gaydon and I.R Hurle, *The Shock Tube in High-Temperature Chemical Physics, Mir,* Moscow, 1966. - 428 p.

[87] V.I. Zvegintsev and A.Yu. Mel'nikov, *One-dimensional models for calculating compressible gas flow with friction through pipeline* [article in this book].

In: Pipelines: Design, Applications and Safety
Editors: M. G. Rivero et al. pp. 207-224
ISBN 978-1-62100-178-2

Chapter 8

ONE-DIMENSIONAL MODELS FOR CALCULATING COMPRESSIBLE GAS FLOW WITH FRICTION THROUGH PIPELINE

V. I. Zvegintsev and A. Yu. Mel'nikov
Khristianovich Institute of Theoretical and Applied Mechanics,
Russian Academy of Sciences Siberian Branch, Russia

ABSTRACT

Two theoretical models for calculating compressible gas flow with friction through pipeline are considered. The first model, most commonly used in practice, is based on the assumption that the flow is adiabatic, and the total enthalpy of the flow presents a conserved quantity. This approach, however, leads to a noticeable inconsistency between the total-pressure value calculated at pipeline outlet and the friction losses in the pipeline. In the second model, an increase of total flow enthalpy due to the work of friction forces is admitted. Calculations show that, in the latter case, an increase in flow stagnation temperature amounting to several ten degrees is possible, especially for high-velocity flows.

Both models were tested for adequacy by comparing the data predicted by these models with available experimental data on the distribution of static pressure over pipeline length. It was shown that no definite conclusion in favor of one of the discussed models can be drawn because of inaccuracies of used experimental data.

INTRODUCTION

The gas flow through pipe is generally considered a comparatively simple problem in industrial aerodynamics, now studied in sufficient detail. Various engineering approaches to calculation of pipe flows through long pipelines with allowance for friction-induced pressure losses, which permit determination of flow velocity, mass rate of the flow, and the loss of pressure, are presently available. Yet, as it will be shown below, at high flow velocities those approaches may yield substantial inaccuracies in predicting flow quantities. In the present publication, we propose a new approach to calculating the flow through a long pipeline with improved accuracy.

THEORETICAL CONSIDERATION

We consider the problem about a flow through pipeline discharged at the open end of the pipe to atmosphere. The gas pressure and the gas temperature, and also the pressure in ambient space at pipeline outlet, are assumed to be specified quantities. It is required to determine the distribution of flow quantities over the pipeline length.

Presently, calculations of gas or liquid flows through pipelines are usually performed in one-dimensional statement of the problem using for calculating the friction and resistance local losses of pressure P_0 the following formula of the Darcy-Weisbach type [1]:

$$\Delta P_0 = \zeta \cdot \frac{\rho V^2}{2} = (\zeta_{fr} + \sum \zeta_L) \cdot \frac{\rho V^2}{2}. \quad (1)$$

In (1), ζ is the hydraulic resistance coefficient, also called the pressure loss coefficient.

In the literature, a specific coefficient of friction Λ, which characterizes the loss of pressure over unit normalized length, is also often used:

$$\Lambda = \frac{\Delta P_0}{\frac{\rho V^2}{2}} \cdot \frac{1}{\frac{L}{D}}. \quad (2)$$

The relation between Λ and ζ is given by

$$\Lambda = \zeta \cdot \frac{D}{L} \quad \zeta = \Lambda \cdot \frac{L}{D} \quad (3)$$

Expressions for frictional losses Λ_{fr} and resistance local losses Λ_L follow from certain theoretical assumptions (e.g., that the flow in the pipeline presents a Poiseuille flow) or experimental data (e.g., from data obtained in the experiments by Nikuradze [2]).

As it is evidenced by Figure 1, the friction coefficient Λ depends on the wall roughness and Reynolds number Re:

$$Re = \frac{\rho \cdot V \cdot D}{\mu}. \quad (4)$$

In (4), ρ is the gas density, V is the flow velocity, D is the pipe diameter, and μ is the dynamic viscosity.

It should be noted here that for a flow through a constant cross-section pipeline we have $\rho V = const$ and, hence, the Reynolds number in such a flow in most cases varies insignificantly (due to viscosity variation only).

An expression for the coefficient of friction that adequetly describes experimental data in a broad range of flow conditions was proposed by A.D. Al'tshul [3, 4]:

$$\Lambda_{fr} = 0,11 \cdot (\frac{h}{D} + \frac{68}{Re})^{0,25}. \tag{5}$$

At high Reynolds numbers ($Re > 10^5$) the friction coefficient depends only on the wall roughness, this dependence being given, for instance, by the following formula reported in [5]:

$$\Lambda_{fr} = \frac{1}{(1,74 + 2 \cdot lg\frac{D}{2h})^2}. \tag{6}$$

Here, D is the pipe diameter and h is the roughness height.

Similar formulas are given in [1]; for instance, the Teplov formula reads [6]:

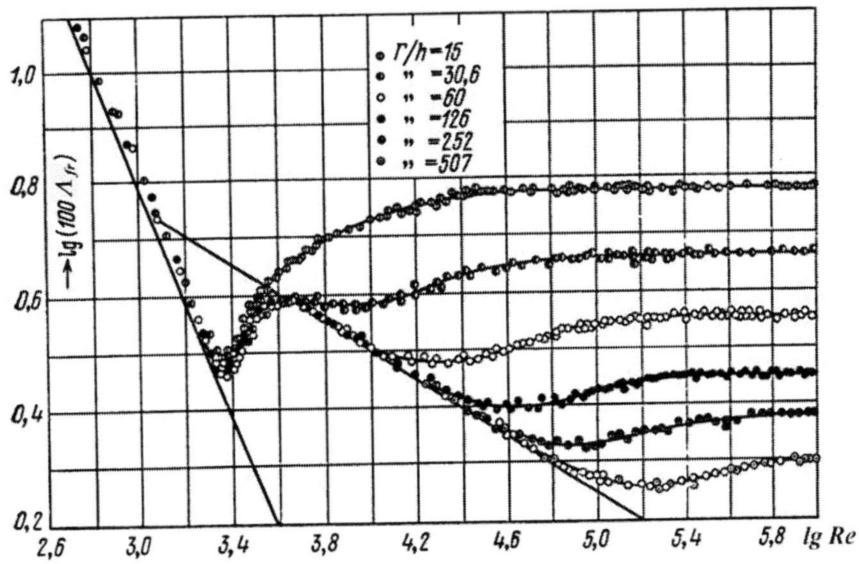

Figure 1. Specific coefficient of friction versus Reynolds number in pipelines with different wall roughness heights as determined from the experimental data by Nikuradze.

$$\Lambda_{fr} = \frac{1}{(1,8 \cdot lg(\frac{8,3 \cdot D}{h}))^2}. \tag{7}$$

As a rule, calculation of flow quantities in a flow through pipeline is accomplished at the stage of determination of total-pressure losses by the above Darcy formula. Here, the gas density ρ and the flow velocity w needed for determination of the dynamic pressure $\frac{\rho V^2}{2}$ are to be taken at the inlet to the pipeline section under consideration. This approach was validated for incompressible gas flows with moderate flow velocities.

As an example, Figure 2 shows calculated data that illustrate the variation of total and static pressures over pipelines of different lengths with identical parameters of the gas flow at the inlet to pipeline (P_o = 0.49 MPa, T_o = 300 K, V = 50 m/s).

The other graph in the same figure, illustrating the variation of flow velocity (Mach number) over pipeline length, shows that, here, the outlet Mach number reaches a value M = 0.6. Yet, even at low pressures the gas velocity can grow from a subsonic one at the inlet to pipeline to a sonic one at pipe outlet. That is why in the majority of actual situations it becomes necessary to consider the case of a compressible gas flow.

There are reported propositions for extending the range of applicability of the available calculation procedures for incompressible flows beyond the traditional area by applying, for instance, a correction to friction coefficient that would allow for gas compressibility (see [7]). However, such propositions are primarily based on empirical data, and they offer only a partial solution to the problem.

Generally, the flow through pipeline can be treated in one-dimensional approximation using the following conservation laws:

Equation of mass conservation:

$$\frac{dG}{G} = \frac{dF}{F} + \frac{d\rho}{\rho} + \frac{dV}{V} \tag{8}$$

Momentum (Bernoulli) equation:

$$\frac{dP}{\rho} + VdV + dL_{fr} + dL = 0 \tag{9}$$

Here, dL_{fr} is the work of friction forces and dL is the work of external forces.

Energy equation:

$$dQ = dQ_{out} + dQ_{in} = dQ_{out} + dL_{fr} = di + VdV \tag{10}$$

Equation (10) implies that the heat supplied to the flow generally comprises an external heat and an internal heat released in the flow due to the work of friction forces $dQ_{in} = dL_{fr}$. The heat supplied to the flow eventually becomes redistributed as a change of flow enthalpy and kinetic energy.

Following some re-arrangements, the momentum equation can be put to the form

$$(M^2 - 1)\frac{dV}{V} = \frac{dF}{F} - \frac{dG}{G} - \frac{dL}{a^2} - \frac{k-1}{a^2}(dQ_{out} + dL_{fr}) - \frac{k}{a^2}dL_{fr} \quad (11)$$

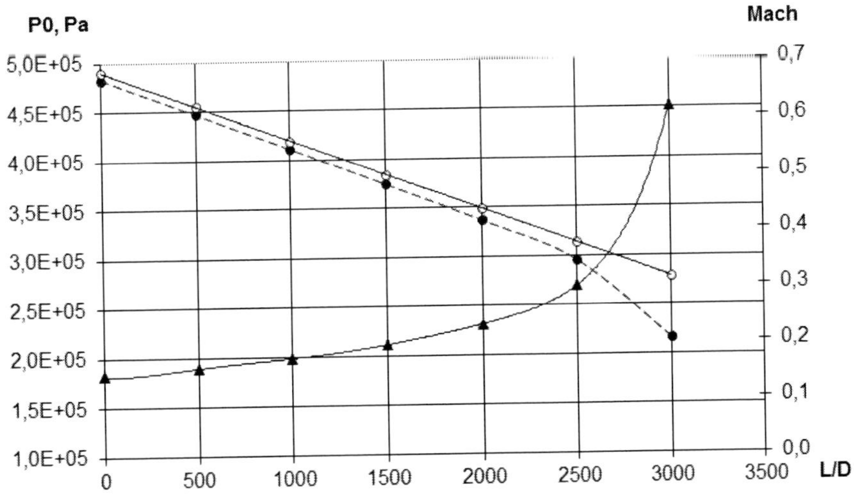

Figure 2. Calculated distributions of total pressure, static pressure, and Mach number over pipelines of different lengths with fixed values of flow quantities at the entrance to pipeline.

Equation (11), initially proposed by L.A. Vulis [8], is called the equation of reverse action. This equation shows how the flow velocity V varies under the geometric action ($\frac{dF}{F}$), under the rate action ($\frac{dG}{G}$), under the mechanical action ($\frac{dL}{a^2}$), under the thermal action ($\frac{k-1}{a^2}(dQ_{out} + dL_{fr})$), and under the work of friction forces ($\frac{k}{a^2}dL_{fr}$). In particular, this equation shows that, for the velocity in subsonic flow to be increased ($(M^2 - 1) < 0$), a reduction in pipe cross-sectional area is required, $dF < 0$, and for reaching further increase of flow velocity in supersonic flow ($(M^2 - 1) > 0$) an increase in pipe cross-sectional area is necessary, $dF > 0$. This principle was used as a guiding one in the invention of the well-known geometric Laval nozzle. The influence of other actions or combinations of such actions on flow velocity can be analyzed in a similar way.

A particular calculation example for compressible flow through a cylindrical pipeline with friction based on solving the equation of reverse action was given in [5]. In the present study, we assume that the problem of interest involves no geometric action ($\frac{dF}{F} = 0$), no rate action ($\frac{dG}{G} = 0$), no mechanical action ($\frac{dL}{a^2} = 0$), and no thermal action ($\frac{k-1}{a^2}(dQ_{out} + dL_{fr}) = 0$). On adoption of those assumptions, only the work of friction forces remains involved in the equation of reverse action:

$$(M^2 - 1)\cdot\frac{dV}{V} = -\frac{k}{a^2}\cdot dL_{fr}\,. \tag{12}$$

Here, $dL_{fr} = \zeta\cdot\frac{V^2}{2} = \frac{\Delta P_0}{\rho}$ is the work of friction forces.

On the assumption of constant values of the specific coefficient of friction $\Lambda = const$ and the stagnation temperature $T_0 = const$ over the pipe length, G.N. Abramovich [5] obtained an analytical solution to Eq. (2) for determining the normalized flow velocity at the outlet from the pipe λ_2 at known velocity of the flow at the inlet to the pipe λ_1 :

$$\frac{1}{\lambda_1^2} - \frac{1}{\lambda_2^2} - ln\frac{\lambda_2^2}{\lambda_1^2} = \frac{2k}{k+1}\cdot\Lambda\,. \tag{13}$$

In (13), $\lambda = \frac{V}{a_\times}$ is the normalized flow velocity, $a_\times = \sqrt{kRT_\times}$ is the speed of sound calculated from the stagnation temperature at the nozzle throat (at M=1), λ_1 is the normalized velocity of flow at the inlet to the pipe, and λ_2 is the normalized velocity of flow at the exit from the pipe.

Using (13), one can determine the value of λ_2 at arbitrary cross-section of the pipe provided that the normalized velocity at the inlet to the pipe λ_1, the pressure loss coefficient ξ, and the adiabatic exponent k are known.

Based on the above consideration, in the present study we have developed an improved algorithm for calculating compressible gas flow through a constant-cross-section pipeline with friction.

Prior to giving a description of the improved algorithm, we will describe the traditional approach most commonly used in practical calculations of compressible gas flows through pipelines.

In this approach, the whole pipeline is divided into a certain number of pipeline sections. At the inlet to each pipeline section, the known quantities are the velocity V_1, the total pressure P_{01}, and the stagnation temperature T_{01}. From these known parameters, we can determine all other flow quantities at the pipe inlet: M_1, a_1, λ_1, static parameters T_{1cm}, P_{1cm}, ρ_{1cm}, and the mass rate of the flow Q_1.

We start the calculations by treating the first pipeline section. Over this pipeline section, the total-pressure losses for friction and the resistance local losses $\zeta = \zeta_{mp} + \sum\zeta_{м}$ are to be determined. The friction losses can be calculated by the Al'tshul formula [4]

$$\zeta = 0.11\cdot\frac{L}{D}\cdot\left(\frac{h}{D} + \frac{68}{Re_1}\right)^{0.25}. \tag{14}$$

In (14), L is the length of the pipeline section, $\frac{h}{D}$ is the relative height of inner-surface roughness, and $Re_1 = \frac{V_1 \cdot D}{\nu}$ is the Reynolds number based on the inlet flow quantities.

Next, we use Eq. (13) to calculate the normalized velocity λ_2 at the exit from the pipeline section.

At fixed value $\zeta = const$, one can find the roots of equation (13) using a simple iteration procedure based on the following iterative formula:

$$\lambda_2^{n+1} = \sqrt{\frac{1}{\frac{1}{\lambda_1^2} + 2 \cdot ln(\lambda_1) - \frac{2k}{k+1} \cdot \zeta - ln(\lambda_2^n)}} \tag{15}$$

As an initial approximation, the value $\lambda_2^1 = 1$ can be adopted. Calculations by formula (15) normally converge in 20-30 iterations.

At supersonic flow velocity at the pipeline inlet ($\lambda_1 < 1$) the velocity at the pipeline outlet never exceeds the sonic velocity ($\lambda_2 = 1$). The flow in which the gas velocity at the pipeline outlet reaches the sonic velocity is called the critical flow. At a given pipeline geometry and at a given value of the pressure difference between the inlet and outlet there exists a certain value of critical velocity at the inlet to the pipeline section ($\lambda_{1\times}$) such that at the outlet there forms a critical flow with attained sonic velocity,

$$\frac{1}{\lambda_{1\times}^2} - 1 - ln\frac{1}{\lambda_{1\times}^2} = \frac{2k}{k+1} \cdot \zeta . \tag{16}$$

If the normalized velocity at the pipeline inlet is lower than the critical velocity ($\lambda_1 < \lambda_{1\times}$), then, at the given length of the pipeline section, the velocity at the pipeline outlet is subsonic ($\lambda_1 < 1$). By increasing the pipe length at fixed conditions at the inlet to the pipeline section, one can obtain the sonic velocity at the pipeline outlet.

Now, the inlet velocity will be the critical one for the given pipe length. On further increase of pipe length, the flow velocity at the pipe outlet will remain sonic; yet, the inlet velocity will start decreasing, and at any time this velocity will be equal to the critical velocity for the given values of the pipe length and resistance coefficient.

From known values of λ_2, P_{02}, and T_{02}, all other flow quantities at the pipe outlet, namely, M_2, a_2, static parameters T_2, P_2, ρ_2, and the velocity V_2 are to be determined. Then, the obtained values can be used as the initial conditions for calculating the flow in the next pipeline section. Such calculations are to be performed, one by one, for all subsequent pipeline sections. At the outlet from the last section, the solution is to be checked for realization of one of the following two possible cases:

Case 1	M≤1	$P_2=P_{amb}$
Case 2	M=1	$P_2 \geq P_{amb}$

If necessary, the initial approximation for the velocity V_1 at the inlet to the pipeline is to be modified, and the calculations are to be repeated.

On the basis of simple calculations, the approach described above enables revealing the main regularities in the behavior of subsonic flows with friction through pipelines under various combinations of determining parameters. The same approach can also be used in analyzing more complicated supersonic flows not considered in the present article. Due to its simplicity, the analytical solution first reported in [5] was cited in many textbooks, and for many years it has been used in calculating high-velocity compressible gas flows.

Nonetheless, a closer examination shows that the above calculation procedure has some shortcomings, and some of the assumptions laid to its basis need to be refined.

First of all, like in the above-considered case of an incompressible fluid, it is required to evaluate the extension (length) of pipeline section over which integration of Eq. (12) can be performed. This matter was not discussed by G.N. Abramovich [5], who possibly implied that the calculation procedure for the normalized inlet and outlet velocities could be considered applicable to arbitrarily long pipelines. Nonetheless, it can easily be shown that, for different partitions of a long pipeline into several pipeline sections the calculation procedure will yield different results, especially at high flow velocities, i.e. in the case where flow compressibility becomes substantial.

Figure 3 shows calculated data on total pressure obtained on dividing the whole pipeline into different numbers of pipeline sections of identical lengths. Here, the flow quantities at the exit from a previous pipeline section were used as initial parameters to be adopted for the flow at the inlet to the next pipeline section. With increasing the total number of pipeline sections, the data obtained converge to a curve that differs considerably from the curve obtained by treating the whole pipeline as a single pipeline section.

The effect due to the total number of partition elements points to the necessity of direct use of differential equation (12), instead of the analytical solution of this equation, in performing critical calculations where an extreme accuracy in predicting flow quantities would appear necessary. Simultaneously, simple solutions remain obviously applicable to situations where obtaining just rough estimates is required.

By way of example, Figure 4 shows the total-pressure, static-pressure, and Mach-number data for a pipeline comprising 20 pipeline sections calculated on the assumption of a constant stagnation temperature. In spite of the identical values of other input parameters, the calculations yield a total-pressure value at pipeline outlet that notably differs from the outlet pressure value in Figure 2.

The experience gained in practical calculations showed that, in the above procedure, partitioning of the whole pipeline into a small number of pipeline sections always resulted in different values of mass rate of the gas flow at the inlet to the pipe and at the outlet. As a result of summation of inaccuracies, the calculated mass rate of the gas flow at pipeline outlet differed from the mass rate of the gas flow at pipeline inlet, and this difference in some cases could appear amounting to 10 – 15%. In this connection, a hypothesis was put forward that the difference between the total mass fluxes at the ends of the pipeline could possibly be eliminated on abandoning the assumption about a constant stagnation temperature in the flow.

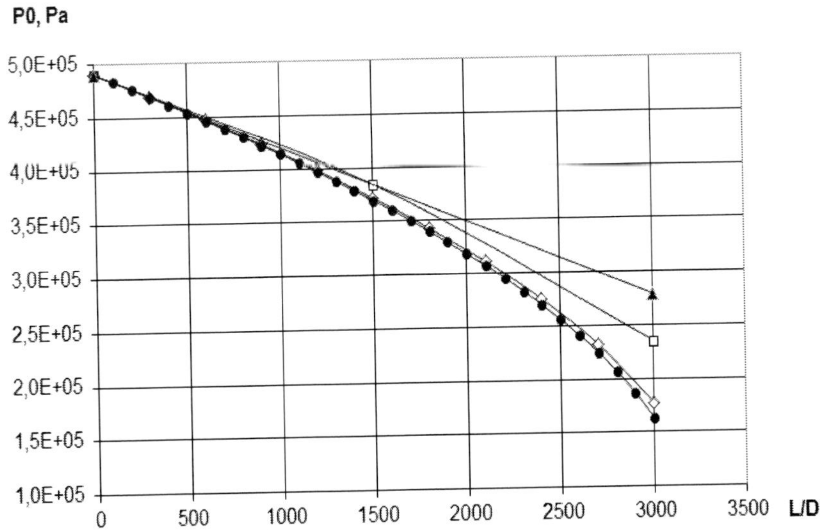

Figure 3. Calculated data on total pressure obtained on dividing the whole pipeline into 1, 2, 20, and 30 pipeline sections.

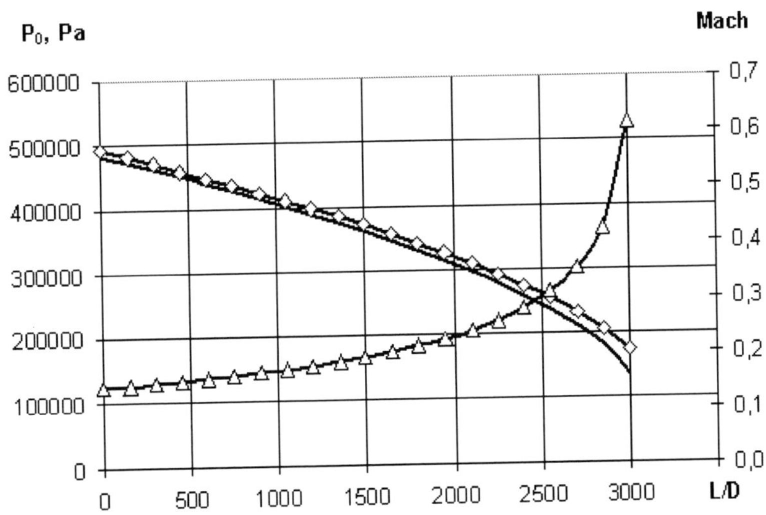

Figure 4. Distributions of total pressure, static pressure, and Mach number obtained on dividing the whole pipeline into 20 pipeline sections. $T_0 = const.$

In the modified statement of the problem, the stagnation temperature at pipeline outlet was calculated from the condition of identical mass rates of the gas flow at pipeline ends. In implementing the latter stategy, prior to using formula (1) we calculated, first, the loss of total pressure ΔP_0 over a current pipeline section and, then, the total pressure at the outlet from this pipeline section $P_{02} = P_{01} - \Delta P_0$. Then, from the known value of λ_2, the flow rate function $q(\lambda_2)$ was calculated as

$$q(\lambda_2) = \left[\frac{k+1}{2}\right]^{\frac{1}{k-1}} \cdot \lambda_2 \cdot \left[1 - \frac{k-1}{k+1} \cdot \lambda_2^2\right]^{\frac{1}{k-1}}. \tag{17}$$

Then, using the obtained value of $q(\lambda)$, from the condition of flow continuity $G_1 = G_2$ we could determine the stagnation temperature at pipeline outlet:

$$T_{02} = \left(\frac{P_{02} \cdot S \cdot q(\lambda_2) \cdot m}{G_1}\right)^2 \tag{18}$$

Here, $m = \sqrt{\frac{k}{R} \cdot \left(\frac{2}{k+1}\right)^{\frac{k+1}{k-1}}}$, and S is the cross-sectional area of the pipe.

Figure 5 shows the data on total and static pressures that were calculated for a pipeline comprising 20 pipeline sections on the assumption of variable stagnation temperature. The obtained pressure and Mach-number values are almost identical to the values in Figure 4; however, here the flow stagnation temperature T_0 at the pipeline outlet has increased by 50 K.

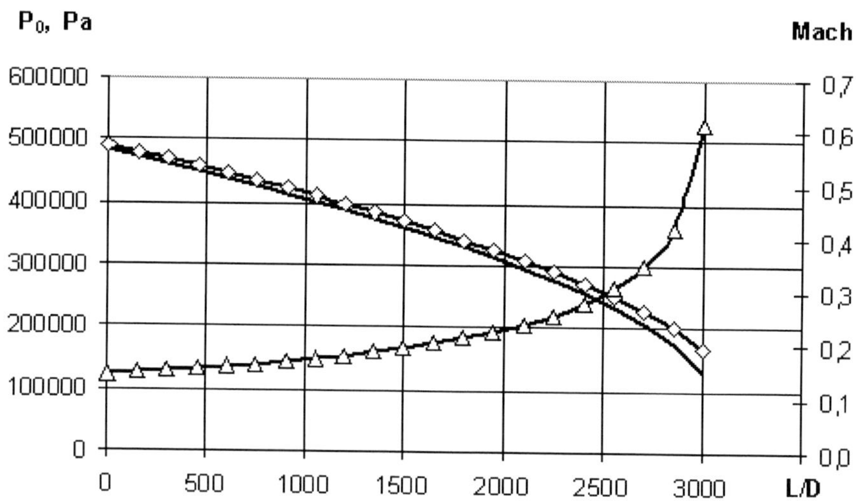

Figure 5. Calculated data on total pressure, static pressure, and Mach number obtained on dividing the whole pipeline into 20 pipeline sections. The data were obtained on the assumption T_0 = *var*.

For improving the accuracy in the calculations based on the assumptions formulated in [5], the calculation procedure can be organized as follows.

The adopted assumptions are

$$dQ = di + VdV = 0,\ dG = 0,\ dF = 0, \text{ and } dL = 0.$$

The starting formulas are

$$\begin{aligned} &\frac{dP}{\rho} + V \cdot dV + dL_{fr} = 0 \\ &dL_{fr} = \Lambda_{fr} \frac{V^2}{2} \cdot \frac{dx}{D} \end{aligned} \quad (19)$$

On combination of both, we obtain:

$$\frac{dP}{\rho} + V \cdot dV + \Lambda_{fr} \frac{V^2}{2} \cdot \frac{dx}{D} = 0 \quad , \quad (20)$$

or

$$\Delta Po + \Lambda_{fr} \cdot \frac{L}{D} \cdot \frac{\rho V^2}{2} = 0, \quad (21)$$

$$dP + \rho \cdot V \cdot dV + \Lambda_{fr} \frac{\rho \cdot V^2}{2} \cdot \frac{dx}{D} = 0 \quad . \quad (22)$$

We integrate equation (22) from some pipe cross-section 1 to some pipe cross-section 2:

$$\int_1^2 dP + \int_1^2 \rho \cdot V \cdot dV + \int_1^2 \Lambda_{fr} \frac{\rho \cdot V^2}{2 \cdot D} dx = 0 \quad (23)$$

Since $\rho \cdot V = \dfrac{G}{F}$ and $V = a_\times \cdot \lambda$, for the difference between the static pressures we readily obtain:

$$P(2) - P(1) = -\frac{G \cdot a_\times}{F} \int_1^2 d\lambda - \frac{\Lambda_{fr} \cdot G \cdot a_\times}{2 \cdot D \cdot F} \int_1^2 \lambda \cdot dx, \quad (24)$$

or

$$P(2) - P(1) = -G \cdot a_\times \cdot \left(\lambda(2) - \lambda(1)\right) - \frac{\Lambda_{fr} \cdot G \cdot a_\times}{2 \cdot D} \int_1^2 \lambda \cdot dx. \quad (25)$$

On the condition that cross-section 2 refers to $x = L$, and cross-section 1, to $x = 0$, we have:

$$P(0) - P(L) = G \cdot a_\times \cdot \left(\lambda(L) - \lambda(0)\right) + \frac{\Lambda_{fr} \cdot G \cdot a_\times}{2 \cdot D} \int_0^L \lambda \cdot dx \quad (26)$$

The difference between the total pressures is

$$dP_0 = \Lambda_{fr} \frac{\rho \cdot V^2}{2 \cdot D} dx = \Lambda_{fr} \frac{G \cdot a_\times}{2 \cdot D \cdot F} \lambda \cdot dx \ , \tag{27}$$

$$P_0(0) - P_0(L) = \Lambda_{fr} \frac{G \cdot a_\times}{2 \cdot D \cdot F} \int_0^L \lambda \cdot dx \, . \tag{28}$$

Here, the total pressure at pipeline outlet should be determined by calculating the integral in (28) via partitioning the whole pipe length L. in a large amount of small intervals.

Simultaneously, with known values of the static pressure and normalized velocity λ at the final cross-section, we can determine the total pressure using the following formula for isentropic flow assuming that the flow at the final cross-section is isentropic:

$$P_0 = P \cdot (1 - \frac{\gamma - 1}{\gamma + 1} \lambda^2)^{\frac{\gamma}{1-\gamma}} . \tag{29}$$

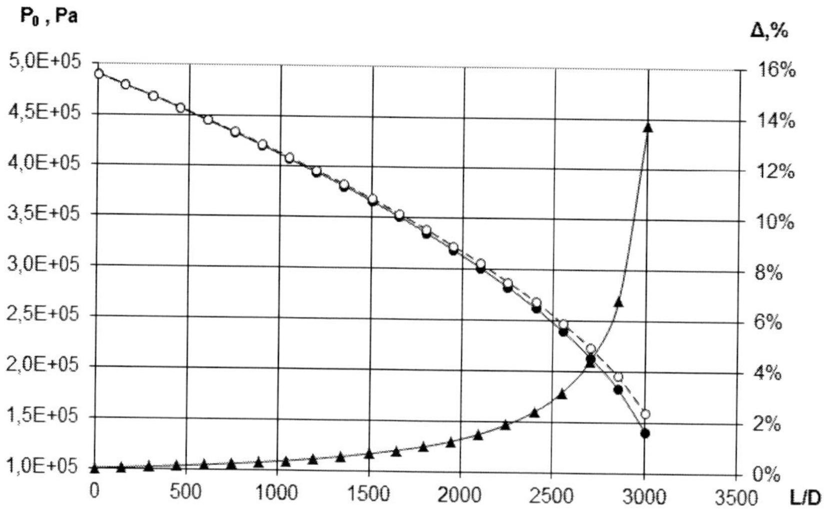

Figure 6. Comparison of the two algorithms for calculating the distribution of total pressure over pipeline length.

Open symbol in Figure 6 shows the distribution of total pressure over pipeline length calculated via numerical integration of Eq. (28). Full symbols in the same figure show the total-pressure values calculated by formula (29).

It is seen that in the considered example there exists a noticeable difference Δ (up to 14% at the pipe outlet) between the total-pressure values calculated by the two algorithms. Despite the various measures taken to improve the integration accuracy, this difference could not be diminished, even on adoption of the assumption of variable stagnation temperature in the flow.

The obtained difference in the values of total pressure determined by the two methods has generated a need in revisiting the previously adopted assumptions used in reducing the general equation (11) to abridged equation (12). Considering the treated case of the gas flow through pipeline, we can state that, here, there is no supply of mechanical work to the flow, there is no rate action, and there is no external supply of heat, so that we have

$$\frac{dG}{G}=0,\ \frac{dL}{a^2}=0, \text{ and } dQ_{out}=0.$$

Simultaneously, there exists a supply of internal heat due to the work of friction forces and, therefore, the net supply of heat to the flow is nonzero ($dQ \neq 0$) even in the absence of an external supply of heat. This assumption entails possible change (namely, increase) of the total enthalpy, or stagnation temperature of the flow, over the pipeline length. In the latter case, the energy equation reads:

$$dQ = dQ_{in} = dL_{fr} = \Lambda_{fr}\frac{V^2}{2}\cdot\frac{dx}{D} = di + VdV. \tag{30}$$

In a pipe flow, a boundary layer always forms on the pipe wall; in one-dimensional case this boundary layer can be allowed for as a reduction of the effective cross-sectional area of the pipe by the displacement thickness. Hence, the geometric action needs to be additionally retained in equation (11) ($\frac{dF}{F}\neq 0$), which also involves the mechanical work of friction forces ($\frac{k}{a^2}dL_{fr}$).

With the above assumptions, the action equation is

$$(M^2-1)\frac{dV}{V}=\frac{dF}{F}-\frac{k-1}{a^2}(dL_{fr})-\frac{k}{a^2}dL_{fr}, \tag{31}$$

or

$$(M^2-1)\frac{dV}{V}=\frac{dF}{F}-\frac{2k-1}{a^2}dL_{fr}. \tag{32}$$

Now, the iteration algorithm can be organized as follows. First of all, given the value of the flow velocity at the inlet to the pipeline, we can determine the mass flow rate G_1. Then, formula (1) is to be used to determine the loss of total pressure dP_0 over the first pipeline section,

$$dP_0=\zeta\cdot\frac{\rho V^2}{2}=\Lambda_{fr}\cdot\frac{dx}{D}\cdot\frac{\rho V^2}{2}, \tag{33}$$

and also the total pressure at the exit from this pipeline section:

$$P_{02} = P_{01} - dP_0 . \tag{34}$$

Then, we calculate the increase of flow enthalpy due to the work of friction forces,

$$di_0 = dL_{fr} = \Lambda_{fr} \frac{V^2}{2} \cdot \frac{dx}{D}, \tag{35}$$

and also the increase of stagnation temperature,

$$dT_{02} = di_0 / C_P . \tag{36}$$

Here, C_P is the specific heat of air at constant pressure (in the present calculations, the value $C_P = 1005$ kJ/(kg·K) was adopted).

At the first calculation step, the change of pipe cross-sectional area over the current pipeline section was assumed to be zero ($\frac{dF}{F} = 0$).

Then, the change of flow velocity over this pipeline section could be determined from the following simple equation:

$$(M^2 - 1)\frac{dV}{V} = -\frac{2k-1}{a^2} dL_{fr} . \tag{37}$$

Next, the outlet velocity V_2 could be calculated as

$$V_2 = V_1 + dV . \tag{38}$$

Now, from known values of stagnation parameters and flow velocity we can calculate the static parameters (pressure, temperature, density) and determine the mass flow rate G_2 at the exit cross-section. As a rule, we have $G_2 \neq G_1$.

For matching the flows, we apply a correction to the exit cross-sectional area:

$$\frac{dF}{F} = \frac{G_2 - G_1}{G_1} = \frac{G_2}{G_1} - 1 . \tag{39}$$

Then, the change of flow velocity over the pipeline section can be determined using the following equation:

$$(M^2 - 1)\frac{dV}{V} = \frac{dF}{F} - \frac{2k-1}{a^2} dL_{fr} . \tag{40}$$

The new value of outlet velocity gives a new value of G_2 and, then, the iteration procedure can be repeated till a constant (to a required accuracy) value of $\frac{dV}{V} = const$ or $\frac{dF}{F} = const$ is obtained.

By way of example, Figure 7 shows the distributions of stagnation pressure and Mach number in the above-considered pipeline. Symbols show the data that were calculated on the assumption of a constant stagnation temperature. The data calculated on the assumption of variable stagnation temperature and variable pipe cross-sectional area are shown with lines.

It is seen that in the latter case the decrease in stagnation pressure is more pronounced in comparison with the former case, in which the heat supply and the variation of pipe cross-sectional area were assumed to be zero (see Figures 5 and 6). The flow velocity reaches the speed of sound much more rapidly (at the distance $\frac{L}{D} = 2700$).

The proposed calculation algorithm was tested for adequacy using the experimental data of [9], where results of accurate measurements of the distribution of static pressure along a pipeline prepared from weldless drawn, technically smooth cylindrical copper pipe of length 1979 mm with inner diameter 24.95 mm were reported, and a detailed description of the experimental facility used was given. In the study of [9], the following measurements were carried out. The mass rate of the air flow through the pipeline was measured with the help of a metering nozzle installed at the entrance to the pipeline. Simultaneously, the inlet stagnation temperature of air flow was measured with 0.1% accuracy.

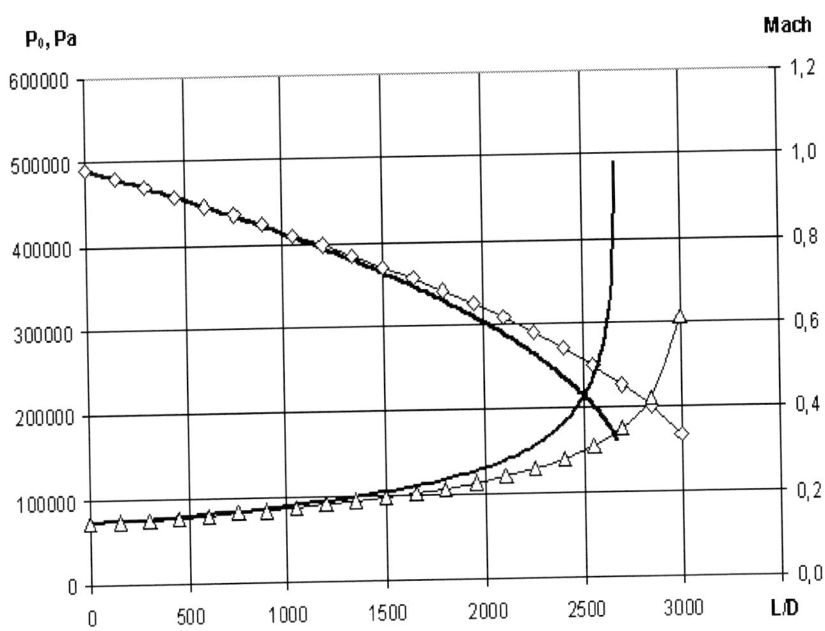

Figure 7. Comparison of calculated data on the distribution of total pressure and Mach number over the pipeline length.

The flow temperature at pipeline outlet was not measured. The total pressure at the inlet to the pipeline was measured with a standard manometer that had scale-division value 0.017 atm. Static-pressure taps were provided at 19 pipe cross-sections over the pipeline length. The accuracy of static-pressure measurements performed at the various pipe cross-sections was ranging from 0.5 to 0.7% depending on the absolute value of measured pressure.

In all experiments, the pipe flow was discharged directly to atmosphere. The examined regimes differed in terms of the total-pressure value and, hence, in the mass rate of the air flow. The range of the mass rate of the air flow was from 0.103 to 0.258 kg/s, and the Reynolds number at the pipeline inlet varied over the pipe diameter from $2.9 \cdot 10^5$ to $7.3 \cdot 10^5$.

In experiments with moderate inlet pressures and moderate mass rates of the air flow (below 0.208 kg/s), the flow at the pipe outlet was subsonic, and the static pressure was strictly equal to atmospheric pressure. In the experiments with enhanced pressures at the pipe inlet the static pressure at the pipe outlet was matched to the sound velocity.

In treating their experimental data, the authors of [9] used the procedure based on the assumption of a constant stagnation temperature. Dividing the pipeline into pipeline sections in between the pipe cross-sections at which static-pressure values were measured, these authors calculated the value of the friction coefficient over each pipeline section and plotted the friction-coefficient values versus Mach number. Substantial reduction of friction coefficient observed at Mach numbers M < 0.9 was attributed to inapplicability of one-dimensional model under transonic flow conditions.

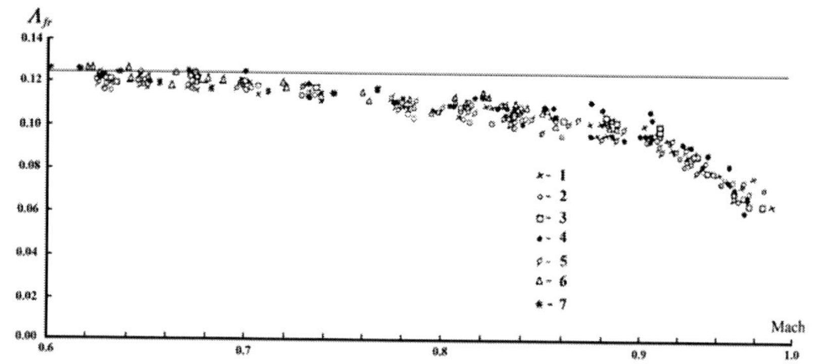

Figure 8. Summary plot of all experimental data on the dependence $\Lambda_{fr} = f(M)$. The designations are explained in Table 1.

Table 1 Designations used in Figure 8

Designation	Mass rate of air flow *G, kg/s*	Reynolds number $Re_D \cdot 10^{-5}$
1	0.255-0.280	7.20-7.30
2	0.247-0.250	7.00-7.10
3	0.236	6.60-6.65
4	0.222	6.25
5	0.209	5.87
6	0.194	5.47
7	0.163	4.58

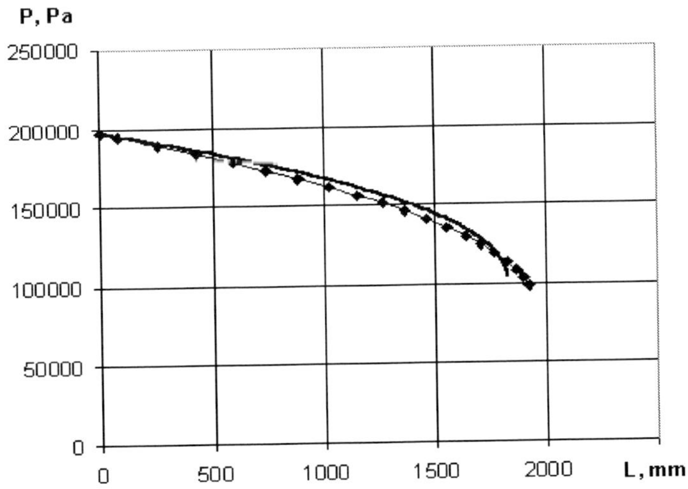

Figure 9. Variation of static pressure over the length of the experimental pipeline. Measured data are shown as points, and the solid line shows calculated data.

The experimental data of [9] were compared to the data calculated by the procedure decribed in the present publication. In the calculations, the friction coefficient was assumed constant, equal to Λ_{fr} = 0.125. The calculated static-pressure values proved to be in a fairly good agreement with the experimental data, although the sonic velocity (M = 1) was attained over a smaller length (shorter approximately by 100 mm, or by 5% of the total pipe length). In the calculations, the stagnation temperature increases to T_o = 317 K, contrary to the normally adopted assumption of a constant stagnation temperature, T_o = 300 K. The increase in flow enthalpy is consistent with the work done by friction forces; i.e. the energy equation is fulfilled here. In the case under consideration, due to the growth of boundary layer on the pipe wall the pipeline cross-sectional area has decreased by 4.8% (the diameter of the one-dimensional flow has decreased from 24.95 mm at pipeline inlet to 24.24 mm at pipeline outlet).

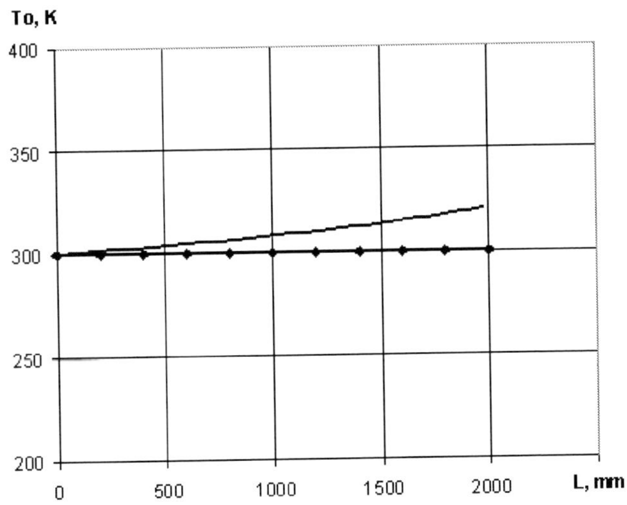

Figure 10. Variation of stagnation pressure over the length of the experimental pipeline. The points show the line of constant stagnation temperature. The solid line shows calculated data.

References

[1] E.I. Idel'chik, Handbook of Hydraulic Resistance, M.O. Shteinberg (editor), third edition, revised and supplemented, Moscow, *Mashinostroenie,* 1992. – 672 p.

[2] I. Nikuradze, *Regularities of turbulent motion in smooth pipelines, in: Problems in Turbulence* / M.A. Velikanov and N.G. Shveikovskii (editors), Moscow, 1936. – pp. 75-150.

[3] A.D. Al'tshul, *Hydraulic Losses due to Friction in Pipelines*, Moscow, 1963. – 256 p.

[4] A.D. Al'tshul, *Hydraulic Resistances,* Moscow, 1982. – 224 p.

[5] G.N. Abramovich, Applied Gas Dynamics, in 2 parts, Pt. 1: Manual for Students of Technical Colleges, third edition, revised and supplemented, Moscow, Nauka, *Glavn. Red. Fiz.-Mat. Literat.,* 1991. – 600 p.

[6] A.V. Teplov, Regularities of pressure flows in pipelines, in: *Similarity Theory and Its Application in Heat Engineering, Proc. of the Moscow Institute of Railway Engineering,* Moscow, 1961. – p. 72-78.

[7] F.S. Voronin, Effect of compressibility on the friction resistance coefficient in turbulent gas flow, *Inzh.-Fiz. Zhurn.,* 1959, Vol. 2, No. 11, pp. 81-85.

[8] L.A. Vulis, On the sonic transition in gas flow, in: *Thermodynamics of Gas Flows,* Moscow, Energoizdat, 1950.

[9] A.F. Gandel'sman, A.A. Gukhman, N.V. Il'yukhin, L.N. Naurits, Investigation into resistance coefficient in near-sonic flow, Pts. I and II, *Zh. Tekhn. Fiz.,* Vol. 24, Iss. 12, 1954. – pp. 2221-2249.

In: Pipelines: Design, Applications and Safety
Editors: M. G. Rivero et al. pp. 225-257
ISBN 978-1-62100-178-2

Chapter 9

PIPE JOINT STRENGTH DESIGN AND SERVICE LIFE OF A PSEUDO HOMOGENEOUS ALLWELD METAL UNDER CONTINUUM FLOW

*Joseph I. Achebo**
Department of Production Engineering, Faculty of Engineering
University of Benin, Benin City, Edo State, Nigeria

ABSTRACT

Effluents emanating from Petroleum reservoirs channeled through steel pipelines comprise of oil, water, gas, and other projectiles suspended in fluid, incessantly flowing in a turbulent manner, in a statistically discernable continuum. Therefore, Pipelines tend to suffer corrosion and wear having endured continuous impact, particularly at various weld joints within their internal walls. These joints have a subtle, but significant difference in composition from the parent metal, and are actually a concoction of the parent metal alloys with other weld elements in varying amounts, hence its pseudo homogenous nature. The structural integrity and design life of these pipelines invariably depends on these welds possessing equivalent strength and toughness potentials as the parent metal. In order to reduce corrosion and wear to the barest, and geared towards meeting increasing demands of operating pressures and loads, the individual effects of each alloying and weld element must be identified, and a predictable, maneuverable protocol attained. Statistical methods such as the Taguchi Method with Grey Rational Analysis could be applied to optimize the welding process parameters so that elements that increase longevity are incorporated to obtain signature uniquely crafted weld chemical compositions. Tensile, toughness, hardness tests, and micro structure analysis are used to assess weld quality and the veracity of the applied statistical methods is thereafter compared with measured values. The mcthods applied proved successfully, that the welding process parameters as well as the various types and proportions of the alloying elements that make up the chemical composition of the weld metal can be satisfactorily optimized and improved.

---

* E-mail: josephachebo@yahoo.co.uk.

Keywords: Correlation analysis, Kendall's coefficient of concordance, Taguchi method with grey relational analysis, welding process parameters, weld strength.

1. Introduction

The oil and gas industry relies heavily on the use of pipelines to convey crude oil directly from underground reservoirs to other facilities, usually for refining, storage, or shipping (Achebo, 2008). A combination of gas, oil, water, and suspended particles (mostly sandstone) are transported from the reservoir to the tank farm through the separator tank by the feeder pipes. During this incessant flow process these particles of indeterminate size and origin are churned together giving rise to random impact between either of two contacting surfaces; being the particle-pipeline surface, and/or a particle-particle surface. Several studies have shown that during the flow of such effluents in the pipeline, the constrictions that occur at welded joints induce a capillary action that increases the pressure and flow rate at these joints (Achebo, 2008:2009).

These flow and pressure conditions make the contacting surfaces to move even more turbulently, contrary to each other, and are consequently forced to cut sharp scratches or grooves, damaging one or both of the surfaces (Achebo, 2008:2009). Welded joints generally have poor strength and toughness capacity compared with the parent metal, the effect of this cutting wear is therefore most significant at these welded joints.

As a result of this wear on the internal walls of the pipeline visible as well as invisible metal loss tends to occur. This leads to the expected consequence of the thinning out of the width of the internal wall. When the internal walls are compromised in this manner they are less able to withstand the high pressures and temperatures inherent in turbulent flow. Thus, whenever pressure surges arise, welded joints which have been most affected by cutting wear could give way, easily leading to pipeline failure.

However, the integrity and life span of a pipeline is of the utmost importance in today's industrial setting (Achebo, 2009). Hopkins (2002) wrote that pipelines are designed, built and operated under well established standards and laws, because the products they convey can pose a significant hazard to the surrounding population and environment, nevertheless the combination of good design, materials, and operating practices, has ensured that transmission pipelines have a good safety record. He emphasized that the prime role of pipeline design is safety.

Transmission pipelines are actually laid by progressively joining composite prefabricated sections. Thus pipelines can extend to infinite lengths. In the 1800s pipe sections were joined by threading the edges and screwing them together. This joining process was difficult to apply when joining pipes with wider diameters, which are ordinarily more susceptible to leaks, especially under high pressure flow conditions.

This uncontrolled leakage problem and its associated hazardous and legal implications has lead to the concept of joining these pipe sections permanently together as opposed to merely screwing them together. Welding is one such process for joining two or more pieces of metal together, and is the trusted method for making such permanent joints, and fixing associated problems such as pipeline wear.

A welded joint is obtained when two clean surfaces are brought into contact with each other and either pressure or heat or both are applied to obtain a bond (Kumar and Kumar,2011). Welding is conducted by making weld deposits in multilayers in the gaps between the steel pipes to be joined. The solidification of the molten weld metals deposited in these gaps, permanently fuses the pipes together (Achebo and Ibhadode, 2008, 2009; Achebo, 2010, 2010*, 2011).

The earliest welding processes geared towards pipeline design were first applied in the 1920s (Hopkins, 2002). Gomez et al. (2008) wrote that weldments experience certain failure modes such as weld failure caused by low strength property, and pipe joint rupture caused by metal loss resulting from wear and erosion corrosion, and pressure surges. Gas transmission pipeline repair by direct deposition of weld metal, or weld deposition repair, is a proven technology that can be applied directly to the area of wall loss (external repair of external wall loss) or to the side opposite to the wall loss (external repair of internal wall loss). There are no apparent limitations to applying this repair technology (Gordon et al. 2004).

The quality of a weld is a key issue in welding, and in many cases a weld must fulfill certain specific quality requirements (Thomsen, 2004). Duane and Miller (1997) said that the design strength of a welded joint or weldment is dependent on either the strength of the weld metal or the parent metal.

This strength is a function of the weld type and the orientation of the weld. Duane and Miller (1997) said that the composition of the weld metal determines the strength of the joint. The authors defined the weld metal strength as the yield and tensile strength of the deposited weld metal, as measured by an all-weld metal tensile coupon extracted from a welded joint made in conformity with the applicable American Welding Society (AWS) filler metal specification.

Gomez et al (2008) wrote that the weld strength for optimum load carrying capacity is assumed to match the strength of the parent metal. Duane and Miller (1997) were of the opinion that matching the weld metal implies that the electrode strength (both yield and tensile strength) is at par with the strength of the parent metal. They further stated that the specifications of the matching weld metal should be the minimum specified yield and tensile strengths equal to, or higher than the minimum specified strength properties of the parent metal.

Thus, when matching the weld metal with the parent metal, the allowable load on both of them should, as nearly as possible, be equal. However, this is not always the case. Welds always have a subtle but significant difference from the parent metal. Welding exposes the electrode metal to the immediate environment and open air, and there is close interaction with the elements of the welding flux itself in the weld pool.

Steel alloys are susceptible to distortion due to their high coefficient of thermal expansion. In some cases, certain steel alloys are quite prone to cracking and reduced corrosion resistance (Kishore et al, 2010). These limitations are even more glaring when these steel alloys are subjected to the welding process (Achebo, 2011).

With the combined effects of the inadvertent introduction of the welding flux elements, the weld, upon solidification, ends up as a random mix, a concoction of wanted and unwanted elements. This pseudo homogenous characteristic of the weld often negatively impacts on the strength qualities of the weld. Considering these limitations, it becomes imperative to optimize the welding protocols and parameters (Achebo, 2011).

Kumar and Kumar (2011) also stated that materials actually fail due to distortions at a molecular level. Where the all weld metal is low in toughness, it lowers its ability to absorb impact energy and resist wear or scratching, as compared to the parent metal. The main goal in trying to arrive at the most appropriate weld specification is in devising welding protocols and parameters that ensure that the steel alloy of the weld metal applied at the joints satisfactorily matches the values attained by the parent metal considering the physical properties of the elements inadvertently introduced (Achebo, 2011*: Achebo and Igbaonogu, 2011).

Other investigators have also done some research in finding new ways of improving weld strength. Ligtenburg (1968), Higgins and Preece (1969), and Clark (1971) observed that transverse fillet welds in tension (where the loading is applied perpendicular to the weld axis) were approximately 60% stronger than longitudinal fillet welds (welds where the line of axis of the applied load is parallel to the axis of the weld). Butler and Kulak (1971) were of the opinion that the increase in loading angle improved the strength yet reduced the weld deformation capacity.

The ultimate capacity of a connection is reached when the ultimate strength and deformation of any element of weld are reached (Butler, Pal, and Kalak, 1972). Kuhlmann et al. (2008), Gunther et al. (2009), and Collin and Johnson (2005), have also investigated the relevance of ultimate strength of fillet welded joints made from various steel grades. Hopkins (2002) observed that most steel pipes used for transporting petroleum products, in the early days, would deform or strain to a great extent before getting to their Ultimate Tensile Strength (UTS) and eventually fail. In recent times, with improved technology and research on the development of steel pipes, pipes do not deform for such a long time span and may only attain about 3% – 5% of their UTS.

The major problem this study intends to address is to produce a weld which possesses a strength that would be able to match the strength of the parent metal. This can be achieved by optimizing the welding process parameters. Process parameters once chosen and applied have a significant effect on the quality and strength of the welded joint. In this study, the Taguchi method with Grey Relational Analysis is used to optimize the process parameters.

The Taguchi method has been found to be a powerful tool used to improve the overall process quality by optimizing the welding process parameters, in such a way that variation is reduced to the barest minimum, and is cost effective as well. Chang et al, (2010) described the Taguchi method as a powerful optimization tool for designing high quality systems that is based on orthogonal array experiments with an optimum setting of process control parameters. Sir Ronald Fisher, a British scientist, studied and proposed a more systematic approach in 1920. He fashioned out the Design of Experiments (DOE) using the Orthogonal Array (OA) in order to maximize the knowledge gained from experimental data (Fisher et al, 1935). The orthogonal array can be thought of as a distillation mechanism through which the engineer's experiment passes (Ealey, 1988). The array allows the engineer to vary multiple variables at one time, obtain the effects that the set of variables has on the average, and determine the dispersion (Clausing and Simpson, 1990). In the 1950s, the OA was later rebranded after detailed research by the researcher Genichi Taguchi, from Japan. The method as it is today is coined the "Taguchi Method" (Eiklenborg et al, 2007). The Taguchi Method is however extensively applied and limited to DOE having single performance characteristics.

Kim and Lee (2009) used the Taguchi experimental design to suggest optimal combinations for process factors of hybrid welding methods to optimize the welding parameters of the resistance spot welding process. Yoon et al (2006) optimized the welding conditions in resistance spot welding of the 7075-T6 aluminum alloy using the Taguchi method (Achebo, 2011*). For multi-variant performance characteristics, an additional statistical tool, the Grey Relational Analysis (GRA) introduced by J. Deng (Hsiao et al, 2007), would be inculcated to provide a clearer horizon for stipulating optimal process parameters. This tool helps transform a multi-variant performance characteristic into the Grey Relational Grade (GRG) which is a single performance characteristic. The optimal solution however obtained from the Grey Relational Grade does not state distinctively which process (independent) parameter significantly affects the control (dependent) parameters, or the result so obtained. Thus, another strong statistical tool being the Analysis of Variance (ANOVA) is employed in classifying the contribution of each process parameter on the average resulting solution (Aneru, 2011). Balasubramanian and Ganapathy (2011) said that the Grey Relational Analysis is a normalization evaluation technique extended to solve the complicated multi-performance characteristics that are optimized effectively. This claim was also supported by Deng (1989) and Lin (2003). Achebo (2011*) cited in his research paper that other researchers who have also adopted this method are Hsiao et al (2008), they optimized the Plasma Arc Welding parameters using the Taguchi Method with the Grey Relational Analysis. Tarng et al (2002) used the Grey based Taguchi Methods to determine the Submerged Arc welding process parameters in hardfacing. Fung (2003) studied the manufacturing process optimization for wear property of fiber reinforced polybutylene terephthalate composites with Grey Relational Analysis. The Taguchi method with Grey Relational Analysis used to optimize the process parameters in this study is shown in section two.

Table 1. Experimental layout using an L_{18} orthogonal array

Expt No.	A Welding current	B Welding time	C Welding speed	D Welding voltage
1	1	1	1	1
2	1	2	2	2
3	1	3	3	3
4	2	1	1	3
5	2	2	2	1
6	2	3	3	2
7	3	1	2	3
8	3	2	3	1
9	3	3	1	2
10	1	1	3	1
11	1	2	1	2
12	1	3	2	3
13	2	1	2	2
14	2	2	3	3
15	2	3	1	1
16	3	1	1	2
17	3	2	2	3
18	3	3	3	1

In this study, the Gas Metal Arc Welding (GMAW) process was used to make weld deposits. This welding process has been widely used for joining both ferrous and non ferrous metals.

2. Application of the Taguchi Method with Grey Relational Analysis

The Taguchi method involves the use of standard orthogonal arrays in its optimization processes. The L_{18} orthogonal array used in this study is shown in Table 1.

Table 2 shows the welding process parameters in their various levels. These process parameters were substituted into Table 1 in their various levels to make weld deposits at the joint gaps of the steel pipes to be joined, eventually producing an average of 18 experimentally iterated results containing the multi response quality characteristics (i.e, root penetration, weld throat width and weld bead appearance). The performance indicators used by Hsiao et al (2008) were adopted for classifying the root penetration, in this study. For the weld root penetration, Table 3 was used as the indicator evaluation criteria.

Table 2.Welding parameters and their levels

Factors	Variables	Level1	Level2	Level3
A	welding current (A)	180	220	260
B	Welding Time (Min)	1.5	2.0	2.5
C	Welding speed (mm/s)	8.5	10.0	12.5
D	Welding voltage (V)	16.0	20.0	24.0

Table 3. Root Penetration Evaluation Levels

Root penetration	Qualitative indicator	Evaluation point
Incomplete Penetration	No adhesion	1
Insufficient Penettation	Depth of weld penetration < 1.0mm	2
Adequate Penetration	Depth of weld penetration 1.0 - 3.9mm	3
Excessive Penetration	Depth of weld penetration ≥ 4.0mm	4
Penetrated Root	Weld spatter due to prolonged heat treatment on the electrode	5

Also for the Weld bead appearance, Table 4 was used as the performance criteria. This is used for evaluating the perceived level of the quality of the bead covering the welded portions of the steel specimens, by the Expert Assessors (Judges).

Table 4. Weld Bead Appearance Evaluation Level

Weld bead appearance	Evaluation Point
Very rough surface	1
Rough surface	2
Smooth surface	3

Table 5. Experimental Result for the Analysis of Welding Characteristics Performance

Expt No.	Weld penetration				Weld throat (mm)				Weld bead appearance			
	1	2	3	4	1	2	3	4	1	2	3	4
1	2	3	1	2	3.0	5.0	3.5	4.0	2	1	3	2
2	1	3	2	3	4.5	2.5	3.0	2.5	3	3	2	1
3	4	2	5	1	5.0	6.0	4.0	4.0	2	3	3	1
4	1	4	3	1	6.0	2.5	4.3	5.0	1	1	2	3
5	4	3	2	2	6.2	4.5	3.5	4.0	2	2	2	1
6	3	4	3	2	2.8	3.7	4.0	2.3	3	2	1	1
7	4	4	2	2	6.3	4.0	5.5	4.5	2	1	1	2
8	3	5	2	3	3.8	5.2	4.8	2.7	3	1	1	3
9	4	3	3	3	4.2	5.4	5.6	3.8	3	1	2	3
10	4	3	2	4	3.5	2.7	4.0	4.2	1	1	2	2
11	5	3	2	3	4.2	4.5	5.2	4.2	2	2	3	1
12	3	4	5	3	5.5	5.0	6.1	3.2	1	1	1	3
13	1	3	4	3	3.5	4.5	5.2	5.3	3	2	2	3
14	4	3	3	5	5.4	5.3	3.8	2.7	2	1	1	1
15	4	4	3	5	4.9	5.4	4.5	3.0	3	3	1	3
16	3	4	3	1	5.2	4.9	5.2	4.5	3	3	2	3
17	4	3	4	2	6.2	6.0	5.1	4.2	3	3	3	1
18	4	3	3	2	4.6	5.0	6.1	7.2	2	3	3	3

When the welding processes have been completed, weld deposits were made on four of the steel specimens which constitute one experimental procedure each. This was repeated 18 times, resulting in a total of 72 specimens used for this study. The three different types of measurements made of the multiresponse quality characteristics are recorded and shown in Table 5.

The next step, was to generate the Signal-to-Noise (S/N) ratios, η for the values in Table 5. This action leads to the generation of Tables 6, 7 and 8. In generating these S/N ratios, Eqs (1-2) were used.

Equation 1 was used for determining the S/N ratios for the root penetration and weld throat measurement.

$$\eta = -10\log_{10}\frac{\bar{y}_i^2}{S_i^2} \tag{1}$$

where S = standard deviation and $\bar{y}$ = average of experimental data for each procedure. The standard deviation is expressed in Eq. (2).

$$S^2 = \frac{\sum_{i=1}^{n}\left(y_i - \bar{y}\right)^2}{n} \tag{2}$$

where,

y_i = evaluation indicator value of root penetration or weld throat width measured for the ith time and n = number of repeated experiments; here n = 4

Equation 1 is a simplified nominal-the-better function, for which the target value is adjusted to the average value.

The weld bead appearance evaluation was done using the higher-the-better type, and the better it is, and its S/N ratio was determined using Eq (3).

$$\eta = -10\log_{10}\left(\frac{1}{n}\sum_{i=1}^{n}\frac{1}{y_i^2}\right) \tag{3}$$

where,

y_i = evaluation indicator value of the bead appearance evaluated at the ith time and n = number of repeated experiment, in this case n = 4.

In Table 6, the signal-to-noise ratio for root penetration was obtained using Eq 1 on Table 5.

Table 6. Signal to noise ratio for weld penetration

Expt no.	y_1	y_2	y_3	y_4	$\bar{y}$	$\bar{y}^2$	$(y_1-\bar{y})^2$	$(y_2-\bar{y})^2$	$(y_3-\bar{y})^2$	$(y_4-\bar{y})^2$	$S_i^2=\frac{1}{n}\sum_{i=1}^{n}(y_i-\bar{y})^2$	$\eta=-10\log_{10}\frac{\bar{y}_i^2}{S_i^2}$
1	2	3	1	2	2	4	0	1	1	0	0.5	−9.0309
2	1	3	2	3	2.25	5.0625	1.5625	0.5625	0.0625	0.5625	0.6875	−8.6709
3	4	2	5	1	3	9	1	1	4	4	2.5	−5.5603
4	1	4	3	1	2.25	5.0625	1.5625	3.0625	0.5625	1.5625	1.6875	−4.7712
5	4	3	2	2	2.75	7.5625	1.5625	0.0625	0.5625	0.5625	0.6875	−10.4139
6	3	4	3	2	3	9	0	1	0	1	0.5	−12.5527
7	4	4	2	2	3	9	1	1	1	1	1	−9.5424
8	3	5	2	3	3.25	10.5625	0.0625	3.0625	1.5625	0.0625	1.1875	−9.4913
9	4	3	3	3	3.25	10.5625	0.5625	0.0625	0.0625	0.0625	0.1875	−17.5077
10	4	3	2	4	3.25	10.5625	0.5625	0.0625	1.5625	0.5625	0.6875	−11.8649
11	5	3	2	3	3.25	10.5625	3.0625	0.0625	1.5625	0.0625	1.1875	−9.4913
12	3	4	5	3	3.75	14.0625	0.5625	0.0625	1.5625	0.5625	0.6875	−13.1079
13	1	3	4	3	2.75	7.5625	3.0625	0.0625	1.5625	0.0625	1.1875	−8.0403
14	4	3	3	5	3.75	14.0625	0.0625	0.5625	0.5625	1.5625	0.6875	−13.1079
15	4	4	3	5	4	16	0	0	1	1	0.5	−15.0515
16	3	4	3	1	2.75	7.5625	0.0625	1.5625	0.0625	3.0625	1.1875	−8.0403
17	4	3	4	2	3.25	10.5625	0.5625	0.0625	0.5625	1.5625	0.6875	−11.8649
18	4	3	3	2	3	9	1	0	0	1	0.5	−12.5527

Table 7. Signal to noise ratio for weld throat widths

Expt no.	y_1	y_2	y_3	y_4	$\bar{y}$	$\bar{y}^2$	$(y_1-\bar{y})^2$	$(y_2-\bar{y})^2$	$(y_3-\bar{y})^2$	$(y_4-\bar{y})^2$	$S_i^2=\frac{1}{n}\sum_{i=1}^{n}(y_i-\bar{y})^2$	$\eta=-10\log_{10}\frac{\bar{y}_i^2}{S_i^2}$
1	3	5	3.5	4	3.875	15.0156	0.7656	1.2656	0.1406	0.0156	0.5469	−14.3863
2	4.5	2.5	3	2.5	3.125	9.7656	1.8906	0.3906	0.0156	0.3906	0.3719	−11.6239
3	5	6	4	4	4.75	22.5625	0.0625	1.5625	0.5625	0.5625	0.6875	−15.1611
4	6	2.5	4.3	5	4.45	19.8025	2.4025	3.8025	0.0225	0.3025	1.6325	−10.8387
5	6.2	4.5	3.5	4	4.55	20.7025	2.7225	0.0025	1.1025	0.3025	1.0325	−13.0213
6	2.8	3.7	4	2.3	3.2	10.24	0.1600	0.2500	0.6400	0.8100	0.4650	−13.4285
7	6.3	4	5.5	4.5	5.075	25.7556	1.5006	1.1556	0.1806	0.3306	0.7919	−15.1220
8	3.8	5.2	4.8	2.7	4.125	17.0156	0.1056	1.1556	0.4556	2.0306	0.9369	−12.5915
9	4.2	5.4	5.6	3.8	4.75	22.5625	0.3025	0.4225	0.7225	0.9025	0.5875	−15.8438
10	3.5	2.7	4	4.2	3.6	12.96	0.0100	0.8100	0.1600	0.3600	0.3350	−15.8756
11	4.2	4.5	5.2	4.2	4.525	20.4756	0.1056	0.0006	0.4556	0.1056	0.1669	−20.8878
12	5.5	5	6.1	3.2	4.95	24.5025	0.3025	0.0025	1.3225	3.0625	1.1725	−13.2010
13	3.5	4.5	5.2	5.3	4.625	21.3906	1.2544	0.0144	0.3364	0.4624	0.5169	−16.1682
14	5.4	5.3	3.8	2.7	4.3	18.49	1.1200	1.0000	0.2500	2.5600	1.2550	−11.6829
15	4.9	5.4	4.5	3.0	4.45	19.8025	0.2025	0.9025	0.0025	2.1025	0.8025	−13.9227
16	5.2	4.9	5.2	4.5	4.95	24.5025	0.0625	0.0025	0.0625	0.2025	0.0825	−24.7276
17	6.2	6.0	5.1	4.2	5.375	28.8906	0.6806	0.3906	0.0756	1.3806	0.6319	−16.6011
18	4.6	5	6.1	7.2	5.725	32.7756	1.2656	0.5256	0.1406	2.1756	1.0269	−15.0402

Table 8. Signal to noise ratio for weld bead appearance

Expt no.	y_1	y_2	y_3	y_4	$\frac{1}{y_1^2}$	$\frac{1}{y_2^2}$	$\frac{1}{y_3^2}$	$\frac{1}{y_4^2}$	$\frac{1}{4}\sum_{i=1}^{n}\frac{1}{y_i^2}$	$\eta=-10\log_{10}\left(\frac{1}{4}\sum_{i=1}^{n}\frac{1}{y_i^2}\right)$
1	2	1	3	2	0.25	1	0.1111	0.25	0.4028	3.9491
2	3	3	2	1	0.1111	0.1111	0.25	1	0.3681	4.3403
3	2	3	3	1	0.25	0.1111	0.1111	1	0.3681	4.3403
4	1	1	2	3	1	1	0.25	0.1111	0.5903	2.2893
5	2	2	2	1	0.25	0.25	0.25	1	0.4375	3.5902
6	3	2	1	1	0.1111	0.25	1	1	0.5903	2.2893
7	2	1	1	2	0.25	1	1	0.25	0.6250	2.0412
8	3	1	1	3	0.1111	1	1	0.1111	0.5556	2.5524
9	3	1	2	3	0.1111	1	0.25	0.1111	0.3681	4.3403
10	1	1	2	2	1	1	0.25	0.25	0.6250	2.0412
11	2	2	3	1	0.25	0.25	0.1111	1	0.4028	3.9491
12	1	1	1	3	1	1	1	0.1111	0.7778	1.0913
13	3	2	2	3	0.1111	0.25	0.25	0.1111	0.1806	7.4328
14	2	1	1	1	0.25	1	1	1	0.8125	0.9018
15	3	3	1	3	0.1111	0.1111	1	0.1111	0.3333	4.7716
16	3	3	2	3	0.1111	0.1111	0.25	0.1111	0.1458	8.3624
17	3	3	3	1	0.1111	0.1111	0.1111	1	0.3333	4.7716
18	2	3	3	3	0.25	0.1111	0.1111	0.1111	0.1458	8.3624

Also, Table 7 contains the signal-to-noise ratio for weld throat widths shown in Table 5. The signal-to-noise ratio shown in Table 7 was obtained using Eq 1.

Table 9. Multi-response signal-to-noise ratio for the welding performance

Expt no.	Root penetration η	Throat width(mm) η	Weld bead appearance η
1	−9.0309	−14.3863	3.9491
2	−8.6709	−11.6239	4.3403
3	−5.5630	−15.1611	4.3403
4	−4.7712	−10.8387	2.2893
5	−10.4139	−13.0213	3.5902
6	−12.5527	−13.4285	2.2893
7	−9.5424	−15.1220	2.0412
8	−9.4913	−12.5915	2.5524
9	−17.5077	−15.8438	4.3403
10	−11.8649	−15.8756	2.0412
11	−9.4913	−20.8878	3.9491
12	−13.1079	−13.2010	1.0913
13	−8.0403	−16.1682	7.4328
14	−13.1079	−11.6829	0.9018
15	−15.0515	−13.9227	4.7716
16	−8.0403	−24.7276	8.3624
17	−11.8649	−16.6011	4.7716
18	−12.5527	−15.0402	8.3624

Lastly, the signal-to-noise ratio of the scores made in respect of the weld bead appearance by the Judges, was obtained using Eq. 2 and this is shown in Table 8. The summary of the multiresponse signal-to-noise ratio for the welding parameters considered in Tables 6 - 8, is shown in Table 9.

2.1. Grey Relational Analysis

In this analysis, the S/N ratio data in Table 9 were normalized using Eq (4). The normalized S/N ratio, x_{ij} for the ith performance characteristic in the jth experiment can be expressed as

$$x_{ij} = \frac{\eta_{ij} - min_j\eta_{ij}}{max_j\eta_{ij} - min_j\eta_{ij}} \tag{4}$$

where: η_{ij} is the S/N ratio for each parameter in each experiment.

$min_j\eta_{ij}$ is the minimum S/N ratio for each parameter in Table 9.

$max_j\eta_{ij}$ is the maximum S/N ratio for each parameter in Table 9.

For root penetration, $\min_j\eta_{ij} = -17..5077$ and $\max_j\eta_{ij} = -4.7712$

For weld throat width, $\min_j\eta_{ij} = -24.7276$ and $\max_j\eta_{ij} = -10.8387$

For weld bead appearance, $\min_j\eta_{ij} = 0.9018$ and $\max_j\eta_{ij} = 8.3624$

The normalized S/N ratios are presented in Table 10. In Table 10, the first row shows the best normalized S/N ratios. The larger normalized S/N ratio corresponds to the better performance and the best normalized S/N ratio is equal to unity.

Table 10. Normalized Signal-to-Noise Ratio for Root Penetration, Weld throat width and Weld bead appearance

Expt No.	Weld Penetration x_{ij}	Weld Throat x_{ij}	Weld bead Appeearance x_{ij}
Ideal sequence	1	1	1
1	0.6656	0.7446	0.4085
2	0.6938	0.9435	0.4609
3	0.9378	0.6888	0.4609
4	1	1	0.1860
5	0.5570	0.8429	0.3604
6	0.3890	0.8135	0.1860
7	0.6254	0.6916	0.1527
8	0.6294	0.8738	0.2212
9	0	0.6396	0.4609
10	0.4430	0.6373	0.1527
11	0.6294	0.2765	0.4085
12	0.3454	0.8299	0.0254
13	0.7433	0.6163	0.8754
14	0.3455	0.9392	0
15	0.1929	0.7780	0.5187
16	0.7433	0	1
17	0.4430	0.5851	0.5187
18	0.3890	0.6975	1

Quality loss function or Difference data series, $\left|x_i^0 - x_{ij}\right|$

The quality loss function values of the Normalized Signal-to-Noise for root penetration, weld throat width and weld bead appearance in Table 10 were determined.

x_i^0 is the Ideal normalized signal-to-noise ratio for the ith performance characteristics, is equal to 1.0. The calculated values for the quality loss function is shown in Table 11.

Table 11. Quality Loss Estimate

Expt no.	Root Penetration $\lvert x_i^0 - x_{ij} \rvert$	Weld Throat Width $\lvert x_i^0 - x_{ij} \rvert$	Weld Bead Appearance $\lvert x_i^0 - x_{ij} \rvert$
1	0.3344	0.2554	0.5915
2	0.3062	0.0565	0.5391
3	0.0622	0.3112	0.5391
4	0	0	0.8140
5	0.4430	0.1571	0.6396
6	0.6110	0.1865	0.8140
7	0.3746	0.3084	0.8473
8	0.3706	0.1262	0.7788
9	1	0.3604	0.5391
10	0.3570	0.3627	0.8473
11	0.3706	0.7235	0.5915
12	0.6546	0.1701	0.9746
13	0.2567	0.3837	0.1246
15	0.6545	0.0608	1
16	0.8071	0.2220	0.4813
16	0.2567	1	0
17	0.5570	0.4149	0.4813
18	0.6110	0.3025	0

2.2. Grey Relational Coefficient ξ_{ij}

The Grey relational coefficient is determined to show the relationship between the ideal (best) and actual normalized S/N ratio. The Grey relational coefficient ξ_{ij} for the ith performance characteristic in the jth experiment can be expressed as

$$\xi_{ij} = \frac{min_i min_j |x_i{}^0 - x_{ij}| + \zeta max_i max_j |x_i{}^0 - x_{ij}|}{|x_i{}^0 - x_{ij}| + \zeta max_i max_j |x_i{}^0 - x_{ij}|} \tag{5}$$

where

$x_i{}^0$ = Ideal normalized S/N ratio for the ith performance characteristic = 1.0

x_{ij} = The normalized S/N ratio obtained

ζ = Distinguishing coefficient which has value $0 \leq \zeta \leq 1$ but usually taken as 0.5

ξ_{ij} = The grey relational coefficient.

$Min_i Min_j \left| x_i^o - x_{ij} \right|$ = the least value of $x_i^o - x_{ij}$ values

$Max_i Max_j \left| x_i^o - x_{ij} \right|$ = the maximum value of the $x_i^o - x_{ij}$ values

Equation(5) is applied to Tables 10-11 to obtain Table 12.

Table 12. Grey Relational Coefficient

Expt No.	Root Penetration ξ_{ij}	Throat Width ξ_{ij}	Weld Bead Appearance ξ_{ij}
1	0.5992	0.6619	0.4581
2	0.6202	0.8985	0.4812
3	0.8894	0.6169	0.4812
4	1	1	0.3805
5	0.5302	0.7609	0.4388
6	0.4500	0.7283	0.3805
7	0.5717	0.6185	0.3711
8	0.5743	0.7985	0.3910
9	0.3333	0.5811	0.4812
10	0.4730	0.5796	0.3711
11	0.5743	0.4087	0.4581
12	0.4331	0.7462	0.3391
13	0.6608	0.5658	0.8005
14	0.4331	0.8976	0.3333
15	0.3825	0.6925	0.5095
16	0.6608	0.3333	1
17	0.4730	0.5465	0.5095
18	0.4500	0.6231	1

For all three characteristics; Root penetration, Weld throat width and Weld bead appearance, in this case, $\min_i \min_j \left| x_i^0 - x_{ij} \right| = 0$ and $\max_i \max_j \left| x_i^0 - x_{ij} \right| = 1$. The calculated values of the grey relational coefficient are shown in Table 12.

2.3. Grey Relational Grade and its Order

Following the above, is to convert the Grey Relational coefficients of each experiment shown in Table 12 into the Grey Relational Grade in Table 13 by using a weighing method.

The overall evaluation of the multiple performance characteristics is based on the Grey Relational Grade which is obtained using Eq (6)

$$\gamma_j = \frac{1}{m}\sum_{i=1}^{m} w_i \xi_{ij} \tag{6}$$

It is assumed that:

$w_1 = w_2 = w_3 = 1$,

where; γ_j = the Grey relational grade for the jth experiment,

w_i =the weighting factor for the ith performance characteristic, and m = the number of performance characteristics (in this case, root penetration, weld throat width and bead appearance, ie, m = 3).

Equation(6) is applied to Table 12 to obtain Table 13.

From Table 13, the best welding process parameters are the ones used to conduct experiment 4.

Table 13. Grey Relational Grade Υ_j and its Order

Expt No.	Grey Relational Grade γ_j	*Order*
1	0.5731	9
2	0.6666	4
3	0.6623	6
4	0.7935	1
5	0.5766	8
6	0.5196	13
7	0.5204	12
8	0.5879	7
9	0.4652	18
10	0.4746	17
11	0.4804	16
12	0.5061	15
13	0.6757	3
14	0.5527	10
15	0.5282	11
16	0.6647	5
17	0.5097	14
18	0.6910	2

Table 14, shows clearly the relationship between the orthogonal array and the Grey Relational Grade.

Table15 is obtained from Table 14 by considering the orthogonal array setup (in Table1), this requires the taking of the respective averages of each level for the process parameters from Table 14. Table 15 shows the optimal levels of each of the welding parameters.

From Table 15, the optimum combination of welding process parameters is $A_2B_1C_1D_3$. This result confirms the prediction made from Table 13, that Experiment 4 has the best process parameters.

Optimal process parameters consist of the Welding current: 220A, Welding time: 1.5 min, Wclding speed: 8.5mm/s, Welding voltage: 24.0V.

The layout of Table 15 is expressed, for further clarity by presenting it in a graphical form. This leads to the construction of Figure 1.

Figure 1 shows the Grey relational grade, where the dashed (center) line is the value of the total mean of the Grey relational grade. Basically, the larger the Grey relational grade, the better the multiple performance characteristics.

Table 14. Experimental setup and Result (Orthogonal Array – Grey Relational Grade Relationship)

Expt No.	Process parameters A	B	C	D	Grey Relational Grade γ_j
1	1	1	1	1	0.5731
2	1	2	2	2	0.6666
3	1	3	3	3	0.6623
4	2	1	1	3	0.7935
5	2	2	2	1	0.5766
6	2	3	3	2	0.5196
7	3	1	2	3	0.5204
8	3	2	3	1	0.5879
9	3	3	1	2	0.4652
10	1	1	3	1	0.4746
11	1	2	1	2	0.4804
12	1	3	2	3	0.5061
13	2	1	2	2	0.6757
14	2	2	3	3	0.5527
15	2	3	1	1	0.5282
16	3	1	1	2	0.6647
17	3	2	2	3	0.5097
18	3	3	3	1	0.6910

Table 15. Response Table for the Grey Relational Grade

Symbol	Process parameter	Level1	Level2	Level3	Maximum – Minimum
A	Welding current	0.5605	*0.6077	0.5732	0.0472
B	Welding time	*0.6170	0.5623	0.5621	0.0549
C	Welding speed	*0.5842	0.5754	0.5814	0.0088
D	Welding voltage	0.5719	0.5787	*0.5908	0.0189

Total mean for the grey relational grade = 0.5805.
* signifies the optimum values.

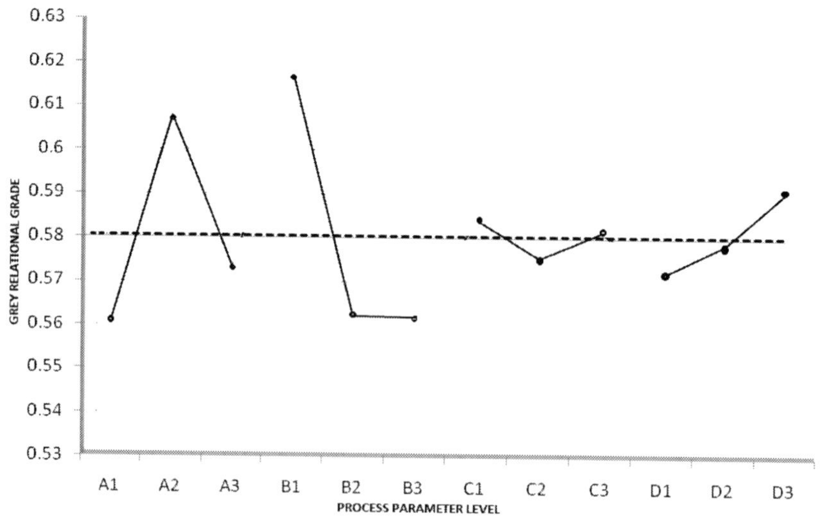

Figure 1. Grey Relational Grade.

2.4. Analysis of Variance (ANOVA)

However, the relative importance of each of these process parameters was determined. In other words, The contribution of each of these welding process parameters to the multiple performance characteristics of the welding method were determined by obtaining the Analysis of Variance (ANOVA) in Table 16.

The determination of ANOVA is outlined as follows.

Sum of squares

i) Sum of squares total SS_T

$$SS_T = \sum_{i=1}^{n} (\gamma_i - \overline{\gamma})^2 \tag{7}$$

and
$$\overline{\gamma} = \frac{1}{N} \sum_{i=1}^{N} \gamma_i \tag{8}$$

where, N = 18

$$\overline{\gamma} = \frac{1}{18}(\gamma_1 + \gamma_2 + \gamma_3 + \ldots + \gamma_{17} + \gamma_{18})$$

$$\overline{\gamma} = \frac{1}{18}(10.4483) = 0.5805$$

$$SS_T = (\gamma_1 - \overline{\gamma})^2 + (\gamma_2 - \overline{\gamma})^2 + (\gamma_3 - \overline{\gamma})^2 + \ldots + (\gamma_{17} - \overline{\gamma})^2 + (\gamma_{18} - \overline{\gamma})^2$$

$$SS_T = 0.14387$$

ii) Sum of squares individual SS_P

$$SS_P = \frac{1}{t} \sum_{j=1}^{k} (\sum \gamma_i)_j^2 - \frac{1}{N} \left[\sum_{i=1}^{N} \gamma_i \right]^2 \tag{9}$$

Where : t = number of times of each level repetition

k = number of levels

$$\frac{1}{N} \left[\sum_{i=1}^{N} \gamma_i \right]^2 = \frac{1}{18} [10.4483]^2 = 6.06483$$

$$SS_A = \frac{1}{6} \left[(3.3631)^2 + (3.6463)^2 + (3.4389)^2 \right] - 6.06483$$

$$SS_A = 0.00717$$

$$SS_B = \frac{1}{6}\left[(3.7020)^2 + (3.3739)^2 + (3.3724)^2\right] - 6.06483$$

$$SS_B = 0.01202$$

$$SS_C = \frac{1}{6}\left[(3.5051)^2 + (3.4551)^2 + (3.4881)^2\right] - 6.06483$$

$$SS_C = 0.00022$$

$$SS_D = \frac{1}{6}\left[(3.4314)^2 + (3.4722)^2 + (3.5447)^2\right] - 6.06483$$

$$SS_D = 0.00110$$

iii) Sum of squares residual SS_R

$$SS_R = SS_T - (SS_A + SS_B + SS_C + SS_D) \quad (10)$$

$$SS_R = 0.14387 - 0.02051$$

$$SS_R = 0.12336$$

Degree of freedom Df

$$\mathrm{df_T} = N - 1 = 18 - 1 = 17$$

$$df_A = n_A - 1 = 3 - 1 = 2$$

$$df_B = n_B - 1 = 3 - 1 = 2$$

$$df_C = n_C - 1 = 3 - 1 = 2$$

$$df_D = n_D - 1 = 3 - 1 = 2$$

$$df_R = \mathrm{df_T} - (df_A + df_B + df_C + df_D)$$

$$df_R = 17 - 8 = 9$$

Table 16. The ANOVA Table

Variation source	df	SS_i	MS_i	Variance ratio	%Contribution
A	2	0.00717	0.003585	0.2615	4.98
B	2	0.01202	0.006010	0.4385	8.36
C	2	0.00022	0.000110	0.0080	0.15
D	2	0.00110	0.000550	0.0401	0.77
Residual	9	0.12336	0.013707		85.74
Total	17	0.14387			100.00

Mean sum of squares, MS

$$MS = \frac{SS_i}{df_i} \quad (11)$$

Variance ratio (F – Statistics or Fisher' ratio)

$$F = \frac{MS_i}{MS_R} \quad (12)$$

Percentage contribution

$$\%Contr. = \frac{SS_i}{SS_T} \quad (13)$$

Using Eqs (7-13), the ANOVA table was created. This is shown in Table 16.

2.5. Confirmation Test $\tilde{\eta}$

Experiments through the Taguchi orthogonal array reveal that the optimal welding parameter combination is $A_2B_1C_1D_3$, which is then employed to predict the Grey relation that represents the welding quality. Only the effects of greater significance (A, B, and D) are taken into account as indicated by the ANOVA. Prediction of the Grey relation of the optimal welding parameters can be made by using Eq (14)

$$\hat{\eta} = \eta_m + \sum_{i=1}^{q} (\bar{\eta}_i - \eta_m) \quad (14)$$

Table 17. The Predicted S/N Ratios and Ultimate Tensile Strength Values of Optimum and Existing Process Parameters

	$A_1C_3D_1$ (*Predicted optimum*)	$A_1B_3C_3D_1$ (*Experimental Test* Results)	$A_2B_2C_3D_3$ (*Initial / existing parameter*)
	0.6545	0.5258	0.6007
Ultimate Tensile Strength	728 *MPa*	467 *MPa*	485 *MPa*

where;

$\hat{\eta}$ = Grey relational grade for predicting the optimal welding parameters

η_m = Grand mean of the grey relational grade

$\bar{\eta}$ = Grey relational grade of the optimal level for the significant factors

q = Number of significant factors (A, B and D) = 3

$\hat{\eta} = 0.5805 + (0.6077 - 0.5805) + (0.6170 - 0.5805) + (0.5908 - 0.5805)$ $\hat{\eta} = 0.6545$

Finally, the confirmation tests were done for the entire optimum combination of $A_2B_1C_1D_3$, the existing process parameters of $A_1B_2C_3D_2$ and an experimental procedure. The summary of the confirmation test is presented in Table 17.

From Table 17, it is shown that the predicted process parameters are 1.09 times, better than the S/N ratios of the existing welding process parameters and the experimental process parameter is 1.14 times less than the existing welding process parameters. The ultimate tensile strength of the predicted process parameters is 1.5, better than that of the existing welding parameters. Whereas, the experimental process parameter is 1.04 times less than the existing welding process parameter. From the results shown, it is clear that with the use of the optimum process parameters, weld strength quality has been greatly increased.

3. Analysis of the Alloying Elements of the Welds Made from the Optimum Process Parameters

Pure metals are ordinarily soft in most cases, and on their own, are not suitable for engineering projects. To increase the strength of these metals, alloying elements are added. Each alloying element is added in minute proportions to the main element to increase its (main element) strength. Each alloying element has its own functional contribution to the performance of the strength property of the main element. Therefore an optimum selection of the proportions of these alloying elements would achieve the required weld strength. The chemical compositions of the eighteen weld specimens are shown in Table 18 with their corresponding impact strengths.

Table 18. Chemical Composition of the Weld Deposits and their corresponding Charpy Impact Strength

C	*Si*	*Mn*	*P*	*S*	*Cu*	*Ni*	*Cr*	*Mo*	*Nb*	*V*	*Ti*	*N*	*Charpy* Impact Test (0-150 J)
0.150	0.420	1.420	0.012	0.006	0.040	0	0.140	0.200	0.020	0.080	0	0.004	100
0.120	0.310	1.810	0.006	0.009	0.080	0	0.090	0.320	0.040	0.120	0	0.020	45
0.20	0.510	1.620	0.008	0.007	0	0.040	0.040	0.420	0.320	0.200	0.320	0.018	65
0.180	0.300	0.920	0.014	0.006	0.020	0.010	0.050	0	0.210	0	0.200	0.009	111
0.130	0.210	1.700	0.009	0.004	0	0.020	0.090	0.090	0.060	0.210	0	0.007	92
0.150	0.150	1.920	0.021	0.007	0.040	0	0.120	0.120	0.040	0	0.030	0.004	42
0.210	0.180	1.630	0.008	0.008	0.110	0.040	0.080	0.100	0	0.310	0.040	0.008	98
0.170	0.320	1.750	0.006	0.004	0	0.050	0.070	0	0	0.320	0.080	0.005	35
0.140	0.450	0.980	0.018	0.007	0.080	0.070	0.080	0.210	0	0.240	0.060	0.004	55
0.120	0.320	1.240	0.021	0.008	0	0.080	0.060	0.180	0	0.230	0.400	0.009	30
0.100	0.280	1.350	0.042	0.009	0.090	0	0.080	0	0.230	0.120	0	0.012	45
0.200	0.260	1.700	0.068	0.006	0	0.090	0.030	0.310	0.430	0.130	0	0.012	105
0.160	0.210	2.320	0.024	0.004	0.020	0.070	0.040	0.320	0.120	0	0.340	0.014	30
0.140	0.190	1.530	0.013	0.007	0.060	0.060	0.080	0	0.002	0.230	0	0.007	48
0.110	0.670	1.720	0.022	0.008	0	0.070	0.090	0.410	0.008	0	0.190	0.008	25
0.160	0.220	1.800	0.008	0.007	0.140	0	0.120	0	0.030	0.310	0	0.006	30
0.170	0.280	1.620	0.007	0.006	0.060	0.050	0.080	0.230	0.320	0	0.210	0.010	68
0.250	0.370	2.100	0.014	0.005	0.210	0.090	0.130	0	0.300	0.180	0	0.015	125

The linear additive (hyperplane) model was used to obtain the coefficients of the alloying elements, containing a system equation of

$$\begin{bmatrix} y_1 \\ y_2 \\ y_3 \\ y_4 \end{bmatrix} = \begin{bmatrix} f_1(x_1) & f_2(x_1) ------ f_p(x_1) \\ f_1(x_2) & f_2(x_2) ------ f_p(x_2) \\ ; & :---------: \\ f_1(x_{18}) & f_2(x_{18}) ------ f_p(x_{18}) \end{bmatrix} \begin{bmatrix} \beta_1 \\ \beta_2 \\ : \\ \beta_p \end{bmatrix} + \begin{bmatrix} \varepsilon_1 \\ \varepsilon_2 \\ : \\ \varepsilon_p \end{bmatrix} \tag{15}$$

The terms used in the design matrix above are f (x) = 1, $f_{k+1}(x) = x_k$ where k = 1, 2, - - -, p.

When this matrix is subjected to further analysis, the following linear equation was obtained

$$y = \beta_1 f_1(x) + \beta_2 f_2(x) ---- + \beta_p f_p(x) + \varepsilon \tag{16}$$

where

ε = uncontrolled factors and experimental errors and $x = [x_1, \quad x_2, \quad x_3 ---- x_{13}]$

Equation (16) is the coefficient containing the alloying elements. It is used to determine the coefficient of variation. Statistical analysis was done to determine the standard error associated with the relationship between the weld metal compositions and the energy absorbed by them, thus:

$$\text{Standard Error} = SE = \sigma\sqrt{\frac{1}{n}} = \frac{\sigma}{\sqrt{n}} \tag{17}$$

However, the term coefficient of variation Cv, is the standard error divided by the coefficient (Furuya et al, 2007) that is,

$$Cv = \frac{\text{Standard error}}{\text{Coefficient}} \times 100\% \tag{18}$$

The relevance of each of the alloying elements is analysed here with the aid of the coefficient of variation. Theory of the coefficient of variation as used by Furuya et al (2007), suggested in their paper that a small coefficient of variation means that the element considered is expected to have a high reliability. However, Achebo and Igboanugo (2011) were of the opinion that low reliability is expected to have occurred as a result of the increased hardness nature of the steel plates.

The coefficient of variation calculated and presented in Table 19 shows that the alloying elements C, P, Cu, Ni, Cr, V, Ti and N have a high reliability of increasing the weld strength. The negative values associated with Cu, V and Ti indicate that even though these elements have been proved numerically to improve weld strength property, they can also in larger quantities be detrimental to improving weld strength. The limitations associated with them are not prominently significant and as a result have not been large enough to reduce the weld load bearing capacity.

Table 19. Result of the Linear Regression Analysis

Element	variable	coefficient	standard error	coefficient of variation (%)
Constant term	β_0	0.1126		
C	β_1	0.6219	0.0091	1.4633
Si	β_2	−0.0365	0.0310	−84.9315
Mn	β_3	−0.0644	0.0825	−128.1056
P	β_4	0.2113	0.0037	1.7511
S	β_5	−8.644	0.0004	−0.0046
Cu	β_6	−0.0417	0.0137	−32.8537
Ni	β_7	0.0586	0.0079	13.4812
Cr	β_8	0.2251	0.0074	3.2874
Mo	β_9	0.0325	0.0351	108
Nb	β_{10}	−0.0447	0.0337	−75.3915
V	β_{11}	−0.0948	0.0273	−28.7975
Ti	β_{12}	−0.0930	0.0267	−28.7097
N	β_{13}	2.8108	0.0011	0.0391

n = sample size of no of observations of the population.

Having considered this, from the values of the coefficient of variation in Table 19, C and Cr indicate that although their content in the weld appear to be adequate, a too high in content value would be detrimental to the ductility of the weld material, by making it brittle and increase the weld hardness. These characteristics would definitely deteriorate the toughness value of weld. This claim was supported by Furuya et al (2007) when they said that an increase in C could deteriorate HAZ toughness. Bergman (2011) was of the opinion that carbon content has a clear effect on impact strength, since the material is more ductile at lower carbon levels. Bergman (2011) said that Cr is an alloying element which gives good hardenability but has the limitation that it has a high affinity for oxygen which could lead to the problems with oxidation. Cr also has a high affinity for Ni (Achebo and Igboanugo, 2011). Ni has been determined to improve toughness in this study. Bergman (2011) said that the effect of Ni on material property depends on the amount of Nitrogen taken up and Ni element could have a positive effect on material, in the case of nitriding (Molinari et al, 1999) but could also lead to the deterioration of mechanical properties if precipitation occurs at grain boundaries (Nayar and Wasiczko, 1990; Achebo and Igboanugo, 2011).

The other alloying elements such as Si, Mn, Mo and Nb have high coefficient of variation but low reliability in improving the weld strength quality. The implication is that the increase in the additions of Mn, Mo and Nb alloying elements would deteriorate the toughness profile of the weld. This claim was buttressed by Achebo and Igboanugo (2011). Also, the increase in the addition of Si alloying element can induce the fluid property of the weldment to create spatter. This element would actually reduce the viscosity of the molten metal formed during the welding process and as a result of this, inadequate weld penetration could occur and oxygen infiltration into the molten metal may cause the oxidization of the weld metal and eventually, a brittle weld is produced.

From the above analysis, the average values of the coefficients (percent by weight) of alloying elements Mn, Mo and Nb should be reduced drastically, as they have not been very effective in improving the weld toughness and strength properties. The alloying element, S is seen not to have any input on the strength of the weld. This observation was supported by the claim made by Furuya et al (2007) that any alloying element content lower than 0.005 mass by weight in percent, should be regarded as 0. Further investigation was made to determine the percent contribution of each of the alloying elements in improving the quality of the weld by using Eq (13). Alloying elements C, P, Cu, Ni, Cr, V, Ti, and N were found to be very reliable when the coefficient of variation was used to measure the reliability of each of the alloying elements. Ti, V, Cu, C, Ni, Cr, and P contributed 0.727%, 0.523%, 0.133%, 0.0588%, 0.044%, 0.038% and 0.0297% to the entire strength performance property of the weld. However, those alloying elements with low reliability such as Si, Mn, Mo and Nb contributed 0.676%, 4.77%, 0.865% and 0.797% to the entire strength performance property of the weld. From the above, it can be found that for higher strength reliability, the mass by weight (% by wt) of P should be increased by at least double the present available content (% by wt). This indicates that the content of P should be increased to about 0.0357% by wt to achieve the desired strength. Mn has contributed most to the strength performance of the weld. This alloying element was determined to have low reliability, therefore, it is highly recommended that even though, it has a low reliability, the presence in the composition of the weld is of great significance. This claim also applies to Si, Mo and Nb. The alloying elements, S and N were found to have a contribution of 0.0000976% and 0.000887%. The contributions of S and N are greatly insignificant to the strength of the weld. As a result, the alloying elements S and N should be removed from the chemical composition of the weld.

Thus, considering the analysis using the coefficient of variation, the weld expected to have the strength comparable to the strength of the parent metal should have the elements C, P, Cu, Ni, Cr, V, Ti, and N. However, having investigated the contributions of each alloying element to the strength performance of the resultant weld, the new chemical composition that should be expected to have the strength required to match the strength of the parent metal should be composed of C, P, Cu, Ni, Cr, V, Ti, Si, Mn, Mo and Nb. It is also suspected that to further increase the strength of the weld, the values of Si, Mn, Mo and Nb should be reduced to an acceptable level that could be determined by further experimental analysis.

4. Mechanical Properties of Steel Weld and Pipe

Test specimens for tensile strength, impact strength, hardness, chemical composition tests, and microstructure tests were made from the all weld metal deposits of the weld made by using the optimum welding process parameters.

Strength is a property that enables a metal to resist deformation under load. The ultimate strength is the maximum strain a material can withstand. Tensile strength is a measurement of the resistance to being pulled apart when placed in a tension load. Impact strength is the ability of a metal to resist suddenly applied loads. Hardness is a property of a material to resist permanent indentation (Gupta, 2009).

Tensile Strength Test

1. Tensile test specimens of dimensions shown in Figure 2 were machined from the all-weld metal deposits.
2. Tensile tests were performed on the specimens using the Avery Extensometer;
3. The tensile test parameters: yield strength, ultimate tensile strength, fracture strength, percent elongation and reduction in area were determined using the extensometer graph and dimensions of specimen before and after test using the following formulae:

i $$\text{Engineering Stress}, \sigma = \frac{\text{Uniaxial tensile load}}{\text{Original Cross sectional Area}} = \frac{F}{A}\ kg/mm^2 \qquad (19)$$

ii. $$\text{Engineering Straine} = \frac{\text{Change in length of Sample}}{\text{Original length of Sample}} = \frac{l - l_0}{l_0} = \frac{\Delta l}{l_0}. \qquad (20)$$

iii 0.2% Offset Yield Strength is determined from the stress-strain graph

iv. $$\%\ elongation = \frac{\text{Final length} - \text{Initial length}}{\text{Initial length}} \times 100\% \qquad (21)$$

v. $$\%\ \text{reduction in area} = \frac{\text{Original area} - \text{Final area}}{\text{Original area}} \times 100\% \qquad (22)$$

where l and l_o represent the final and original lengths of the tensile test specimen.

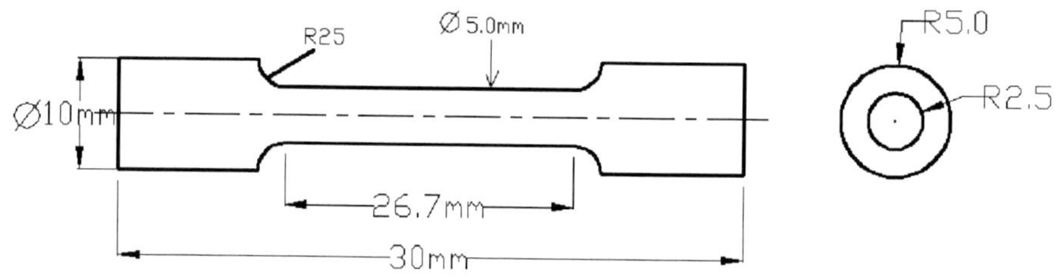

Figure 2. Tensile Strength Test specimen.

V-Notch Impact Test

1. Impact test specimens of dimensions shown in Figure 3 were prepared from the all-weld metal deposits;
2. Impact tests were performed on the specimen using the Avery Dennison Charpy-Izod Impact testing machine of a capacity of 0 – 150J on the scale;
3. The impact test results which indicate the energy absorbed as the specimen is fractured were read off the scale.

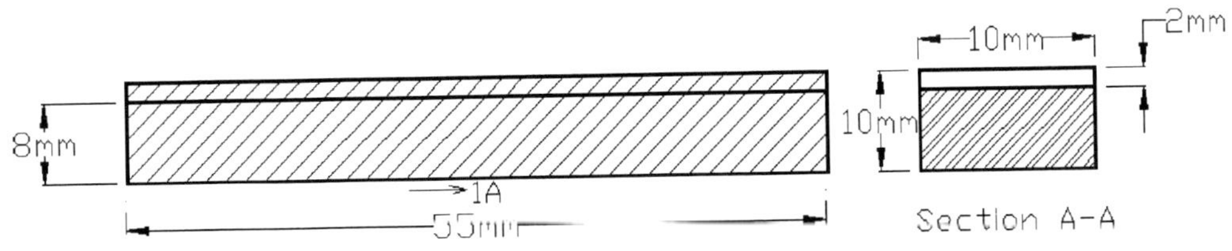

Figure 3. Notch Impact Strength Test specimen.

Hardness Test

1. Test samples made from the all-weld metal deposits were machined using a lathe machine;
2. The machined samples were further smoothened using a 400-grit emery paper to produce mirror-like weld samples;
3 The Avery Brinell hardness tester was used to make an impression or indentation on the smoothened samples for 10 seconds each;
4. The Avery pre-calibrated microscope was used to determine the diameters of the indented samples;
5. Equation (23) was used to determine the Brinell hardness number (BHN) of the samples;

$$BHN = \frac{2P}{\pi D\left(D - \sqrt{D^2 - d^2}\right)} \tag{23}$$

where, D = diameter of the ball,10 mm
d = diameter of the impression, mm
P = applied load, 3000 kg

Chemical Composition Analysis

1. The samples were made from the all-weld metal deposits and machined using the lathe machine;
2. The machined samples were smoothened with a 400-grit emery paper and a 0.5 μm emery cloth which has a finer textured surface was used to further smoothen the samples to very high precision level of smoothness;
3. The samples were placed on an ARL1640 model spectrometer made in Switzerland. As the nib of the spectrometer struck the polished surfaces of the samples, the chemical compositions immediately are displayed on a digital screen and printed out with the use of a printer.

Weld Microstructure

- The test samples were sectioned using a power hacksaw and a new sample measuring 20 x 20 x10 mm was removed;
- The all weld metal deposits were machined using the Lathe machine into test samples and then ground;
- The Avery grinding machine consisting of four compartments used to grind the specimens. The first, second, third and fourth compartments contain 240, 320, 400 and 600-grit emery papers respectively;
- The sectioned samples were made to rub the 240, 320, 400 and finally 600 grit-emery papers in the presence of silicon carbide solution;
- These polished samples were further polished by the Econet II Polisher. This machine contains 1.0 μm and 0.5 μm emery cloths. The samples were initially polished using 1.0μm emery cloth and finally polished using 0.5 μm emery cloth under the influence of silicon carbide solution;
- These well polished samples were etched using 2% NaOH solution and taken to the weld metallurgical microscope;
- The weld metallurgical microscope contains a black and white photographic film. Photographs of the treated sectioned welds were taken, the films developed and the microstructural views were printed out.

From the microstructural views of both the weld metal and that of the pipe material (see Figures 4 and 5). There appears to be no spores present. The weld metal appears to be more affected by the flux material. This indicates that the welding process was effective, as there was no trace of gas entrapment in the weld. Summarily, the weld metal and the pipe material are of comparatively high quality.

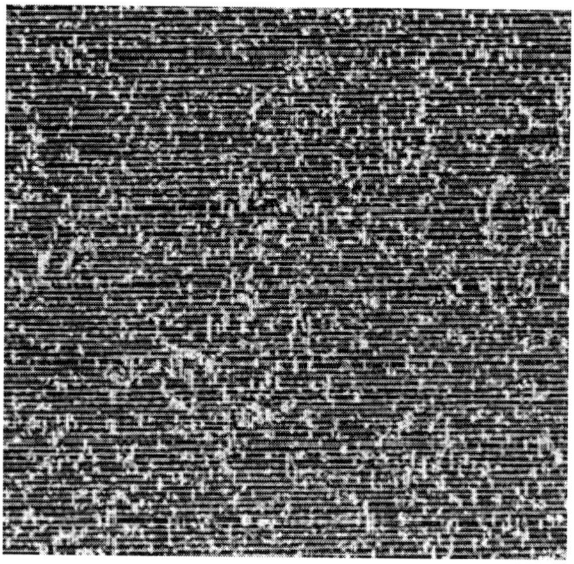

Figure 4. Microstructural View of the Weld Metal.

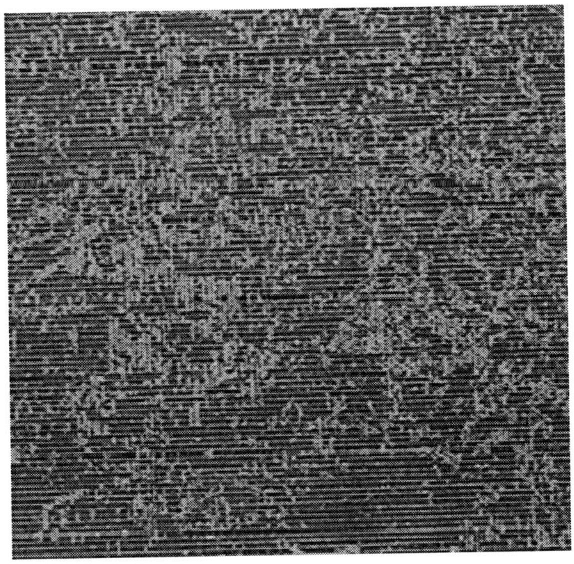

Figure 5. Microstructural View of the Pipe Material.

Table 20 shows the mechanical properties of both the weld and pipe material.

Table 20. The Mechanical Properties of the Weld Metal and the Pipe Material

Mechanical Test		Weld Property	Pipe (Parent Metal) Property
V-Notch Impact Test		111(J)	90(J)
Tensile Test			
a.	0.2% Yield strength	478 MPa	482 MPa
b.	Ultimate Tensile Strength	728 MPa	716 MPa
c.	Percent Elongation	20%	15%
d.	Percent reduction in area	48%	37%
Hardness Test			
Brinell Hardness Number		320	350

From Table 20, it can be seen that the weldmetal yielded before the pipe material. This could be attributed to the fact that the pipe material is harder than the weldmetal as indicated by the Brinell Hardness Number. The ultimate tensile test result of the weld is more than that of the pipe and the ability of the weld to absorb more impact energy reveals that it could be more ductile than the pipe material. However, the values of the mechanical properties of the weld are closely matched to that of the pipe. The effect of the alloying elements on percent elongation and percent reduction in area were studied by other investigators. Okazaki et al (1993) suggest that elongation and reduction of area, which are tensile properties could decrease with the addition of nitrogen, Ni and oxygen, but can be effective, for improving strength without lowering ductility. They also stressed that 2% proof strength and ultimate tensile strength increase linearly with increase in Ni and O in titanium alloying elements. From the opinions of Bergman (2011), Molinari et al (1999) and Okazaki (1993) it can be seen that Ni has a positive effect on mechanical properties of material depending on the content available in the weld. Its improvement on ultimate tensile strength shows that it has a great significance on charpy impact energy. But, if the alloying element content is significantly above 0.0586% by weight, it could affect the weld specimen elongation and reduction in area, which is a function of the weld's ductile property (Achebo and Igboanugo, 2011).

To find out whether there is a correlation (relationship) between the weld properties and that of the pipe. The excel software package was used to find the linear regression using the values in Table 20, which displays properties between the weld and that of the parent metal. The analysis is shown in Table 21.

Table 21. Least Square Model for determining SSE and SST

y	x	$\bar{y} = -5.462 + 1.010x$	$\left(y - \bar{y}\right)^2$
90	111	106.648	277.156
482	478	477.318	21.921
716	728	729.818	190.937
15	20	14.738	0.069
37	48	43.018	36.216
350	320	317.738	1040.837
		1689.278	$SSE = 1567.136$

From the linear regression analysis, ȳ = - 5.462 + 1.010x was determined

SST = $(1689.278)^2$ = 2,853,660.161

$$MST = \frac{SST}{k-1} = \frac{2,853,660.161}{5-1} = 713,415.040$$

$$MSE = \frac{SSE}{n-k} = \frac{1567.136}{10-5} = 313.427$$

where, SST is the total sum of squares; MST is the mean total sum of squares; MSE is the mean sum of errors

$$F_{test} = \frac{MST}{MSE} = \frac{713,415.040}{313.427} = 2,276.176$$

$$F_{table} = F^{k-1}_{n-k,\alpha=0.05} = F^{4}_{5,\alpha=0.05} = 5.19$$

Since the $F_{cal} > F_{table}$, there is a strong correlation between the properties of the weldmetal and the pipe material properties.

Coefficient of determination, r^2

$$r^2 = 1 - \frac{SSE}{SST} = 1 - \frac{1567.136}{2,853,660.161} = 1 - 0.00055 = 0.99945$$

r^2 was found to be 0.99945. The values of F ratio and r^2 show that there is a good correlation between the weld property and that of the pipe (parent metal). Therefore the strength quality of the weld compares well with that of the parent metal (pipe material).

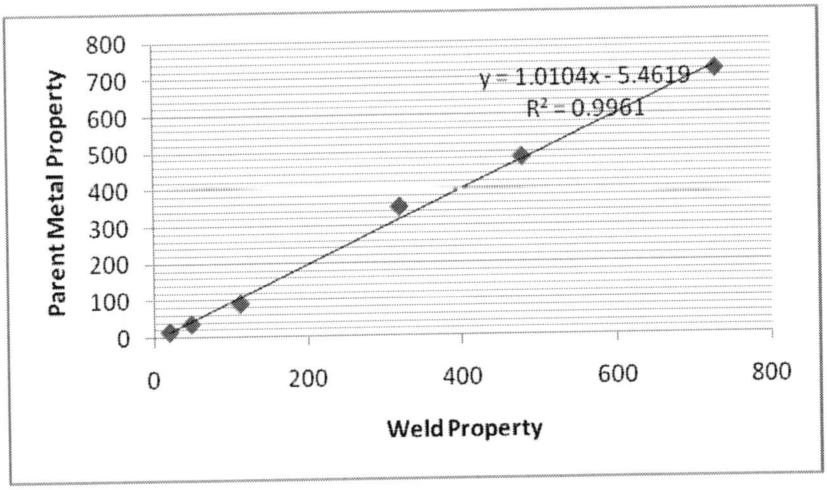

Figure 6. Correlation Analysis Curve.

From the scatter graph of Figure 6, we can see that there is a very strong positive correlation between the parent metal property and the weld property indicating a strong relationship between Weld Property and the parent material property. From the bivariate plot, r^2= 0.9961

5. Expert Assessment of Weld Strength Quality Using the Kendall's Coefficient of Concordance

A questionnaire containing some structured questions was given to five Judges who have professional Welder's certification and are Professors of Manufacturing Engineering. Presented before them were the weld joint specimens, microstructural views of the welds and parent metal and the result values obtained from the mechanical tests carried out on the weldments.

The Questionnaire contains the following questions:

1. How would you score, the appropriateness of the value set for the weld throat measurement used with the optimum welding process parameters?
   (a)Excellent ----, (b) Very good ----, (c) Good ----, (d) Not very good ----, (e) Bad ----
2. How would you score, the ultimate tensile strength of the weld made by using the optimum welding process parameters?
   (a)Excellent ----, (b) Very good ----, (c) Good ----, (d) Not very good ----, (e) Bad ----
3. How would you rate the bead appearance of the weld made using the optimum welding process Parameters?
   (a)Excellent ----, (b) Very good ----, (c) Good ----, (d) Not very good ----, (e) Bad ----
4. In your opinion how would you rate the overall quality of the weld?
   (a)Excellent ----, (b) Very good ----, (c) Good ----, (d) Not very good ----, (e) Bad ----
5. In your opinion has the strength of the weld been improved by welding with the optimum welding process parameters?

(a)Excellent ----, (b) Very good ----, (c) Good ----, (d) Not very good ----, (e) Bad ----

6. From the analysis of the study how would you rate the viability of this approach, comparing the strength of the weld deposit with the strength of the parent metal?
   (a)Excellent ----, (b) Very good ----, (c) Good ----, (d) Not very good ----, (e) Bad ----
7. In your opinion how would you rate the impact energy absorbed by the weld specimen?
   (a)Excellent ----, (b) Very good ----, (c) Good ----, (d) Not very good ----, (e) Bad ----
8. In your opinion, how would you assess the weld microstructural view of the weld specimen?
   (a)Excellent ----, (b) Very good ----, (c) Good ----, (d) Not very good ----, (e) Bad ----
9. How would you rate the ease of slag removal from the welded joint?
   (a)Excellent ----, (b) Very good ----, (c) Good ----, (d) Not very good ----, (e) Bad ----
10. How would you rate the ductility of the weld when compared to that of the parent metal?
    (a)Excellent ----, (b) Very good ----, (c) Good ----, (d) Not very good ----, (e) Bad ----

The responses from the questionnaires administered to the five Judges and their expert assessment were tabulated in Table 22.

Table 22. Expert Assessment on Weld Strength Quality

	J u d g e s							
Questions	1	2	3	4	5	t	$\sum_{i=1}^{n}\left(R_i - \overline{R}\right)^2$	$\left(t^3 - t\right)$
1	4	3	4	3	5	8	2.80	504
2	5	5	3	3	4	8	4.00	504
3	4	5	5	4	3	6	2.80	210
4	4	4	4	3	3	8	1.20	504
5	3	3	5	5	3	8	4.80	504
6	4	4	4	5	4	6	0.80	210
7	3	3	3	4	4	8	1.20	504
8	5	5	4	5	5	9	0.80	720
9	4	3	4	4	4	8	0.80	504
10	4	5	5	5	4	9	1.20	720
	40	40	41	41	39	78	20.40	4884

Scoring Criteria

Excellent [5], Very Good [4] Good [3] Not Very Good [2] Bad [1]

The Kendall's coefficient of concordance (W) is obtained by first determining the sum of squares statistic over the row sum of observations, R_i

$$S = \sum_{i=1}^{n}\left(R_i - \overline{R}\right)^2 \tag{24}$$

S = sum of squares statistic over the row sum of observations, R_i
The Kendall's W statistic is expressed in Eq. (25)

$$W = \frac{12S}{p^2\left(n^3 - n\right) \quad pT} \tag{25}$$

where n is the number of questions contained in the questionnaire, p the number of judges. T is a correction factor for the observations (Siegel, 1956; Siegel and Castellan, 1988; Zar, 1999) T_j is calculated for the jth row, and then used to find Kendall's coefficient of concordance, W.

$$T_j = \sum_{k=1}^{m_j}\left(t_k^3 - t_k\right) \tag{26}$$

J = 1, 2, - - - k

Where k is the number of trials, m_j is the number of groups of matches in row j. A group of matches exists when at least two samples in row j have the same rating. For each group of matches in a row j, t_i is the number of observations in group i. The Kendall's coefficient of concordance, W is within the range of $0 \le W \le 1$. The relationship between Kendall's W and the Friedman's χ^2 statistic is obtained from

$$\chi^2 = p(n-1)W \tag{27}$$

This is compared with the table value which is expressed as

$$\chi^2_{(0.05,\, n-1)} \tag{28}$$

Using the value that represents Eq. (24) in Table 22, we have

$$S = \sum_{j=1}^{m}\sum_{i=1}^{n}\left(R_i - \overline{R}\right)^2 = 20.40$$

Substituting this value of S into Eq. (25)

$$W = \frac{12\ x\ 20.40}{5^2\left(10^3 - 10\right) - 5\left(4884\right)} = 0.74$$

Incorporating the Friedman's χ^2 statistic we have

$$\chi^2_{cal} = 5\left(10-1\right)0.74 = 33.38 \approx 33.40$$

The table value of $\chi^2_{table(0.05,\,9)} = 16.92$

The this case, the calculated value, χ^2_{cal} is greater than the table value, χ^2_{table}. This indicates that the expert evaluation made by the Judges are in agreement that the weld strength of the pipe joints are well compared to those of the parent metal (pipe).

The Judges' individual assessment was investigated by using

$$F = \frac{12\sum_{j=1}^{k} R_j^2 - 3N^2k(k+1)^2}{Nk(k+1) + \dfrac{Nk - \sum_{i=1}^{N}\sum_{j=1}^{\delta_i} t_{ij}^3}{(k-1)}} \tag{29}$$

where N is the number of rows (scores), p = k is the number of columns (Judges), R_j is the sum of ranks in the jth column, t_{ij} is the size of the jth set of tied ranks in the ith group and ∂_i is the number of sets of tied ranks in the ith group. From the analysis carried out, it was found that Judges 1, 2, 3, 4 and 5 have F- values of 37.5, 37.5, 36.45, 36.45 and 38.52 respectively. These values were compared with the table value, $F^{k-1}_{n-k,\alpha=0.05} = F^{4}_{5,\alpha=0.05} = 5.19$. Since the calculated F values are greater than the table value, It was concluded that the individual assessment of the Judges suggest that the independent opinions of the Judges agree that the quality of the weld is good and compares well with those of the parent metal.

CONCLUSION

The challenges and inhibitions caused by corrosion and wear occurrence in steel pipes cost industrial facilities billions of dollars. Corrosion and wear occur mostly at the welded joints, because the strength of these welded joints is usually lower than that of the parent metal. As a result, initial pipe failures occur at these joints. Therefore, to avoid these failures, joints need to have comparable strength with their parent metal. To achieve a uniform service life of both the pipe materials and their welded joints, the welding process parameters and the types and proportions of the alloying elements that make up the chemical composition of the weld metal should be optimized and improved. This study has been able to show these improvements.

This study further determined the mechanical properties of both the weldments and the pipe (parent metal). Statistical analysis shows that there is a correlation between the properties of the weld metal and the pipe material allowing predictability and maneuverability of process parameters to meet specific needs. Also expert assessment on the strength quality of the weld metal was conducted by five Professors of Manufacturing Engineering who possess certifications in welding technology. Kendall's coefficient of concordance was used to analyze these assessments, and it was found that the Judges unanimously agreed that through the application of optimum process parameters, the strength of the improved weld metal compares well with the parent metal of the pipeline. This chapter has successfully elucidated the possibilities of improving weld strength by optimizing the welding process parameters.

REFERENCE

Achebo, J.I. (2008) Application of Modified Langelier Saturation Index Model to Pipeline Corrosion. *JEAS,* 4; p. 44 – 46.

Achebo, J. I. & Ibhadode, A. O. A (2008) 'Development of a New Flux for Aluminium Gas Welding'Materials and Product Technologies, Edited by Z. Y. Shen; M. N. James; W. D. Li, and Y. X. Zhao. Trans Tech Publications Ltd, Switzerland, Vol. 44 - 46 of *Advanced Materials Research,* p 677- 684.

Achebo, J.I. (2009) Evaluation of Wear Severity in Pipeline, *JEAS*, 4(1): p. 74 – 76..

Achebo, J. I. & Ibhadode, A. O. A (2009) 'Development of Optimum Welding Flux Composition using the Bend Strength Test' Materials and Product Technologies, Edited by A. O. A. Ibadode. Trans Tech Publications Ltd, Switzerland, Vol. 62 - 64 of *Advanced Materials Research,* p 322 .

Achebo, J.I. (2010*) Numerical Investigation of Keyhole Shape during CO_2 Laser Welding of 5456 Aluminum Alloy, *International Journal of Engineering Science and Technology,* Vol. 2 (6): p. 2034 - 2039 .

Achebo, J.I. (2010) A Parametric Analysis of CO_2 Laser Heat Absorption Profile of 5083 Aluminum Alloy, *International Journal of Engineering Science and Technology,* Vol. 2 (6): p. 2029 – 2033.

Achebo, J.I. (2011) 'A Multiphysics Analysis of Aluminium Welding Flux Optimization Methods' Chapter 11, In *Advances in Computer Science and Engineering.* InTech Open Access Publishers Rijeka, Croatia, p. 215-236

Achebo, J.I. (2011*) Optimization of GMAW protocols and Parameters for Improving Weld Strength Quality Applying the Taguchi Method, Presented at *the 2011 World Congress on Engineering: International Conference of Manufacturing Engineering and Engineering Management,* Imperial College, London , July 6 – 7, 2011.

Achebo, J.I. & Igboanugo, A.C. (2011*) Influence of Alloying Elements on HAZ Toughness of Multilayer Welded Steel Joints* submitted to the Nigerian Institution of Welding.

Achebo, J. I. & Oghoore, O. (2010) Computational Analysis of Condensed Vaporized AlloyingElements of 5456 Aluminum Alloy, Trans Tech Publications Ltd, Switzerland, Vols. 118 – 120 of *Advanced Materials Research,* p 855- 859.

Aneru, S. A; Aigbogun, C. J; Ovabor, K. & Awolumate, O. (2011) *The Study of the Methodology for Optimizing Plasma Arc Welding Parameters Using Taguchi Method with Grey Relational Analysis*. B. Eng Thesis, Department of Production Engineering, University of Benin, Benin City, Edo State, Nigeria.

Balasubramanian, S. & Ganapathy, S. (2011) Grey Relational Analysis to determine Optimum Process Parameters for Wire Electro Discharge Machining (WEDM). *International Journal of Engineering Science and Technology,* Vol. 3, No. 1, p. 95 - 101.

Bergman, O. (2011) *Effect of Nitrogen Uptake during Sintering on the Properties of PM Steels Pre-alloyed with Chromium*. Swedish Institute for Metals Research, Stockholm.

Butler, L. J. & Kulak, G. L. (1971) Strength of Fillet Welds as a Function of Direction of Load. *Welding Research Supplement.* Vol. 36, No. 5, p. 231s-234s.

Butler, L.J; Pal, S. & Kulak, G. L. (1972) Eccentrically Loaded Welded Connections. Journal of the Structural Division, *ASCE,* Vol. 98, ST5, May, p. 989-1005.

Chang, C; Yang, J; Ling, C & C. Chou, C (2010) "Optimization of Heat Treatment Parameters with the Taguchi Method for the A 7050 Aluminim Alloy". *IACSIT International Journal of Engineering and Technology,* Vol.2, No. 3, , p. 269-272.

Clark, P.J. (1971) Basis of Design for Fillet Welded Joints under Static Loading. *Proceedings of Conference on Improving Welded Product Design.* The Welding Institute, Cambridge, England, Vol. 1, p. 85-96.

Clausing, D. and Simpson, B. H. (1990) Quality of Design. *Quality Progress* 23, p 41-44.

Collin P. & Johnson, B. (2005) Design of Welds in High Strength Steel. *4th European Conference on Steel and Composite Structures.* Maastricht.

Deng, J. (1989). Introduction to Grey System. J. Grey System, *1* (1), p. 1–24.

Duane, K. & Miller, P.E. (1997) *Use Undermatching Weld Metal where Advantageous. Welding Innovation,* Vol. XIV, No. 1.

Ealey, L. A. (1988) *Quality by Design: Taguchi Methods and U.S. Industry,* ASI Press, Dearborn, Mich.

Eiklenbory, M; Ioannou, S; King II, G & Vilcheck, M. (2007) Taguchi Methods for Achieving Quality, *Project Report at Engineering Management, School of Engineering,* San Francisco State University.

Fisher, R.A. (1935) *The Design of Experiments.* Edinburgh: Oliver & Boyd, RE, RI.

Fung, C.P. (2003). Manufacturing Process Optimization for Wear Property of Fiber-reinforced Polybutylene Terephthalate Composites with Grey Relational Analysis. *Wear, 254,* 298–306.

Furuya, H; alhara, S. & Morita; K.(2007) A new Proposal of – HAZ Toughness Formulation by Chemical Compositions, *Welding Journal,* Vol. 86, No.2, p. 44s – 50s.

Gomez, I; Kanvinde, A; Kwan, Y. & Grandin, G. (2008) *Strength and Ductility of Welded Joints Subjected to Out-of-Plane Bending.* American Institute of Steel Construction.

Gordon, R.; Bruce, B.; Harris, I.; Harwig, D.; Porter, N.; Sullivan, M. & Neary, C. (2004) *Internal Repair of Pipelines: 12 – Month Technical Progress Report.* EWI Project No. 46211GTH Edison Welding Institute, Columbus, OH 43221, p. 1 – 95.

Gunther, H-P, Hildebrand, J; Rasche, C; Versch, C; Wudtke, I; Kuhlmann U; Vormwald, M; & Wemer F. (2009) *Welded Connections of High Strength Steels for Building Industry.* IIW XV 1315-09.

Gupta, J. (2009) *Mechanical and Wear Properties of Carburized Mild Steel Samples.* Master of Technology Thesis, National Institute of Technology, Rourkela.

Higgins, T.R. & Preece, F.R. (1969) Proposed Working Stresses for Fillet Welds in Building Construction. *Engineering Journal,* AISC Vol. 6, No. 1, p. 16-20.

Hopkins, P. (2002) The structural integrity of Oil and Gas Transmission Pipelines, in *Comprehensive Structural Integrity,* Vol. 1, Elsevier Publishers (Final Draft for Editors).

Hsiao, Y. F, Tarng, Y. S. & Huang, W. J. (2008). 'Optimization of Plasma Arc Welding Parameters by Using the Taguchi Method with the Grey Relational Analysis', *Materials and Manufacturing Processes,* 23: 1, p. 51 - 58.

Kim, H. R.& Lee K. Y.(2009) Application of Taguchi Method to determine Hybrid Welding Conditions of Aluminum Alloy. *Journal of Scientific and Industrial Research,* Vol. 68, p. 296 - 300.

Kishore, K; Gopal Krishna, P. V; Veladri, K. & Ali Syed Qasim (2010) Analysis of Defects in Gas Shielded Arc Welding of AISI 1040 Steel Using Taguchi Method. *ARPN Journal of Engineering and Applied Sciences,* Vol. 5, No. 1, p. 37 - 41.

Kuhlmann, U; Gunther, H.P & Rasche, C. (2008) *High Strength Steel Fillet Welded Connections. Steel Construction,* Design and Research, Vol.1.

Kumar, S.R.S. & Kumar, A.R.S. (retrieved online on 19/05/2011) Design of Steel Structures. Indian Institute of Technology, Madras.

Ligtenberg, F.K. (1968) *International Test Series Final Report,* IIW Doc. XV-242-68, International Institute of Welding.

Lin, C.L. (2003) Use of the Taguchi Method and Grey Relational Analysis to Optimize Turning Operations with Multiple Performance Characteristics. *Materials and Manufacturing Processes*, 19, p. 209–220.

Molinari, A; Bacci, T; Campestrini, P; Pellizari, M; & Tesi, B. (1999) *Plasma Nitriding of Fe-Cr-Mc sintered steels, powder metallurgy,* Vol.42, No.2 p.119-125.

Nayer, H-S. & Wasiczko, B. (1990) Nitrogen Absorption Control during Sintering of Stainless Steel Parts. *Metal Powder Report,* Vol.45, No.9 .

Okazaki, y; ito, y, ito, A & Teteishi, T.(1993)Effect of Alloying Element on Mechanical Properties of Titanium Alloys for Medical Implants. Material Transactions, *JIM,* Vol.34, No.12, p.1217-12 .

Siegel, S. (1956) *Nonparametric Statistics for the Behavioral Sciences,* McGraw-Hill, New York.

Siegel, S & Castellan, N. J; Jr (1988) *Nonparametric Statistics for the Behavioral Sciences* (2nd ed.), McGraw-Hill, New York.

Tarng, Y.S.; Juang, S.C. & Chang, C.H. (2002). The Use of Grey-based Taguchi Methods to determine Submerged arc Welding Process Parameters in Hard Facing. *J. Materials Processing Technology,* p. *128*, 1–6.

Thomsen, J.S. (2004) *Advanced Control Methods for Optimization of Arc Welding,* Ph.D. *Thesis Department of Control Engineering, Aalborg University,* Fredrik Bajers Vej 7, 9220. Aalborg East, Denmark. p. 63 – 76; 183 – 215.

Yoon, H; Byeong Hyeon, M; Chil Soon, L; Hyoung, K. D; Kyoun, K. Y & Jo, P. W.(2006). Strength Characteristics on Resistance Spot Welding of Al Alloy Sheets by Taguchi Method. *International Journal of Modern Physics* B, 20, , p. 4297 – 4302.

Zar, J. H. (1999) *Biostatistical Analysis* (4th ed.), Prentice Hall, Upper Saddle River, New Jersey.

INDEX

A

B

C

D

E

I

J

K

L

O

P

T

U

V

W

X

Y

Z